Preparing for Contact

George Michael

Preparing for Contact

When Humans and Extraterrestrials Finally Meet

RVP Press
New York

RVP Publishers Inc.
95 Morton Street, Ground Floor
New York, NY 10014

RVP Press, New York

Illustration cover: Diverstudio

Library of Congress Control Number: 2014953059

ISBN 978 1 61861 322 6

www.rvppress.com

For my friends George Hathaway, Charles Kellett,
Thomas Malloy, and Louis Namm
and my sisters Sandra and Susan

Table of Contents

Acknowledgments ix
Introduction xi

CHAPTER 1: Early Speculations 1
CHAPTER 2: Astronomy and the Plurality of Worlds 13
CHAPTER 3: Astrobiology 26
CHAPTER 4: The Structure of Extraterrestrial Civilizations 50
CHAPTER 5: The Search for Extraterrestrial Intelligence 62
CHAPTER 6: Preparing for Indirect Contact 78
CHAPTER 7: The Ancient Alien Hypothesis 96
CHAPTER 8: The UFO Controversy 112
CHAPTER 9: The Government's Response to UFOs 135
CHAPTER 10: Close Encounters 158
CHAPTER 11: UFO Conspiracy Theories 174
CHAPTER 12: Preparing for Direct Contact 209
CHAPTER 13: Exploring the Cosmos 230
CONCLUSION: A Race Against Time 261

Suggested Reading 277
Endnotes 283

Acknowledgments

I have long had an abiding interest in the prospect of extraterrestrial life, advances alien civilizations, and UFOs. Numerous scientists, investigators, and journalists have made notable contributions to these fields which have guided my research including Isaac Asimov, Philip Coppens, Paul Davies, Frank Drake, John Gribbin, Albert A. Harrison, Stephen Hawking, Ray Kurzweil, Carl Sagan, Michio Kaku, Leslie Kean, Seth Shostak, and Douglas A. Vakoch.

In particular, I greatly appreciate those individuals who took the time to converse directly with me and offer their insights on these topics, including Steven J. Dick, Michael A.G. Michaud, Stanton Friedman, and Nick Pope.

I want to express my gratitude to René van Praag and Astrid Bosch for their confidence in this project and giving me the opportunity to publish my book with RVP Press. I am also grateful to Christie Rears for her efforts in proofreading my manuscript.

Finally, I thank my friends George Hathaway, Charles Kellett, Thomas Malloy, and Louis Namm and my sisters Sandra and Susan for their support and encouragement. It is they to whom I dedicate this book.

Introduction

The prospect of extraterrestrial alien intelligence both intrigues and tantalizes people. Some biologists are skeptical of the notion insofar as they find it highly unlikely that the evolutionary process that led to *Homo sapiens* could be replayed on another planet. Other scientists think that the topic is moot because the vast distances between stars and the low probability that extraterrestrial civilizations would exist nearby and contemporaneously with our species, conspire to make the prospect of interstellar communication practically nil. Some of the more doctrinaire religious people, most notably fundamentalist Christians, reject the extraterrestrial hypothesis, since it would seemingly violate the precepts of their theology: The existence of another intelligent species would undercut the notion of humanity's "uniqueness" and special relationship with the creator. In the realm of popular culture, aliens are an enduring staple, as they simultaneously conjure up our worst fears of planetary subjugation in the style of H.G. Wells' *War of the Worlds* and also our yearnings for divine deliverance as depicted in Steven Spielberg's *E.T.*

The extraterrestrial life debate is ongoing. On the one hand, pessimists argue that there were so many contingencies that were necessary to give rise to intelligent life on Earth that the process is unlikely to be replicated elsewhere in the cosmos.[1] Unlike Charles Darwin's theory of evolution, there is still no universally accepted theory of life's origin. Furthermore, there is nothing in the laws of physics that singles out "life" as a favored destination.[2] Optimists, on the other hand, note that life emerged relatively quickly on Earth after the planet formed and had time to cool down. Fossil evidence indicates that single-celled organisms existed a mere half a billion years after the formation of the Earth. What is not promising, though, is that multi-cellular life did not emerge until about 3.5 billion years later. Furthermore, intelligent life, in the form of *Homo sapiens*, would come much later, and actual "civilizations" are of even more recent vintage. Currently, the Sun is at middle age; moreover, environmental conditions on Earth are always changing, and it is not certain that the planet will be habitable for humans for the duration of the Sun's life. The inference is that although life in the universe could be fairly widespread, there may not be adequate time for intelligent life to develop, survive, and build enduring civilizations capable of communicating with other sentient beings.

A number of obstacles must be surmounted for intelligent life to emerge, yet more and more people are agreeable to the extraterrestrial hypothesis. Whereas once scientists were highly skeptical about the existence of life elsewhere in the cosmos, today the pendulum has swung in the opposite direction as both scientists and journalists now often proclaim that the universe is teeming with life. As the astrophysicist Paul Davies noted, for scientists to openly profess a belief in extraterrestrial life during the 1960s and 1970s was tantamount to career suicide, noting that, "[o]ne might as well have expressed a belief in fairies."[3] Why, he asks, is it now scientifically respectable to search for life beyond the Earth? To be sure, several new developments have endowed the field of astrobiology (the study of life on other planets) with greater credibility, including the discovery of exoplanets (planets outside of our solar system) and the dawning realization that life can survive in a much wider range of physical conditions than what was once recognized. To Davies, the change in sentiment is due more to fashion rather than discovery. As he points out, new theories in physics such as antigravity, string theory, the multiverse, dark matter, and dark energy can make speculation about extraterrestrial life seem quaint in comparison.[4] He concedes that the origin of life may have been more "law-like" than "chance-like," suggesting that it will emerge more or less automatically wherever conditions permit. Nevertheless, he argues that these sentiments are fundamentally philosophical and not scientific.[5]

Recent astronomical discoveries, though, lend credence to the extraterrestrial life hypothesis. According to a "cosmic census" extrapolated from the National Aeronautics and Space Administration's (NASA) Kepler telescope, at least 50 billion planets reside in the Milky Way galaxy.[6] Moreover, approximately fifty exoplanets have been identified as not only similar to Earth in mass, but also located in the habitable zones of their respective solar systems.[7] And new discoveries of so-called extremeophiles on Earth—organisms that can live in severe environments from the very cold to the very hot and even the very toxic—suggest that life could exist elsewhere in the universe. If these exoplanets share similar properties to Earth, the likelihood that they could host extraterrestrial civilizations greatly increases. Proponents of convergent evolution argue that similar environments will produce similar adaptations, even in unrelated organisms, thus it is not unreasonable to assume that intelligent life could emerge on other planets. These findings would seem to augur well for the SETI (Search for Extraterrestrial Life) enterprise, which uses scientific methods to scan the galaxy for electromagnetic transmissions from extraterrestrial civilizations.

Yet despite over fifty years of effort, SETI has yielded no evidence of radio transmissions emanating from alien civilizations. Some observers infer that this "eerie silence" confirms the Fermi paradox—that is, if extraterrestrial aliens do indeed exist, they should have visited Earth by now. This enigma has bewildered those who advance the position that alien civilizations exist in the universe. Proponents of the so-called Rare Earth hypothesis posit that the existence of Earth-like planets is extremely uncommon, which would explain why no signals have been detected.

Perhaps extraterrestrial aliens have already visited Earth. Although the prospect of alien visitation seems outlandish to many people, a sizable number of seemingly credible witnesses have come forward over the years to make the case that there is something real behind UFO (unidentified flying object) sightings rather than hoaxes and misidentifications that can be attributed to earthly sources. In fact, from the time that the modern era of UFO sightings commenced in 1947, the U.S. government took a keen interest in the topic.

Project Blue Book—as it came to be known—was a U.S. Air Force investigative program that lasted until 1970. Although the official program is now defunct, the U.S. government has on occasion demonstrated an interest in UFOs. Something as earth-shaking as the prospect of extraterrestrial visitations, so the argument goes, is too sensitive for the government to release to the public. Such speculation notwithstanding, smoking gun evidence for the extraterrestrial hypothesis remains elusive.

For a variety of reasons, some people do not look to the prospect of extraterrestrial aliens with equanimity. If representatives of an extraterrestrial civilization ever visited Earth, they would certainly be far more technologically advanced than we are insofar as they would have mastered interstellar travel. In that vein, in 2010, the distinguished British astrophysicist, Stephen Hawking, opined that making contact with aliens was "a little too risky." Drawing upon human history, he noted that when a more technologically advanced civilization encountered a less advanced one, the results have often been catastrophic for the weaker party. In contrast, optimists argue that any civilization so technologically advanced that it is capable of interstellar travel would not find anything on Earth worth plundering. For if extraterrestrial aliens made the long journey to Earth they would be motivated primarily by curiosity and altruism rather than aggression.[8]

Considering the potential perils, we must consider the consequences of both direct and indirect contact with extraterrestrial aliens. Although initially exciting, detecting an alien message could create a number of quandaries. For instance, governments might fear that their authority could be challenged if it is confirmed that another species could be in a preeminent position vis-à-vis humans. Once a message is received, we will have to decide if we should remain silent or respond.

People have long wondered if we are alone in the cosmos.[9] Chapter 1 looks at early speculations on the prospect of extraterrestrial life. For centuries, our ancestors pondered if they were alone in the universe. On the one hand, the multitude of celestial bodies suggested that life could be common throughout the cosmos. On the other hand, humans, as the only intelligent species on Earth, understandably felt a sense of uniqueness. In myths and religions, anthropomorphic gods intervened in human affairs. Astronomy seemed to confirm a special place for the Earth, which was conceptualized as the center of the universe. But the Copernican Revolution removed the centrality of the Earth in the cosmos, and in doing so, ushered in new interest in the prospect of life beyond our planet. During the eighteenth and nineteenth centuries, astronomical discoveries of numerous planets and moons in our solar system led a great number of secular Western intellectuals to believe that life existed on other planets. Over time, even Christian scholars began to incorporate the possibility of extraterrestrial life into their doctrines.

In the twentieth century, major achievements in science added new vigor to the extraterrestrial life debate. With exciting discoveries in astronomy, chemistry, and physics, scientists were able to observe in much greater detail the characteristics of the planets in our solar system. Over time, however, the prospects for life—or at least intelligence—existing outside of Earth in our solar system grew dim, as probes on Mars found no evidence of life and Venus was determined to be too inhospitably hot to support life. But in recent years, images taken from the Hubble telescope have bolstered the plurality of worlds doctrine, and by extension, the prospects for the existence of extraterrestrial life. Numerous so-called exoplanets have been confirmed to exist in other star systems. To date, most exoplanets detected have been "hot Jupiters," that is, huge bodies that have short-period orbits very close to their stars. In recent months, some Earth-like planets have even been identified.[10] As more and more exoplanets with terrestrial features are discovered, astrobiologists are more sanguine about the prospect of detecting life outside of Earth. Concomitant with the discovery of new celestial bodies, however, astronomers have come to understand the perilous nature of the cosmos. For instance, events such as supernovas have the potential to wipe out life over broad swaths of space. These topics are explored in Chapter 2.

As astronomers discovered that the observable universe contained far more celestial bodies than originally conceived, more and more people became open to the possibility of life existing beyond Earth. A new and exciting field of science—astrobiology—endeavors to explain the nature of life outside of the Earth and is the focus of Chapter 3.

What makes the possibility of extraterrestrial life exciting is the prospect of alien intelligence. As exoplanets are discovered at the rate of about one a week, scientists now assume that these worlds are plentiful in the universe. Recent astronomical evidence suggests that small worlds—approximately the size of Earth—might be common in our galaxy. Given the sheer number of star systems in the Milky Way, some scientists believe that there could be extraterrestrial life on other planets that preceded life on Earth by millions, if not billions, of years. In fact, one study suggested that roughly 30 percent of the stars harboring life in our galaxy are, on average, 1 billion years older than our Sun.[11] Furthermore, Charles Linewater of the Australian National University, calculated that three-quarters of all Earth-like planets in the galaxy are older than ours.[12] Proponents of convergent evolution argue that similar environments will produce similar adaptations, even in unrelated organisms, in which case it is not unreasonable to assume that intelligent life could emerge on other planets. As a consequence, there could be numerous advanced alien civilizations, whose structures are speculated on in Chapter 4.

If these civilizations exist, we might be able to establish contact with some of them. There is a strong possibility that their level of technological sophistication would be far superior to ours. When using the example of Earth, the evolutionary period from the emergence of simple hunting and gathering bands to complex societies is relatively short on both geological and astronomical time scales. If intelligent beings emerge elsewhere in the cosmos, it is likely that they have contemplated the prospect of life existing beyond their respective planets and would be interested in communication with alien civilizations. With that assumption, the SETI enterprise has used various scientific methods to scan the galaxy for electromagnetic/radio transmissions from alien civilizations and is the focus of Chapter 5.

If the SETI enterprise is someday successful, it could have serious implications for our institutions and our conception of our place in the cosmos. The detection of messages from an extraterrestrial civilization would be the culmination of the Copernican principle, which implies that there is nothing particularly special about our solar system, the Earth, and man, for we would know for certain that intelligence is not unique to our planet. This finding could prove to be a conundrum for some religious traditions—especially Christianity—insofar as they assume a special relationship between *Homo sapiens* and a divine creator. At first, detecting an alien message could be exciting, but over time, it could lower our collective self-esteem if the sender of the message exhibits technological superiority. Finally, after receiving the message, we will have to decide what to do next. Should we attempt to respond, or would it be more prudent to remain silent? Chapter 6 looks at some of the potential consequences of indirect contact.

Although rejected by the vast majority of the SETI community, some researchers insist that extraterrestrial aliens have already visited our planet. Chapter 7 explores the ancient alien hypothesis. Throughout history, witnesses have recorded numerous anomalies in the sky. In concordance with the prevailing belief systems at the time, these UFOs were often imputed to the gods of their various belief systems. Although the ancient alien hypothesis has gained currency in popular culture, so far, it has not found acceptance in the scientific community.

The field of UFO studies, or Ufology, has yet to attain respectability in the research community. While SETI is practiced as a professional science, Ufology is often derided as "the province of amateurs, or of professional scientists who are operating outside of their fields."[13] Nevertheless, over the years numerous sightings of anomalous aerial objects have been reported from seemingly credible people including high-ranking military officers, civilian pilots, and even public officials. Chapter 8 looks at the contemporary UFO era, which commenced not long after World War II.

With the advent of the Cold War, the U.S. government understandably took reports of UFO sightings seriously for fear that they might have been Soviet in origin. Chapter 9 examines the U.S. government's response to UFOs. On December 30, 1947, the Pentagon gave an order to commence a study of UFOs. The next year, the "Sign" project was established within the Air Material Command with operations located at Wright Field in Ohio. Sign was later renamed "Grudge," and in 1951, "Project Blue Book," which became a repository for UFO cases and a place for people to call and file reports of sightings. Other agencies took an interest in UFOs as well. Project Blue Book was officially terminated in early 1970, thus ending all official government investigations into the topic of UFOs. As a consequence, the UFO question shifted to the margins, although the U.S. government still demonstrates an occasional interest in the topic.

There is a dearth of evidence to make a solid case for the extraterrestrial hypothesis. With the multitude of UFOs sightings that have been reported over the years, however, many people are open to the idea. More controversial, is a small subset of UFO sightings in which some individuals claim to have had intimate contact with beings that are presumably extraterrestrial in origin. The veracity of their claims is contested. While some testimonies appear to be highly suspicious, others have a quality of verisimilitude. Amazingly, some researchers have even conjectured that aliens could originate from alternative dimensions rather than distant planets. In a strange intersection of science and mysticism, there is speculation that the powerful hallucinogenic drug DMT (dimethyltryptamine) could be used as a possible gateway to make contact with alien entities. At first blush, this suggestion may sound preposterous, but Rick Strassman, a physician who conducted a government-sanctioned study of DMT, found that many of his subjects reported encounters with bizarre figures, which they perceived as residing in different dimensions.[14] Chapter 10 discusses the various theories surrounding the alien abduction phenomenon.

Despite the fact that millions of Americans when surveyed express a belief in the extraterrestrial hypothesis to account for UFOs, mainstream researchers are far more skeptical. What accounts for this disparity of opinions? Some researchers maintain that the U.S. government has extensive knowledge on the true nature of UFOs, but withholds this information. The presence of alien visitors would be so momentous that the government would be loath to release such sensitive information to the general public, for if the government could not protect its skies from alien intruders, it might call into question its power and competence. For this reason, the government would have a strong reason to keep silent on the issue. Practically from the start of the contemporary UFO era, insiders have accused the government of an official cover-up. Chapter 11 looks at various conspiracy theories concerning UFOs and government secrecy, most notably the incident at Roswell in 1947.

Someday, aliens could land on Earth, if they have not already done so. Even if extraterrestrial life exists elsewhere in the galaxy, many people believe that they would have no way of ever making direct physical contact with Earth. After all, the laws of physics—at least as we currently conceive them—would seem to rule out the prospect of interstellar travel. But perhaps for a civilization that is millions of years more technologically advanced than ours, this could well be within the realm of possibility. On the one hand, optimists, including Carl Sagan and Frank Drake, once imagined that we might gain access to an "Encyclopedia Galactica," which would answer all of the important questions in science, engineering, and social sciences. Pessimists, on the other hand, point out that any extraterrestrial civilization with which we might come into contact would almost certainly be far more advanced than we are and could pose an existential threat to us. Moreover, the fact that intelligence is more likely to arise in predator species is disquieting as well.[15] Considering the potential perils, it behooves us to consider the consequences of direct contact with extraterrestrial aliens, which is covered in Chapter 12.

According to some estimates, roughly 400 billion stars reside in the Milky Way alone. What is more, over the past two decades, astronomers have confirmed that exoplanets are common in the galaxy. Based on these facts, it is not unreasonable for SETI scientists to assume that there could be many extraterrestrial civilizations in the galaxy far more advanced than ours. But alas, we have no concrete proof yet for this assumption. And despite the myriad of UFO sightings and reports of alien abductions, conclusive evidence remains unestablished. It may be necessary for humans to spread out in the cosmos to initiate first contact with alien life forms. At first, the quest to explore might be taken out of curiosity or a cosmic sense of wanderlust. But in the long term, space exploration will be necessary for the survival of our species insofar as the Sun has a finite lifespan. Assumedly, practical interstellar space travel could only be viable for very advanced civilizations that have learned how to harness enormous amounts of energy. At our current level of technological development, interstellar travel is beyond our reach. But as our technology advances, perhaps humans will someday make the ultimate journey and explore planets outside of the solar system. Chapter 13 discusses potential methods for exploring the cosmos and the physical and psychological rigors for those who would brave the journey.

The prospect of contact with extraterrestrial aliens is exciting and cited by some people as potentially the most important event in human history if it should ever occur, for it would prove that we are not alone in the cosmos. By inference, it will demonstrate that another civilization was able to work through the myriad of problems associated with advanced technology (e.g., the threat of nuclear annihilation, overpopulation, environmental degradation, and global warming) and survive. But is anyone really out there? Some skeptics ask, with all the multitudes of stars and planets, why does the cosmos remain silent? Are we alone after all? Judging by the experience of Earth, the development of intelligent life is a race against time. Fossil evidence indicates that life on Earth emerged relatively quickly after only about half a billion years after the planet's creation. However, it would take 3 billion more years for multi-cellular life to form. The emergence of *Homo sapiens* and civilization would come even later. Although humans have come a long way from their single-celled ancestors, their time on Earth is finite. By a conservative estimate, our planet will be uninhabitable in a billion years. But this will not automatically mean the end of humankind. Conceivably, before this time, humans could attain immortality through technology, but any major calamity—whether natural or man-made—could present a setback from which our species may not have adequate time to recover. If we can navigate past these perils, however, as a species, we could survive long enough to discover life beyond our planet. The conclusion discusses these hurdles we must surmount to reach the stars. Contact with extraterrestrial intelligence would have a transformative effect on us all.

It is hoped that this book will help bridge the gap between SETI and Ufology. At the present time, there is little to no dialogue between the two fields. As Michael Swords and Robert Powell noted, a strange paradox in the study of extraterrestrial life is that it is deemed scientifically appropriate to search for radio transmissions emanating from advanced alien civilizations as long as one does not consider the possibility that these aliens could actually master interstellar travel and have already visited Earth.[16] Generally, SETI scientists deride Ufology as a pseudoscience populated with frauds. By contrast, some UFO researchers derogate SETI as the "Silly Effort To Investigate," whose practitioners dogmatically refuse to consider the extraterrestrial hypothesis for UFOs. Why bother to search for faint radio signals, they reason, if aliens are already here?[17] For their part, scientists seek to marginalize the field of Ufology, lest it discredit the SETI enterprise by association.[18] This sentiment is understandable because Ufology certainly includes some charlatans in its ranks. Nevertheless, there is serious research that takes place in Ufology, and as such, should be included as part of a comprehensive search for alien intelligence. UFO researchers could certainly benefit from the knowledge of astronomy, astrobiology, and speculations on future technology insofar as these disciplines could provide insight on the nature of extraterrestrial beings that could be visiting Earth. By drawing upon both the SETI and UFO fields we can gain a better understanding of the possibilities of extraterrestrial life.

Chapter 1

Early Speculations

The extraterrestrial life debate has a long pedigree. For centuries, our ancestors wondered if we were alone in the universe. To some early observers, the multitude of celestial bodies implied that it was almost a certainty that life existed beyond Earth. But as the only intelligent species on the planet, our forbearers understandably felt a sense of choseness by a creator. This sentiment was reflected not only in myths and religions in which anthropomorphic gods intervened in human affairs, but also in science, more to the point—astronomy—which conceptualized the Earth as the center of the universe.

According to one theory, humans first entertained the notion of life beyond Earth after primitive religions failed to give meaning to those aspects of their environment that had no simple explanations.[1] Moreover, as they learned more about the planet, they began to ponder their place in the cosmos. From about 500 B.C., learned persons in China, Greece, and elsewhere began to accept the fact that the Earth was a globe.[2] Some of the early Greek philosophers were among the first to advance the principle that the Earth was unexceptional in the universe. For instance, Thales of Miletus (624–546 B.C.), the founder of the Ionian school of philosophy, taught that the Sun and the Earth were composed of the same materials. If the laws of chemistry were the same throughout the universe, it would follow that life could take hold on other planets. Based on this premise, as far back as the fifth century B.C., Anaxagoras (510–428 B.C.) hypothesized that the Moon was an Earthlike world inhabited with beings. An early proponent of the heliocentric (sun-centered) theory of the solar system, he argued that the "seeds of life" from which all living things were derived, were dispersed throughout the universe.[3]

Building on these theories, Democritus (460–370 B.C.) theorized that all matter consisted of differing combinations of atoms. His belief in innumerable worlds derived from his atomistic doctrine. Taking this assumption to its logical conclusion, he posited that other worlds were formed from the agglomeration of atoms not unlike the Earth. To Democritus, the cosmos was random and purposeless. No higher deity intervened in a universe that was infinite and homogenous throughout.[4] Likewise, Epicurus (341–270 B.C.) spoke of an infinite number of worlds resulting from an infinite number of atoms.[5] Inasmuch as these worlds were composed of the same elements found on Earth, it was not unreasonable to assume that life could inhabit them. Therefore, representatives of the Epicurean school of materialist philosophy taught that many habitable planets similar to Earth existed in the cosmos.

This atomist, or materialist tradition in philosophy, took hold outside Greece and was invoked to support the plurality of worlds doctrine. For instance, in his book *De rerum natura* (On the Nature of Things), the Roman poet Titus Lucretius Carus (99–55 B.C.) supported the idea of Epicurus that other worlds must exist elsewhere, and in doing so, spread the atomist doctrine throughout Europe. Furthermore, he added that nothing about our part of the universe or the Earth that was unique, thus life could flourish on other planets as well as reflected in his famous poem "On the Nature of Things," in which he wrote: "Nature is not unique to the visible world; we must have faith that in other regions of space there exist other earths, inhabited by other peoples and other animals."[6]

Still, most of the Classical Greco-Roman philosophers maintained that the Earth occupied a special place in the universe. Despite these early astronomical models that favored the plurality of worlds, considered opinion accepted the centrality of Earth view as exemplified by Aristotle (383–322 B.C.). Like his mentor Plato (428–347 B.C.), he rejected the plurality of worlds doctrine.[7]

Aristotle posited that a motionless Earth occupied a "natural place," around which the planets, sun, and all other stars revolved.[8] According to the Aristotelians the universe was teleological (purposefully directed) in nature in which a supreme being governed by design. According to this view, the heavens were unchanging and had attained perfection.[9]

The Aristotelian idea was reinforced in the second century A.D., when Ptolemy (Claudius Ptolemaeus, 90–160 A.D.) posited that the Earth was the center of a universe. The Aristotelian view and Ptolemaic model dovetailed with the Christian theology, which viewed Earth and man as distinct in God's scheme. As such, the early Christian church endorsed the concept of a geocentric (Earth-centered) universe.[10] Adding a secular spin to the geocentric model, William of Ockham (of Ockham's razor fame) (1287–1347)[11] altered Aristotle's doctrine of natural place by positing that the elements of each individual world would return to its natural place within their own worlds without the intervention of God.[12] Democritus' plurality of worlds remained a minority view for centuries, as the geocentric model remained the dominant paradigm.

A remarkable rebirth of learning took hold in Europe in the twelfth and thirteenth centuries, which called into question old dogmas. By the fourteenth century, some critics of Aristotle argued that he was wrong to place limits on God by confining Him to a finite universe. In the fifteenth century, Nicholas Copernicus (1473–1543) boldly asserted that the Sun, not the Earth, was the center of the universe. Copernicus did not invent the heliocentric conception of the universe, for he knew that Aristarchus of Samos (310–230 B.C.) had proposed the idea long before.[13] Nevertheless, Copernicus resurrected the idea and gave it new force backed by new findings in astronomy. Furthermore, he still retained a special status for Earth insofar as its parent star was relegated as the center of the universe. By doing so, he still maintained the exceptionality of our solar system. Despite the limitedness of his assertion, it set in motion a shift in attitudes that would have far-reaching consequences in science and philosophy. The Copernican revolution in astronomy erased the distinction between the celestial and terrestrial realms that was preserved by the pagan philosophers.[14] More than any other concept, it was the heliocentric system that gave rise to the plurality of worlds theory.[15]

By downgrading the status of Earth in the cosmos, Copernicus unwittingly had launched a revolution that would reverberate beyond the field of astronomy, for it would call into question the very intellectual foundations of Europe. According to Michael J. Crowe, it was primarily Copernicus who was responsible for the entry of extraterrestrials in our discourse. Although Copernicus did not explicitly invite aliens into his heliocentric system—"in fact, he did nothing more than open the door an inch"—it was still enough to initiate the invasion.[16] He transformed the Earth into just another planet, which in turn suggested that other planets may be earth-like in characteristics, perhaps even hosting life.[17] The infinitization of the universe emerged out of the fundamental changes Copernicus made in the arrangement of the solar system.[18] Despite this, the plurality of worlds doctrine remained a minoritarian view, which would not gain widespread currency until the seventeenth century.

Between 1500 and 1750 the idea of extraterrestrial life went from being a radical idea accepted only by a few scientists and philosophers, to a doctrine that was taught in college classrooms, championed by preachers, and celebrated by poets. New and exciting developments in astronomy made possible by the invention of the telescope in the seventeenth century were the key factors that brought about this new perspective. Concomitantly, the evolution of religious and philosophical sentiments during this period added new vigor to the extraterrestrial life debate.[19] In fact, what initially troubled the church authorities was not so much the heliocentrism of the Copernican model, but rather the prospect that alien beings inhabited worlds other than Earth.[20] Based on this assumption, Giordano Bruno (1548–1600) postulated that the universe was full of planets that contained intelligent life, a position that eventually cost him his life.[21] In 1600, the Roman Catholic church ordered him burned at the stake for his heretical views.[22]

Over the years, influenced by their conception of an infinite God, philosophers and theologians in the Middle Ages gradually accepted a universe that was infinite in scope.[23] In fact, as far back as the thirteenth century, Thomas Aquinas (1225–1274) reasoned that God was quite capable of creating numerous worlds.[24] But it was the French theologian William Vorilong (ca. 1390–1463) who was widely recognized as the first author to raise the question of whether the notion of the plurality of worlds could be reconciled with the Christian doctrines of divine incarnation and redemption.[25]

The Age of Enlightenment ushered in a sea change in how men viewed the cosmos, and by the end of the seventeenth century, the tide of informed opinion shifted toward the plurality of worlds doctrine.[26] Gradually, an increasing number of people began to think of the universe as containing many planets, many of which were thought to be inhabited.[27] As George Basalla noted, scientific theories of advanced life in the cosmos emerged from a trio of ideas that first appeared in the religious and philosophical thought of antiquity, and later, the Middle Ages. The first idea was the universe was very large, if not infinite in extent. The second was that humans were not alone among intelligent beings in the universe. The third was that there was an essential difference between the superior beings of the celestial world and the inferior ones who resided on Earth.[28] As the Renaissance unfolded, the plurality of worlds doctrine became increasingly rooted in science, rather than theology.

The leading scientists and philosophers of the period voiced support for the plurality of worlds doctrine. For instance, Johannes Kepler (1571–1630), who is best known for discovering the laws of planetary motion,[29] also postulated that our nearest worlds were populated with humanoid beings. Although Kepler frequently invoked God in his astronomical work, he maintained that Earth and its inhabitants were not unique in the universe.[30] In his book *Somnium* (Dream), he went so far as to claim that three groups of intelligent beings resided on the Moon. On the near side of the Moon lived the Subvolvans. The Privolans occupied the far or hidden side of the Moon. And in between them, in the high mountainous region, stayed an unnamed savage race of thieves who raided the civilized settlements of the Subvolvans. Although no cities were visible on the Moon, he argued that they were buried beneath craters.[31] Some literary critics cite *Somnium* as the earliest known work of science fiction.[32] A true visionary, Kepler believed that someday space travel would enable humans to settle other worlds in the cosmos.[33]

Sir William Herschel (1738–1822), the pioneer of stellar astronomy and discoverer of the planet Uranus, also claimed that life inhabited the Moon. In 1776, he claimed to have observed large areas of plant life on the Moon. Furthermore, he conjectured that many of the visible circular formations on the surface were towns erected by the "Lunarians." His imagination running wild, he later professed to have found canals, pyramids, turnpike roads, and more vegetation.[34] Astonishingly, he also believed that the Sun was inhabited with life, a position that even Isaac Newton shared.[35]

Although Isaac Newton (1642–1727) insisted that the will of God was necessary for the formation of all ordered systems, the laws that he uncovered made it increasingly clear that there was less need for a divine agency in the cosmos as told in the Bible.[36] Newton's model of the universe was static, but infinite. Although he wrote extensively about the motions of celestial bodies, Newton was silent about the possibility of other planets outside of our solar system.[37] Richard Bentley (1662–1742) was the first person to address the topic of extraterrestrial life within the framework of the system that Newton had created in his *Principia*. He noted that favorable location of the Earth in the solar system—which today we call the habitable zone—as evidence of God's beneficent design.[38]

More and more, philosophers began to question the choseness of man. In *Principa Philosophiae* (Philosophical Principles, 1644), René Descartes revived atomism with a mechanistic model in which atoms in motion once again formed the basis of cosmology.[39] Descartes rejected the void of the ancient atomists and argued that moving particles, or atoms, filled every portion of the universe. In his model, the Sun is a star and every star is at the center of a whirlpool of celestial matter. Prefiguring the nebular hypothesis, which will be discussed in the next chapter, he reasoned that the universe contains as many planetary systems as there are stars.[40]

The Copernican model gave rise to the principle of mediocrity, that is, there is nothing particularly special about the Earth or our area of the cosmos. If the Earth was not at the center of the universe, then perhaps neither was it unique as a locus of intelligent life.[41] Applying the principle of mediocrity and anticipating the theory of convergent evolution, Christian Huygens (1629–1695) argued that those planets that share a similar orbit to that of the Earth's around the Sun, must also share similar life forms as Earth's. He speculated that intelligent extraterrestrial beings would probably not be exact replicas of humans. Nevertheless, he suggested that they would be upright in posture and have hands that would enable them to operate instruments.[42] He took the principle of mediocrity to new heights, modeling his extraterrestrial civilizations on countries and cultures in Europe.[43]

Despite efforts to invoke God as the designer of the cosmos, an inexorable secular shift in attitudes continued apace, as was reflected in Bernard le Bovier de Fontentelle's (1657–1757) book *Conversations on the Plurality of Worlds* (1686). Early in his book, he dismissed the teleological claim that God created all of nature for the exclusive benefit of mankind. Although he did not explicitly rule out God's role in the cosmos, he tended to replace Him with the laws of nature. He speculated that humans could one day develop the technology to make lunar flight possible. Likewise, he believed that inhabitants on the Moon might develop flight technology and visit the Earth. He once mused that the inhabitants of the Moon might have already arrived, perhaps making him the first proponent of the ancient astronaut hypothesis. Foreshadowing contemporary theories of extraterrestrial intelligence, Fontentell believed that humans would be in an inferior intellectual position vis-à-vis aliens. His contemporary Christian Huygens differed with Fontenelle on the prospect of life on the Moon. Inasmuch as Huygens could not detect any water or surrounding atmosphere on the Moon, he concluded that it could not sustain life. Drawing upon the nascent studies in naturalism, Fontentell surmised that just as on Earth, where the greater the geographic distances between peoples usually implies the greater physiological differences between them, so in our solar system, a planet's proximity to the Sun would account for distinctive features between life forms.[44]

Until the end of the end of the seventeenth century, the Roman Catholic church violently opposed the heliocentric model. But emerging worldviews on the plurality of life in the universe increasingly gained currency in European consciousness. During the eighteenth and nineteenth centuries, astronomical discoveries of numerous planets and moons in our solar system led a great number of secular western intellectuals to believe in life beyond Earth. Still, throughout this period, the debates on the plurality of worlds struggled with the Christian worldview.[45] In *The Age of Reason* (1793), Thomas Paine bluntly stated that "to believe that God created a plurality of worlds at least as numerous as what we call stars, renders the Christian system of faith at once little and ridiculous and scatters it in the mind like feathers in the air. The two beliefs cannot be held together in the same mind." As Paine reasoned, with millions of world under his care, would God send His Messiah to save humanity on one small world? Or did the redeemer hop from one world to the next saving the souls of all sentient beings in the universe, and in doing so, suffer an endless succession of deaths?[46] Rejecting Paine's thesis, William Whitewell, a philosopher, scientist, and Master of Trinity College, Cambridge, argued in *Of a Plurality of Worlds: An Essay* that it was the idea of the plurality of inhabited worlds, not Christianity, which should be rejected.[47]

But over time, even Christian scholars began to incorporate the possibility of extraterrestrial life into their doctrines. In fact, some new Christian sects, including the Mormons, the Seventh-day Adventists, and the Swedenborgians, even embraced the plurality of worlds doctrine.[48] Consistent with the Christian worldview, it seemed plausible that an omnipotent God could create life throughout the universe.[49] Reverend William Whewell (1794–1866) eventually concluded that the belief in extraterrestrial life could be reconciled with the central doctrine of Christianity.[50] Previously, Christian proponents of the plurality of worlds hypothesis had argued that if vast reaches of space were vacant of life, then it could indicate that God's efforts had been wasted in creating them. This claim was based on the assumption that God would never act in such a manner that His efforts would be wasted. Whewell responded to this argument by pointing out that the geological record suggested that for the great majority of Earth's history the planet lacked intelligent life. By analogy, he reasoned that this same principle applied to space, which suggested that God occasionally acts wastefully. Based on telescopic observations of the planets and the Moon, he concluded that life probably did not exist on any of them. Still, he conceded that Mars, since it once might have had conditions comparable to Earth a long time ago, might be populated with dinosaurs and other creatures. Not unlike contemporary astrobiologists, he argued that the propinquity of Venus and Mercury to the Sun made them unlikely abodes for life because their surface temperatures would be too hot. Conversely, Jupiter and Saturn were too far away from the Sun to be candidates to support life.[51]

As Steven J. Dick noted, the rise of the popularity of aliens in literature was closely tied to late-nineteenth-century science, especially evolutionary theory, astronomy, and the plurality of worlds doctrine.[52] Perhaps most important was Charles Darwin's *The Origins of the Species*, which ushered in the conceptual framework that gave impetus to the idea of the physical evolution of the universe.[53] According to Darwin (1809–1882), life emerged on Earth in a kind of primeval soup, containing ingredients that were haphazardly put together. Although he did not mention the topic of alien life, his research presented scientists with a powerful theory for the extraterrestrial life debate.[54] Taking up his theory of evolution, Richard A. Proctor argued in his books *Our Place Among Infinites* and *Science Byways* (both published in 1875) that in due time all planets would develop life, a position that would prefigure contemporary astrobiology.[55]

And so the theory of evolution added momentum to the extraterrestrial thesis. In the late nineteenth century, the French astronomer Camille Flammarion, who would exert a strong influence on the twentieth-century debate, argued that natural selection applied elsewhere in the cosmos. As a consequence, it was folly to assume that humans were the highest life form in the universe.[56] He concluded that no preeminence had been bestowed upon the Earth and that other planets were inhabited as well.[57] Nevertheless, he maintained a teleological approach, arguing that the planets were created purposefully for the formation of life.[58]

If life existed outside of Earth, a natural place to begin looking for it would seem to be Mars. In 1858, Angelo Secchi (1818–1878), a Jesuit monk in Italy, began mapping the red planet and observed some vaguely linear features—*canali*—meaning channels. At the time, the general opinion was that the canals were cracks in the Martian crust formed during the planet's solidification. Some observers, however, argued for their artificiality.[59]

In 1877 Secchi's compatriot, Giovanni Schiaparelli (1835–1910), produced an improved map of Mars. An Italian civil engineer and the director of Milan's Brera Observatory, Schiaparelli created charts displaying the planet Mars marked with straight dark lines, which he also called "canali." For three decades, he entertained the idea that they were part of an advanced civilization's irrigation system that addressed a chronic water shortage. The seasonal melting of the Martian ice caps was central to his interpretation of the landscape of the red planet. The melting of snow and ice, he argued, caused great inundations.

For Schiaparelli, Mars was essentially an aquatic planet, not unlike the Earth. Flooding was the main source of water for all of the seas, oceans, lakes, swamps, and canals on Mars. Rain was very rare. In the English-speaking world, the view is that Schiaparelli was cautious in interpreting the canals of Mars, as he was loath to ascribe any intelligent agent to them. However, in a paper published in 1895 in Italian, he let go of these inhibitions. In it, he suggested that the canals might have been designed by Martian engineers.[60] He went on to speculate about the political structure of Martian civilization, musing that a "collective socialism" developed from the needs of citizens to direct their energies to taming the harsh environment of the planet. To build such intricate and massive canals, Martian society had cultivated mathematics, meteorology, physics, hydrography, and structural technology of a high degree of perfection. International rivalry and war were unheard of in the Martian paradise. Just two months before his death, however, he conceded that natural forces could account for the dark lines that he observed on Mars.[61]

It was in the English-speaking world where the intelligent life on Mars thesis gained the most currency. Confusion in translations contributed to this development. In Italian, "canali" refers either to natural channels or artificial water courses.[62] Over time, the sobriquet became liberally construed in English as "canals" with a hint of artificiality.[63] Englishmen and Americans began referring to canali as canals instead of channels. As a consequence, they began to think of them as built by intelligent beings.[64]

Although most astronomers did not detect the straight lines on Mars, an American astronomer, Percival Lowell (1855–1916), was much intrigued by Schiaparelli's story. The scion of a wealthy Boston family, in 1894 Lowell used his considerable resources to build the Lowell Observatory in Flagstaff, Arizona. All totaled, Lowell determined that 183 canals had been constructed deliberately by Martians. Lowell did not go so far as to claim that he actually saw canals on Mars; rather, he believed that he had found the strips of fertilized land which bordered them.[65] According to his planetary timetable, intelligent beings conquered nature earlier on Mars than on Earth. This assumption suggested that the canal network was the physical manifestation of a much older, superior civilization.[66]

Similar to Schiaparelli, Lowell argued that the harsh Martian environment forced its inhabitants to work together and innovate, or else perish. Their survival depended on the conservation of the planet's scant water supply in an efficient irrigation system. Fear was the greatest catalyst for the construction of the Martian canals.[67] Ominously, Lowell predicted that the Earth was destined to become a desert planet like Mars, an assumption that is actually consistent with contemporary astronomical predictions.[68] Unlike Schiaparelli, Lowell did not believe that Mars had a socialist government. A lifelong opponent of socialism, he interpreted Martian society through the lens of Herbert Spencer's conservative social philosophy. Given the harsh conditions on Mars, only the fittest had survived. Nevertheless, he believed that a benevolent oligarchy consisting of elite technicians governed Mars. These technocrats assigned citizens their social roles and tolerated no dissent. Lowell believed that Mars had lessons for the Arizona socialists of his era of whom he was critical. He argued that society would remain weak and inefficient until all citizens pulled together and healed the bridge between capital and labor. The common man, he explained, lacked the wisdom and foresight to act on his own and was therefore better off under the leadership of his intellectual superiors.[69]

Described as "the Carl Sagan of his time," Lowell was convinced that the canals were real, but by the time of his death in 1916, few astronomers shared his view, its popularity with the general public notwithstanding.[70] Eventually, astronomers came forward to debunk his theory. E.W. Maunder was the first astronomer to assert that Lowell's theory was illusory.[71] However, it was the observations of the Greek-born astronomer Eugene M. Antoniadi (1870–1944) that contributed most to solving the riddle of the canals on Mars. Using a 33-inch refractor telescope, he determined that the canals were an optical illusion.[72] Although his theory of artificial canals on Mars has been discredited, Lowell left a lasting legacy in the sense that he gave voice to the idea that Mars could have once been a planet on which water flowed and life flourished.[73]

In retrospect, many of the early commentators on extraterrestrial life appear naïve in their belief that life was common in the universe. Some of these observers, however, anticipated the astronomical discoveries of the twentieth century, which cast doubt on the prospects of abundant extraterrestrial life in the universe. For example, in 1853, the English philosopher William Whewell (1794–1866) wondered if perhaps not every planet was a suitable habitat for life. He correctly anticipated astronomic findings of the twentieth century, conjecturing that the larger planets of the solar system could be composed of "water, gases, and vapor" that would make them unsuitable for life.[74] Likewise, Alfred Russell Wallace (1823–1913) shocked his contemporaries by arguing that probability suggested that humanity's place in the universe was special and unique after all. He reasoned that the numerous conditions necessary to give rise to intelligent life would be unlikely to be replicated elsewhere in the cosmos.[75] With new powerful telescopes, astronomers were able to observe in much greater detail the characteristics of the planets in our solar system. Over time, the prospects of intelligent life existing outside of Earth in our solar system grew dim. But concomitantly, new advances in astronomy would bolster the plurality of worlds doctrine and, by extension, the prospects for the existence of extraterrestrial life beyond our solar system.

CHAPTER 2

Astronomy and the Plurality of Worlds Doctrine

ASTRONOMY DATES BACK TO THE earliest human observations of the skies that were made with the naked eye. Prior to electrification, there was little light pollution, so people could see vivid images of celestial bodies. Gazing into the night sky, our ancestors were curious about their place in the cosmos. Filled with awe, they found their pantheon of gods in the stars. On a more practical level, the positions of the stars guided their seafaring and agricultural planning. An important milestone was reached in 1608, when a German spectacle maker, Hans Lipperhey, applied for a patent for a device that magnified objects up to three-times their actual size. A year later, Galileo Galilei became the first person to use such an instrument for the purpose of astronomy. The invention of the telescope led to new discoveries that informed the extraterrestrial life debate, for it revealed that the size of the universe was larger than what earlier observers assumed it to be. By the middle of the eighteenth century, interest in the stars and dim nebulous patches scattered throughout the skies gradually increased. In *An Original Theory or New Hypothesis of the Universe* (1750), Thomas Wright presented the first exegesis on the "island universe" theory by suggesting that some of these nebulous patches were actually collections of stars in what came to be called galaxies.[1] Finally, while studying the spectrum of some spiral nebulae in 1909, Vesto M. Slipher discovered that these patches were not nebulous stars or solar systems in the early stages of development; rather, they were gigantic star-filled galaxies located at enormous distances from the Milky Way.[2] As a result of his discovery, our conception of the universe was tremendously expanded.

For centuries, astronomers assumed that the universe was in a steady state. Isaac Newton's theory of gravity, however, raised questions about this assumption. In 1692, using Newton's theory, Reverend Richard Bentley determined that if gravity were truly attractive, then the stars in the universe should come together. As a result, the universe would collapse due to gravity and ultimately form a giant fireball.[3] To account for the seeming stability, by fiat, Albert Einstein designated the "cosmological constant" as a hypothetical factor consisting of repulsive anti-gravity to balance gravity in the universe.[4] Eventually, the steady state universe gave way to the theory of an expanding universe. In 1922, a Russian cosmologist, Alexander Friedmann hypothesized that the universe was expanding contrary to Einstein's static universe model. Through his astronomical observations, Edwin Hubble discovered in 1929 that the universe was in fact expanding, which lent more credence to Friedmann's theory. Extrapolating backwards, in 1931, Georges Lemaître theorized that all matter in the universe must have once been condensed into a single point—"a primeval atom"—that eventually erupted in a violent explosion in which the fabric of both time and space came into existence in an event that came to be known as the Big Bang. Still, some astronomers, such as Thomas Gold, maintained that the universe remained in a "steady state" and that matter emerged spontaneously, thus keeping the density of the cosmos constant as it expanded.[5] But in 1960, two astronomers working at Bell Labs—Arno Penzias and Robert Wilson—discovered the background radiation believed to have been a remnant of a cataclysmic event that created the universe.[6] The Big Bang is now the recognized theory explaining the origins of the cosmos.[7]

Each astronomical new discovery seemed to confirm the Copernican principle, that is, the Earth was unexceptional and occupied no special place. By inference, this gave credence to the extraterrestrial life hypothesis. After all, if life could emerge on Earth, why could it not arise on other planets as well? Although the existence of planets outside of our solar system had yet to be confirmed, it seemed reasonable to assume that they must be out there.

As the discipline of astronomy progressed, the concept of life beyond Earth developed a more scientific foundation. With sophisticated astronomical instruments, scientists learned more about stellar processes. For example, in 1814 Joseph von Fraunhofer invented the spectroscope, and in doing so, discovered that spectrum analysis could be used to determine the chemistry of astronomical objects.[8] When light passes through a substance, some of it is absorbed. This spectrum is discrete, rather than continuous, which means that light is strongly absorbed only at certain sharply defined wavelengths. Which wavelengths are absorbed depends on the type of material doing the absorbing. Each element or compound has an absorption spectrum determined by its atomic or molecular structure, and possesses a distinct fingerprint that can be detected through spectroscopy. Through this method, astronomers were able to determine the chemical composition of stars and planetary atmospheres.[9]

By tracing the earliest stars, astronomers gained insight into the origins of the elements necessary for life. So-called first-generation stars were born shortly after the Big Bang. Most of these stars resided in the compact central regions of galaxies and were composed almost entirely of hydrogen with a little bit of helium and virtually nothing else. When a large star explodes—in an event called a supernova—it produces enough heat and pressure to cause the light elements to fuse into the heavier elements, such as carbon, oxygen, sulfur, and iron.[10] Inasmuch as these heavy elements—the building blocks of life—were not present in appreciable amounts in first-generation stars, we can assume their planets were lifeless.[11] But over time, these heavy metals spread farther and farther out from the center of the galaxy, thus expanding the Galactic Habitable Zone—that is, the region in which it might be possible for life to take hold in the galaxy.[12] Because life is composed in large part of the heavy elements created in supernovae, in a sense we are stardust.[13]

The conventional wisdom is that only second- or third-generation stars, like our Sun, could be planets on which intelligent life evolved because only these planets have the necessary heavier elements, or metals, in sufficient abundance.[14] Spectroscopy provides the clearest evidence ever obtained for the chemical homogeneity of the universe. This implies that the laws governing terrestrial chemical reactions also apply throughout the cosmos.[15] Astronomical observations, space probes, and laboratory analyses of carbonaceous meteorites suggest that the basic building blocks of life are widespread in the Milky Way and beyond.[16] To date, organic chemicals have been detected in at least 50 galaxies.[17] Earth-based life is chemically-derived from a few select ingredients. Four elements—hydrogen, oxygen, carbon, and nitrogen—account for more than 95 percent of the atoms in the human body and all other known life on Earth. We are made up of the most common ingredients in the cosmos.[18] There is nothing to suggest that life is chemically special. Consequently, it would seem to follow that life on other planets would be composed of a similar mix of elements.[19]

Presumably, planets are the most fertile ground for the emergence of life, although some scientists speculate that other celestial bodies, perhaps comets, could host organisms as well.[20] So how did planets form? According to the nebular hypothesis, first proposed by Emanuel Swedenborg (1688–1772) and later refined by Immanuel Kant (1724–1804) and Pierre-Simon Laplace (1749–1827), planets were formed out of the rotating disks of gas and dust surrounding stars. After the clouds settle into thin sheets, particles draw together to form planetesimals (proto-planets), which in turn attract additional matter within their gravitational fields to form larger bodies, eventually becoming planets.[21] In this model, the Sun and the planets were formed at roughly the same time. As a result of the centrifugal forces generated by rotation, matter was drawn to the periphery of the contracting nebula, thus forming an equatorial belt of gas and dust. Over time, the rings were believed to have condensed to form planets.[22]

An alternate theory of planetary formation—the Chamberlin-Moulton hypothesis—proposed in 1905 by the geologist Thomas Chrowder Chamberlin (1843–1928) and the astronomer Forest Ray Moulton (1872–1952), theorized that solar systems were formed out of the collisions or close encounters of stars in space.[23] Similarly, Sir James Jeans (1877–1946) postulated that the original material from which the planets were derived was ejected by the Sun after it had already reached its present form after a near-collision from a passing star. The gravitational tidal forces of the interloper star sucked out a filament of material from the surface layers of the Sun. After a period of time, some of this material condensed and formed planets.[24]

In the middle of the nineteenth century, James Clerk Maxwell (1831–1879), a Scottish physicist who is best known for his classic theory of electromagnetism, discovered a fundamental flaw with the nebular hypothesis. Invoking the principle of angular momentum (the tendency of a rotating body to resist braking), he noted that the majority of this property is concentrated in the orbital motions of the giant planets—Jupiter and Saturn. In his mind, the Kant-Laplace hypothesis lacked any mechanism to explain how this angular momentum could be transferred from the proto-sun to the orbiting ring. Many years later, however, the Swedish physicist Hannes Alfvén (1908–1995) demonstrated that the Sun was indeed able to transfer its angular momentum to the planets through a magnetic field mechanism. His theory was later refined by Fred Hoyle (1915–2001). As a consequence, the nebular theory was rescued.[25] As this was believed to be a universal process, it followed that there would be a multitude of planetary systems in the universe, which in turn increased the probability of extraterrestrial life.

For centuries, our view of the cosmos from Earth was obscured by the presence of the atmosphere. But in 1990 the Hubble Space Telescope, which was launched aboard the Space Shuttle *Discovery*, began producing startling images of the cosmos from space. Primary among them are pictures of exoplanets, or planets that reside outside of our solar system. The first exoplanet was discovered by the radio astronomers Aleksander Wolszczan and Dale Frail in 1992.[26] Currently, exoplanets are discovered primarily through indirect methods, such as measuring very slight "wobbles" in stars due to the mutual gravitational pull with the celestial bodies that orbit them. As far back as 1844, Friedrich Wilhelm Bessel (1784–1846) had measured variations in the motion of the stars Sirius and Procyon, and concluded that they were caused by unseen celestial companions. It later transpired, though, that these companions were stars and not planets. Despite this, Bessel introduced the method of detecting "wobbles" in astronomy that would prove vital in the subsequent search for exoplanets.[27]

In recent years, another search method has been employed. Because of their size and brilliance, stars are relatively easy to find. Planets are much smaller and are obscured by stars that outshine them by as much as a million times.[28] Nevertheless, an alternative exoplanet search strategy—the transit method—measures the very slight diminution of brightness of the star when a planet passes between the star and our telescopes. When the star is approaching the planet, its light will be wavelength-shifted toward the blue. Conversely, when the star is moving away from the planet, the light will be shifted toward the red. The tiny difference between the two cases enables astronomers to infer the presence of a planet in orbit around a star.[29]

As of June 2014, 1,732 exoplanets have been confirmed.[30] So far, most have been "hot Jupiters"—that is, huge planets that have short period orbits very close to their stars.[31] In recent months, however, some Earth-like planets have been detected. New discoveries suggest that small worlds—approximately the size of Earth—might be common in our galaxy. In fact, less massive planets are now reliably inferred to outnumber the giant planets that dominated the discovery of exoplanets in the past.[32] To date, most of the exoplanets that have been discovered have highly elliptical orbits, which presumably would not make them good candidates for complex life—half of the planet would be scorched for part of the year, while the other half would freeze. However, fifty-four exoplanets have been found in the Goldilocks, or habitable, zone in their respective solar systems, which holds out promise that they could host life.[33] Even more exciting, in November of 2013, an academic journal—*Proceedings of the National Academy of Sciences*—published a study which concluded that as many as 40 billion Earth-like planets resided in the Milky Way.[34] Extrapolating from the data received over a three-year period from the Kepler telescope, Erik Petigura estimated that roughly one out of every five sunlike stars in the galaxy has a planet the size of Earth orbiting it in the habitable zone where surface temperatures should be compatible with liquid water. To maintain life, presumably, these planets would have to be rocky rather than gaseous in mass. Nevertheless, there is strong reason to believe that our solar system is not the sole province of rocky planets.[35]

The sheer multitude of stars in the Milky Way (perhaps up to 400 billion) and the vast number of galaxies in the universe (perhaps up to 1 trillion), lead some scientists to conclude that the existence of alien life is all but a certainty. Furthermore, over the past two decades, scores of extra-solar planets have been discovered—about one a week—which suggests that they are common in the universe. Nevertheless, there are many variables that militate against the emergence of advanced life in the cosmos. In fact, some scientists go so far to argue that Earth is the only planet on which intelligent life exists.[36]

The center of the Milky Way would probably not be suitable for the development of advanced extraterrestrial civilizations. The clustering of stars in the region would make for a perilous neighborhood. For instance, the radiation produced by a supernova occurring within a mere 30 light years of a planet would probably destroy all of its surface life.[37] The energy emitted from a supernova explosion is so great that its brightness can surpass the entire galaxy that contains it.[38] According to calculations, when the Milky Way was less than 1 billion years old, the frequency of supernovae explosions was approximately 100 times greater than today, which probably would have snuffed out any nascent life.[39] Perhaps even worse, the x-rays produced by the massive black hole at the galactic center would fry the complex molecules associated with life in that region of the galaxy.[40] Thus, a substantial area of the galaxy may not be conducive to life.

The Galactic Habitable Zone is a region around a thin disk of stars where planetary systems and planets like the Earth are likely to be found.[41] Despite the vastness of our galaxy, according to Charles Lineweaver and his colleagues, the Galactic Habitable Zone contains no more than 10 percent of all the stars ever formed in the Milky Way.[42] Fortunately for us, the Sun is located in the suburbs, far from the galactic center. To some observers, this reinforces the Fermi Paradox, that is, if intelligent extraterrestrial beings exist, they should have visited Earth by now. If advanced life could only be expected to take hold in a limited region, it would make it that much easier for an advanced civilization to explore on the habitable zone portion, rather than the entire galaxy, looking for other life-bearing planets.[43]

Most of the remaining stars residing outside of the center of the galaxy would not be good candidates for supporting life. Stars that are much bigger than our Sun are unlikely to provide the conditions suitable for habitable planets.[44] The largest stars have a mass about 100 times greater that of the Sun.[45] For evolution to run its course, planets would have to orbit stars that are not too large. The main sequence (the period in which a star radiates a steady amount of heat) of large stars would probably not be long enough to allow for the development and evolution of advanced life on their orbiting planets. The large stars burn their nuclear fuel to quickly and explode before life can take hold and evolve.

The small stars—those about 100 times less massive than our Sun—are more plentiful than large stars, but are also probably not good hosts of life either.[46] For instance, red dwarfs, because they radiate less heat, would have very narrow bands of habitable zones that would be relatively close to the star. As a consequence, this would increase the probability that most of their planets in their solar systems would be excluded as candidates suitable for life. Furthermore, red dwarf stars have frequent flares of activity and would release large amounts of dangerous ultraviolet radiation, x-rays, and other harmful particles that would be particularly damaging because of the propinquity of the star to the planet that happened to reside in the habitable region.[47]

Roughly fifty to seventy percent of all star systems in the Milky Way are binary and would probably be inhospitable for life, as the orbits of two stars that are too closely spaced would prove unstable for their planetary systems.[48] Between collisions and gravitational interactions, the planets would likely nudge themselves away from stable orbits.[49] And as stars age, they emit more heat, which steadily expands their habitable zones. A planet in a binary system would have to contend not only with the steady expansion of its parent star's habitable zone, but also the encroachment of the second star's as well.[50] This would be bad news, as Michael Hart pointed out, because a steady habitable zone would presumably be required for the long timeframe necessary for advanced life to evolve.[51]

Despite these somber facts, in September 2012, the first exoplanet in a binary system was discovered. Although the planet—known as Kepler-47c—is a gas giant and thus probably not suitable for life, it resided in its parent star's habitable zone, which could be good news for the prospect of extraterrestrial life in that it demonstrates that binary systems can have orbiting planets.[52]

Periodic mass extinctions as indicated in the Earth's fossil record suggest a galactic culprit may be responsible. The physicists Richard Muller and Robert Rohde found a distinct 62-million-year cycle in the pattern of marine extinctions.[53] Perhaps as our solar system passes through the Milky Way's spiral arm, the Earth is subjected to a number of exogenous forces stemming from that part of the galaxy.[54] For instance, approximately every 100 million years, the Earth experiences an impact from a meteorite that is capable of annihilating all life-forms bigger than a dog.[55] The most powerful explosion in the universe—the gamma-ray burst—emits more energy in a few seconds than the Sun will emit in its lifetime. Such explosions, which might be produced when neutron stars merge with each other, occur about every hundred years in the Milky Way.[56] Even if one occurred in a distant region in the galaxy, it could possibly sterilize all life on a planet. If these gamma ray bursts were more common when the galaxy was young, then life would not have very long to evolve and produce a technological civilization.[57] Consequently, intelligent life could be a relatively recent development on an astronomical scale.

To be in the ideal zone, an Earth-like planet would have an Earth-like orbit around a Sun-like star.[58] But just because a planet may be habitable does not mean that it will be inhabited.[59] Dirk Schulze-Makuch, an environmental scientist at Washington State University, created a rating system called the Planetary Habitability Index to measure the likelihood of life on exoplanets. Some of the criteria for the Planetary Habitability Index include a solid surface that is protected by an atmosphere. Some energy source, such as sunlight or geothermal heating, is also necessary along with a chemical environment that allows for the formation of complex molecules. Finally, a liquid solvent is required, but not necessarily water.[60] A small planet does not have enough gravity to hold onto an atmosphere. Conversely, a large planet, with its massive gravity, would have too dense an atmosphere, which would create crushing pressure at the surface.[61] The best candidate for life would be a so-called "rocky" planet neither too big nor too small in size that resides in the habitable zone in its solar system.[62]

There are seemingly unique properties of our solar system and planet that enabled life to take hold. The so-called "Rare Earth hypothesis" posits that the existence of such planets is extremely uncommon, as a number of unlikely events and conditions would be necessary to give rise to them. The Earth in particular seems especially endowed for life. The high metallicity of the Earth, with its iron core, produces a magnetic field that protects life on the planet from harmful cosmic rays.[63] Orbiting within the Sun's habitable zone, the Earth avoids overly hot or cold temperatures, enabling the planet to retain liquid water on its surface—assumedly a sine qua non for the emergence of life. The Earth is fortunate to have a large moon. Most contemporary astronomers believe that the Moon formed from a collision between the Earth and a Mars-sized object about 100 million years after the formation of the solar system.[64] The impact occurred late in the accretion process and nearly ripped the Earth apart.[65] The Moon may have been essential for the evolution of life on Earth. Among other things, the tidal forces it exerts ensure that the Earth has plate tectonics, which are vital to sustaining surface temperatures in the range required for the existence of our kind of life. Moreover, it influences the Earth's tilt, which is responsible for the cycle of the seasons and is vital to the evolution of life. Finally, tides played a significant part in the migration of life from the sea and on to the land.[66]

Considering the multitude of factors that were necessary for the development of intelligent life on Earth—astronomical, geological, and biological—the astrophysicist John Gribbin makes the case that we are alone in the universe.[67] Although he believes that life is common in the universe, he maintains that intelligence is extremely rare and is likely to have arisen only on Earth.[68] If that is the case, humankind occupies a unique position in the cosmos after all.

According to the Copernican principle there is really nothing special about humans or our place in the universe. This premise is invoked to support the weak Anthropic Principle, which posits that the reason why the laws of physics allow us to observe life on Earth, and perhaps someday on other planets, is because if it were otherwise, it would not be possible for anyone to ponder that concept. By contrast, the strong Anthropic Principle avers that the probability of physical laws of the universe naturally being fine-tuned to allow for life is so remote that it suggests that some kind of divine agency was responsible.

In the nineteenth century, the astronomer John Herschel (1792–1871) argued that Darwin's theory of evolution did not sufficiently account for a guiding and controlling agency. New discoveries in physics have drawn some contemporary scientists to entertain the same conclusion. The parameters of the universe seem so precise that some scientists believe that a creator was responsible for their establishment. The laws of physics that permit the emergence of life lie within a very narrow range of values. Lee Smolin once estimated that the probability of picking a set of random parameters that would generate a universe favorable to life would be 1 in 10^{229}. This would seem to rule out an appeal to good luck.[69] As Freeman Dyson once opined, "In some sense, the universe knew we were coming."[70] In a similar vein, Fred Hoyle once suggested that some intelligence might have preceded us and created our universe as a structure for carbon-based life. According to his theory, our universe shows deliberate fine-tuning in its fundamental parameters. For instance, the ratios of respective strengths of the four fundamental forces are seemingly calibrated to allow for the physical and chemical processes that allow for life in our universe.[71] His theory dovetails well with the theory of intelligent design, which seeks to reconcile Darwin's theory of evolution with Creationism.

How does one account for the fact that the universe seems to reside in so many Goldilocks zones? According to Stephen Hawking and Leonard Mlodinow, universes can arise naturally from physical law; hence, the hand of God is not necessary for a universe to come into existence.[72] Proponents of the multiverse theory propose that there could be an infinite number of universes, the majority of which would not be suitable for life. In the froth of quantum bubbles that emerge, the vast majority of them remain microscopic and fail to blossom into a full-blown cosmos. However, a tiny fraction of them will inflate, grow, and attain the status of a universe.[73] Inasmuch as there are an infinite number of universes, some would just so happen to be so finely calibrated to permit life to take hold.[74] M-theory (discussed in greater detail in Chapter 10) postulates that ours is not the only universe; rather, it is just one of a possible infinite number of universes collectively known as the multiverse. Conceivably, these different universes could be governed by different physical laws.

In recent years, the multiverse concept has gained currency in theoretical physics. Some leading theoretical physicists speculate that our universe is but one of an endless number of parallel, but non-interacting universes.[75] In his book, *The Hidden Reality: Parallel Universes and the Deep Laws of the Cosmos*, Brian Greene explored the possibility that other realms could coexist with our own. There are several different possible versions of the multiverse and they are not necessarily mutually exclusive. One type posits that space extends infinitely encompassing an unlimited number of universes. In the Inflationary model, universes pop continuously, not unlike bubbles in a bathtub filled with water and soap. As mentioned earlier, another version draws upon M-theory, postulating that our universe exists on a three-dimensional brane, which floats in a higher-dimensional expanse populated by other branes, or other parallel universes. The "Many Worlds" version draws upon quantum mechanics, positing that an infinite number of parallel universes "coexist in an amorphous, but mathematically precise probabilistic haze." That is, a new universe is created for every possible reality that can occur in the quantum realm and follows its own unique trajectory. Finally, as Greene notes, it will probably be possible to one day create realistic, but simulated universes. In fact, over time, these artificial universes might tend to be the most numerous of all the variants.[76]

Even though the parameters of physics allowed for life to take hold on Earth, it does not necessarily follow that life can be found elsewhere. Generally, astronomers are sanguine about the prospects for life in the universe, while biologists are more skeptical. Members of the latter group are quick to point out that the various contingencies that gave rise to the evolution of *Homo sapiens* were far from certain and might not be replayed elsewhere on seemingly habitable planets. Astronomers counter with the facts that the sheer number of stars in the universe, buttressed with recent discoveries of exoplanets, imply that it is hubristic to think that we are alone in the cosmos.[77] Despite the seemingly insurmountable obstacles precluding the emergence of life, Steven Dick, a SETI scientist and former Chief Historian for NASA remained optimistic, explaining that "If it happened here, it can happen elsewhere, and there are plenty of 'elsewheres,' as the Kepler spacecraft is now showing. Yes, intelligence takes a long time to develop, but solar-type stars have long lifetimes."[78] Likewise, Michael G. Michaud, a former U.S. State Department official who played an active role in U.S.–Soviet space cooperation, was skeptical about the Rare Earth hypothesis:

> While the Rare Earth hypothesis deserves consideration, it has been challenged many times. Until the 1990s, we had no proof that planets existed anywhere but in our own solar system. Now astronomers are finding thousands of extrasolar planets in a surprising variety

> of types. If even one in a thousand of such planets had the potential to evolve life, the Rare Earth theory would have to be questioned. The time limit foreseen for Earth life is driven by the evolution of our Sun. Astronomers have pointed out that smaller stars, which are much more numerous, have much longer lifetimes. Intelligent life would have more time to evolve. Some astronomers have proposed that habitable planets may have appeared in our galaxy billions of years before the Earth did.[79]

Over the past two decades, scores of extra-solar planets have been discovered, which suggests that they are common in the universe. Powerful telescopes on the Moon could one day be used to identify extrasolar planets capable of supporting life.[80] New hope for finding extraterrestrial life was rekindled in December 2011, when NASA announced that it had found an exoplanet similar to Earth in several key aspects. Orbiting a star very much like our Sun, the planet is likely to have water and land. The surface temperature of the planet was estimated to 72 degrees.[81] Sometime within the next 20 years, NASA is scheduled to launch its Terrestrial Planet Finder system, which should be able to measure exoplanets as small as the Earth in the habitable zones abound other solar systems.[82] Advances in science and technology could give us more clues if life is out there in the cosmos. Most promising in this regard is the fledging field of astrobiology, the focus of the next chapter.

Chapter 3

Astrobiology

As scientists unlock the mysteries of the universe, more and more people are open to the prospect of life existing beyond Earth. Surveys indicate that roughly between 60 to 80 percent of Americans believe intelligent life exists on other planets.[1] Several arguments have been adduced in favor of the existence of extraterrestrial life. First, the laws of physics and material of matter are presumably the same throughout the cosmos. Through the use of spectroscopes, astronomers can determine the chemical features of stellar bodies that confirm this assumption. Not only is the universe composed of the same atoms, but these atoms are generally present in approximately the same proportions.[2] Hence, the elements that compose life on Earth appear to be common in the universe, which suggests that the requisites for life are available on other planets as well. According to the principle of uniformity of nature, the physical processes that produced life on Earth could be replicated elsewhere. Second, the universe is extremely large. A typical galaxy like the Milky Way contains a few hundred billion stars and the number of galaxies that can be observed is approximately 80 billion. The principle of plentitude assumes that whatever is possible in nature tends to become realized. If there is no impediment to the formation of life, then it will form. Third, the Copernican principle posits that the Earth does not occupy any special position in the universe.[3] Finally, the discovery of exoplanets over the past two decades confirms the existence of possible habitats for alien life.[4] On average, one exoplanet is discovered each week, which suggests that other worlds are quite common in the galaxy. What is more, in recent years, a number of these discovered exoplanets have earth-like characterizes and reside in the habitable zones of their star systems. All of these findings would seem to augur well of the extraterrestrial life hypothesis.

Biologists, however, tend to be far more skeptical. On Earth, many contingencies occurred that gave rise to complex life. If the videotape of life were to be replayed at some moment from the distant past, it is far from certain that it would unfold in consonance with its original trajectory. Even the SETI scientist Seth Shostak conceded that intelligent life could be extraordinarily rare, noting that *Homo sapiens* are the consequence of many forks in the evolutionary road. If any of the other roads had been taken, our species would not have appeared.[5] On the one hand, if a species has a mutation rate that is too slow in the face of a rapidly changing environment, it will not adapt quickly enough to adverse situations and die out. On the other hand, if a species mutates too rapidly, it could evolve out of it ecological niche "and either starve, get eaten, or otherwise become a square peg in a round hole."[6] Nevertheless, as Shostak pointed out, even if only one in a million of these exoplanets harbored intelligent life, that would still leave 10,000 civilizations in our galaxy.[7]

Research in the fledging field of astrobiology holds much promise is the search for extraterrestrial life and is the focus of this chapter. Astrobiology can be defined as the study of the origins, evolution, distribution and future of life in the universe.[8] This new scientific discipline is interdisciplinary in that it draws upon several fields, including astronomy, chemistry, physics, planetary science, and evolutionary biology.[9] But what exactly is life? For most observers, the answer seems self-evident, but it can be challenging for scientists and philosophers to define the concept in unequivocal terms. According to the Merriam-Webster Dictionary, life is "an organismic state characterized by capacity for metabolism, growth, reaction to stimuli, and reproduction."[10] But arguably, this definition is in some cases too restrictive. For example, a mule is obviously a living organism, but insofar as it cannot reproduce, it would not meet some definitions of life. Conversely, viruses are capable of replicating, but only inside the living cells of an organism, and as such, are generally not considered to be living things. As will be discussed in greater detail in the next chapter, some futurists predict that computers will someday become self-aware and exceed humans in intelligence. Would such supercomputers constitute life? In *Astrobiology: A Brief Introduction*, Kevin Plaxco and Michael Gross define life as any self-replicating, evolving chemical system.[11] If the right elements are available under the right environmental conditions, then the assumption is that life has a good chance of emerging. Among the most important elements for the creation of life is carbon.

If organic elements can be found on other planets, then perhaps life could take hold as well. As Steven J. Dick argued in *The Biological Universe*, just as the principles of physics and chemistry hold true throughout the universe, so do the principles of biology.[12] Carbon has many qualities that make it an excellent compound for life. As discussed in the previous chapter, the massive explosions of supernovae produced the heavy elements, including carbon, oxygen, and iron which are presumably the building blocks of life.[13] Carbon atoms are able to link up to one another to form complex chains with other carbon atoms and atoms of other compounds. There are conceivable alternative to carbon, perhaps, silicon. Some scientists speculate that on a planet much hotter than Earth, silicon compounds could combine to form life. And even on worlds colder than Earth, silicon would be able to bond to other atoms[14] At low temperatures, some silicon compounds have high stability and can generate as much complexity as carbon compounds. But at room temperature, silicon compounds are not nearly as stable as comparable carbon compounds. Furthermore, in the presence of liquid water, many of the silicon compounds dissociate. However, in ultraviolet light, these compounds are much more stable than many carbon compounds. In short, silicon-based biochemistry may be appropriate in low-temperature, non-aqueous environments with high ultraviolet fluxes.[15] But silicon is not well-suited for forming long molecular chains as carbon is at Earth-like temperatures. With the exception of very low temperatures, there is a much wider variety of complex compounds that can be formed with carbon than from the alternatives.[16] Silicon lacks the ability to form the extraordinary range of complex molecules that carbon can.[17] To be viable, life needs flexibility, which silicon lacks.[18] Thus it seems unlikely that life could take hold in elements other than carbon. As Plaxco and Gross succinctly put it, "if aliens ever visit us, the smart money says we should welcome them with carbon-based cakes and not with silicon-based rocks."[19] Assuming that carbon is a sine qua non of life, carbon-based life began could have begun as far back when the universe was about 6 billion years old (roughly 11.7 billion years ago). Based on this assumption, intelligent life could have emerged relatively recently on a cosmic timescale, perhaps as early as 3 billion years ago.[20]

When looking for signs of extraterrestrial life, astrobiologists use a "follow the water" strategy. Water is an anomalous liquid whose characteristics seem critical to its biological role. It is well-suited as a basis for the complex chemistry necessary to create life. Evidence of oxygen and water can be obtained spectrosopically from great distances even with our technology.[21] Astronomers believe that comets delivered much of the water that resides on Earth more than 4 billion years ago.[22] Water has unique characteristics that make it an ideal medium in which life can evolve. No terrestrial form of life can survive without water. The chemistry that we normally associate with life would only be possible against a liquid background.[23] No simple molecule can mimic all of water's biological functions.[24] Water provides an effective solvent in that it enables chemicals to dissolve and chemical reactions to take place. In the fluid environment of water, organic molecules can be dissolved "so that they can interact in the chemical dance of life as compounds enter and leave cells."[25] Water is indispensable for life, as it dissolves nutrients which can then be circulated to every part of an organism.[26] Moreover, the water molecule has a magnetic polarity that encourages other molecules to line up in a certain way that are crucial for life. The human body contains about 70 percent water, which highlights the important solvency properties for terrestrial life.[27]

Water is also critical in regulating the climate on Earth, for when it freezes and becomes ice, it assumes a less dense property that causes it to float. This is vitally important because during an ice age, the frozen water produces a skin at the top of the ocean, which provides the water beneath it with an insulating layer. If ice sank to the bottom of the ocean, the top layer would be uncovered and turn into ice. As this process repeated, eventually the oceans would turn entirely into solid ice.[28]

Finally, water is believed to be cosmologically plentiful because its atoms—hydrogen and oxygen—are the first and third most abundant atoms respectively in the universe.[29] Although other liquids might serve these functions, water remains fluid over a wider range of temperatures than the alternatives that would most likely be found on other worlds (for example, ammonia, methane, and ethane).[30] Conceivably, other substances could serve as a solvent, such as liquid ammonia—which is, after all, the fourth most common molecule in the universe. However, unlike water, ammonia does not expand when frozen. As a consequence, during periods of cold weather, frozen ammonia would sink to the bottom of the oceans, insulated from the summer's warmth. Prevented from seasonal melting, with each passing year, more and more of the ammonia ocean would be locked up as a solid until eventually, the planet would freeze over. It is unlikely that a frozen ocean could support life, as it would lack the ability to transport nutrients and modulate temperatures on the planet.[31] But water in and of itself does not guarantee the emergence of life. Paul Davies took issue with the "follow the water" strategy, calling it a "fallacy." Although water is necessary condition, it alone is not a sufficient condition for life. As he explains, the reason life on Earth inhabits almost all aqueous niches is because Earth has a contiguous biosphere. Life has invaded all of these niches but did not arise there ex nihlio.[32] Other factors besides water are necessary for complex life to emerge and flourish.

Oxygen appears vital to the development of complex life. About 4 billion years ago, the Earth's atmosphere contained primeval gases composed of either the nebula out of which the Sun and the planets were formed, or gases emitted from the body of the Earth. The present healthy mix of oxygen in the Earth's atmosphere did not originally exist on the planet but was probably a byproduct of life.[33] The early atmosphere is believed to have contained much less hydrogen and more carbon dioxide. As a result, it was a more oxidizing atmosphere, but also worked well to produce the materials of life.[34] Due to the loss of hydrogen by way of photolysis (i.e., chemical decomposition induced by light or radiant energy), the Earth's atmosphere began to oxidize from the day that it was formed. As Plaxco and Gross pointed out, it is quite possible that life on Earth captured a very narrow window of opportunity between the end of the sterilizing impacts and the oxidation of the primordial atmosphere.[35] Early in its geological history, Earth's atmosphere was rich in carbon dioxide, which produced a strong greenhouse effect and kept the surface of the planet warm. This raises the question, why did not a runaway greenhouse effect take place on Earth like it did on Venus? James Lovelock argued that life on Earth regulated the composition of the atmosphere by removing carbon dioxide gradually as the Sun warmed, which kept temperatures on Earth comfortable for life.[36]

The creation of the Earth's atmosphere was a dynamic process. The composition of the atmosphere changed in tandem with other variables. Oxygen is produced almost entirely by green plant photosynthesis (i.e., the process by which plants convert light from the Sun into chemical energy).[37] The invention of photosynthesis led to the advent of our oxygen atmosphere and had wide-ranging consequences for the anaerobic (i.e., not requiring oxygen for growth) bacteria that had theretofore had dominated the biosphere. Indeed, for most life forms the oxygen atmosphere was catastrophic, as it killed off many branches of the tree of life.[38] But the atmospheric change was drawn out over time, which provided sufficient time for some life forms to adapt to the oxic (i.e., containing oxygen) conditions and eventually benefit from the new opportunities that it created. Among the opportunities was the chance to colonize land. If it were not for the formation of a protective blanket of ozone high in the Earth's atmosphere, the living organisms would have fried upon treading on land. Likewise, in primitive times, the top of the ocean served as an ultraviolet absorbing blanket.[39] Finally, oxygen-producing photosynthesis paved the way for oxygen consuming metabolisms, which enabled the surviving life forms to burn carbohydrates and allowing for a much more efficient way to produce energy.[40] A great selective advantage was conferred upon those organisms that evolved mechanisms to cope with the presence of free molecular oxygen in the atmosphere. The capability to oxidize foodstuffs enabled these organisms to extract more energy from their nourishment.[41] After the Earth cooled and became habitable, it took only a few hundred million years for single-celled organisms to arise. The oldest known fossils of Prokaryotes (cells without a membrane-bound nucleus) date back to 3.5 billion years ago. Life in this primitive form remained stuck for almost 2 billion years. More complex eukaryotic cells (those that contain a distinct membrane-bound nucleus) did not arise until at least a billion years later, and the birth of multicellular life forms required the formation of an oxic atmosphere, which took another billion years after that.[42] For 3 billion years, life did not evolve much beyond blue-green algae, which suggests that large life forms with specialized organs are difficult to evolve, more so than even the origin of life.[43]

Why did it take so long for complex life to emerge on Earth? The early plant life produced oxygen in the atmosphere, in effect terraforming the Earth. Early in Earth's life, oxygen levels remained very low in the atmosphere, which postponed the emergence of animal life.[44] Oxygen is mainly a product of life, not a precondition for it.[45] In fact, the first life forms did not require oxygen; as mentioned earlier, for some, exposure to oxygen meant certain death. Prior to Cambrian explosion, all life on Earth consisted of simple single-celled organisms.[46] The increase in oxygen levels might have ushered in a massive extinction.[47]

The astronomer Mario Livio identified two key phases in the development of a life-supporting atmosphere. The first involved the release of oxygen from the photo-dissociation (i.e., a chemical reaction in which a chemical compound is broken down by photons) of water vapor. This phase lasted about 2.4 billion years on Earth. The second phase involved the increase in oxygen and ozone levels to about 10 percent of their present values. This provided a shield against harmful ultraviolet radiation for life on Earth.[48] These processes took eons to unfold. Thus it reasonable to assume that intelligent life will only be able to develop on those planets whose stars have a main sequence (the period in which a star radiates a steady amount of heat) of at least 10 billion years. Keep in mind, even a planet in the Goldilocks zone of such a star will not remain habitable throughout the entire main sequence. Currently, our Sun is in near the middle of its main sequence. The vast majority of stars in the Milky Way are near the end of their lives.[49]

The rise in oxygen content in the atmosphere may have ignited the Cambrian explosion—the relatively rapid emergence around 542 million years ago of most major animal organisms on Earth. Perhaps an arms race between predators and prey led to a flourishing of diverse life forms during this period.

As Michio Kaku explained, the first multi-celled organisms capable of devouring other organisms forced an accelerated evolution of the two, with each one racing to outmaneuver the other.[50] Animals have lived on land for only 300 million years.[51] Throughout Earth's history, somewhere between 5 and 50 billion species have lived on the planet; approximately 40 million live today.[52] The overall trend in the diversity of life on Earth has been upward, as more and more species exist than ever before.[53] Despite our knowledge of biological evolution, scientists have yet to determine the origins of life. For much of our history, it was assumed that life could arise spontaneously from inanimate matter, such as maggots from putrefying meat or lice from sweat. For higher animals, the dominant view found in Genesis 1 and the Sumerian creation myths was that each species was created by a form of divine fiat. But some philosophers wondered if lower forms could emerge seemingly spontaneously, could it not be possible that the earliest self-producing higher organisms developed from simpler antecedents? Anaximander (610–546 B.C.), a pre-Socratic philosopher, theorized that life emerged initially in the sea and that humans were distantly related to fish. Foreshadowing Darwinian natural selection, Empedocles (490–430 B.C.) thought that it might be possible for higher organisms to have evolved from lower organisms and adapted to their environments though some natural process.[54]

Finally, in 1665 the Italian physician, Francesco Redi (1626–1697), put the spontaneous generation of life hypothesis to an experimental test. When putrefying meat was covered with a fine gauze, he found that maggots never developed. Later, he discovered that maggots were the larval forms of flies, which deposited their eggs in the meat. When the meat was covered, the flies were unable to deposit their eggs—hence, no maggots emerged. But a decade after Redi disproved the spontaneous generation theory at the level of the house fly, a Dutch scientist, Antonie van Leeuwenhoek (1632–1723), discovered microorganisms. Leeuwenhoek found that even seemingly pure water abounded with microorganisms. He believed that these tiny seeds of microorganisms—animalcules—were present everywhere and would grow once they had access to some nutrient source. As germs develop from microorganisms themselves, there was no need to invoke spontaneous generation. Inadvertently, van Leeuwenhoek extended the debate on the spontaneous generation of life theory for another two centuries; however, by the early eighteenth century, this doctrine was in doubt. Finally, Louis Pasteur (1822–1895) put the theory to rest in 1857, when he demonstrated that microorganisms were always derived from preexisting microbes. In 1861 he discovered that sterile organic media not exposed to germs would never develop microbial cultures.[55] The implication was that life as we know it always came forth from life that had existed before.[56] Although Charles Darwin (1809–1882) formulated a compelling theory on how life evolves, he failed to conclusively determine how life emerged in the first place. He proposed that life emerged on Earth in a kind of primeval soup containing ingredients that were haphazardly put together. Early in its history, the Earth was forbiddingly hot for life to take hold. Moreover, the Earth's surface was relentlessly pummeled with meteorites which effectively sterilized the planet. After the planet cooled and the meteorite bombardments subsided, life emerged relatively quickly. With the passage of time, various sugars, amino acids, and other compounds accumulated in a primordial soup until at some time and place, perhaps a small lake, a threshold was reached and reproduction began.[57] The evidence discovered so far, implies that life was simmering in Earth's waters perhaps only 400 million years after the planet came into existence.[58] From radioactive dating experiments, mound-like artifacts of microscopic life have been discovered in fossils in the Pilbara hills of Western Australia that date back nearly 3.5 billion years.[59] Complex life, however, did not emerge on Earth until about after 4 billion years after the planet was created. About 370 million years ago, the first plants invaded land. Animals followed over the next few tens of millions of years.[60] But even with all the proper conditions, how did life emerge in the first place? The chemical theory of the origin of life was advanced independently by the Russian biochemist A.I. Oparin (1894–1980) and the British biochemist J.B.S. Haldane (1892–1964). They reasoned that the origin of life could be explained entirely in terms of the general laws of physics and chemistry as applied to primitive Earth conditions.[61] Both theorized that life arose from the slow evolution of chemical systems of increasing complexity.[62] In an effort to determine the origins of life, in 1951 Melvin Calvin (1911–1997) (who would later win a Nobel Prize in chemistry) and his colleagues at Berkeley conducted an experiment to determine if life was a byproduct of carbon chemistry. They put together a mixture of carbon dioxide, hydrogen, and water in a closed flask and applied a jolt of energy to it from the university's 60-inch cyclotron (an early atom smasher). The cyclotron's spark was intended to approximate a burst of cosmic rays from outer space, thus simulating a bolt from the blue that could have given life on Earth a jump start.[63] Nothing conclusive resulted from the experiment.

In 1955 Stanley Miller and Harold Urey conducted a similar experiment to Calvin's that mixed together ammonia, hydrogen, methane, and toxic chemicals, which they believed were present on the early Earth, in a flask.[64] A small electrical current was applied to the solution, which sought to approximate the composition of the early Earth's atmosphere and ocean. A tube connected the solution with a water-cooled condenser, which mimicked rain and set up a simple hydrological cycle.[65] It was believed that the electrical current was sufficient to break apart the carbon bonds within ammonia and methane and then rearrange the atoms into amino acids, which are the precursors to proteins.[66] After a week, they found evidence of amino acids forming spontaneously in the flask. Their experiment was hailed on the front page of *The New York Times* as a breakthrough that would lead to a deep understanding or our origins; however, much of the early optimism was misplaced.[67] Although the results of Miller and Urey's experiments were exciting, they still did not produce actual life. As Seth Shostak observed, manufacturing the building blocks may be easy, but constructing the edifice of life is much tougher.[68] It took hundreds of millions of years for the self-replicating molecule known as DNA (deoxyribonucleic acid) to develop, probably deep in the oceans.[69] For many years scientists had speculated on the nature of DNA, which was first proposed by the Swiss biologist Friedrich Miescher (1844–1895) in 1869. In a lecture titled "What is Life?" presented at Trinity College in Dublin, Ireland, in 1943, Erwin Schrödinger (1887–1961), the co-founder of quantum mechanics and Nobel laureate, speculated on how genetic information might be encoded in something resembling an aperiodic crystal—that is, a regular structure with information-bearing variations. His idea was not unlike our current knowledge on the structure of DNA. He conjectured that the evolution of highly complex organisms from a pool of simple, lifeless chemicals was kept in line with the immutable laws of thermodynamics. Subsequently published as a book— *What is Life?*—his lecture was hugely influential. It was the first time that a prominent scientist raised the question of how the physics of the universe fundamentally constrained its biology.[70] In 1953, James Watson and Francis Crick published their paper on the double helix structure of DNA, which finally unlocked many of the secrets of life. The complexity and specific features of DNA that are universally found in all forms of life is highly suggestive of a common origin for all life.[71] DNA is crucially important for three reasons. First, it carries a complete set of instructions for building an entire organism. Second, it enables an organism to make an exact replica of itself. Finally, if for some reason, the copy is not an exact copy—a mistake or mutation—the new DNA carries a revised set of instructions that will grow into an altered kind of organism.

These mutations form the basis of evolution. Darwinian evolution is built around the two principles of variance and selection. As organisms experience random genetic mutation, those that are best adapted to their environments will survive and multiply.[72] Mutation is essentially a random process engendered by radiation or chemicals in the environment. Those mutations that interfere with an organism's ability to survive or reproduce will usually die out. However, those mutations that improve survival capabilities will tend to become more numerous over time. A certain balance is necessary for a species to prevail over a long period. Too many mutations lead to birth defects and are counterproductive; too few and a species is unable to adapt to changing environmental conditions when they arise.[73] But mutation is just one part of the evolutionary process—natural selection operates as well. In essence, natural selection is the process by which biological traits become more or less common in a population depending on their tendency to impart reproductive success. Finally, a minoritarian school of evolution advanced primarily by the French biologist Jean-Bapiste Lamark (1744–1829) and the Soviet agronomist Trofim Lysenko (1898–1976) posits that genetic traits can be acquired through the lifetime experiences of an organism and passed down to its descendants. An ongoing debate continues if this model has any merit, but in the main, Darwinism is considered the dominant theory of evolution.

The universality of DNA codes suggested to Watson and Crick that an "infective" theory was responsible for the origins of life on Earth. As knowledge of the complexity of life emerged in the 1950s, some biologists, and even astrobiologists, became increasingly concerned about the probability of life having emerged by chance.[74] Actually as far back as the late nineteenth century, Lord Kelvin and Hermann von Helmholtz (1821–1894) proposed what became known as the panspermia hypothesis to explain the origins of life. There is indeed strong evidence that the building blocks of life arrived on Earth by way of meteorites that bombarded the planet during its early history.[75] In December 2010, scientists discovered amino acids in a meteorite believed to have formed when two asteroids collided. The shock of the collision heated the meteorite to more than 2,000 degrees Fahrenheit (1,093 degrees Celsius), which was hot enough to kill all organic molecules such as amino acids. Nevertheless, the amino acids were found anyway. This finding suggested that some of life's key ingredients could have formed in space and then been delivered to Earth by meteorite impacts.[76]

Another early proponent of the panspermia thesis, Svante Arrhenius (1859–1927), argued that hardy dormant spores could survive in the cold dark vacuum of space for long enough to be transported between the stars. According to his theory, bacterial spores could escape the upper atmosphere of their home planet and be launched into interstellar space by the pressure of light. As photons have momentum, he reasoned that they could accelerate something as small as bacterial spores to high speeds.[77]

Besides the seeds of life, perhaps diseases had extraterrestrial origins as well. The British astrophysicists Sir Fred Hoyle (1915–2001) and Chandra Wickramasinghe maintained that a number of viruses—including the common cold—originated from outer space and arrived on Earth by way of passing comets whose dust settled in our upper atmosphere.[78] They theorized that some of the epidemics and plagues that have ravaged humans throughout history have been caused by microbes that arrived from space. For example, Wickramasinghe looked at the Black Death which spread in Europe from 1334 to 1350 and claimed that the disease did not follow a route traveled by people; rather, it travelled a route as if carried by the winds. Hoyle and Wickramasinghe also argued that evolution proceeded in bursts of activity stimulated by the arrival of new genes from the sky.[79] Their theories, however, are not supported by any compelling scientific evidence.[80] Critics point out that it would be extremely unlikely that life-bearing germs could survive both the harsh environments of outer space and the destructive forces encountered on entering a planet's atmosphere.[81] For example, Carl Sagan (1934–1996), argued that the overall prognosis of the panspermia thesis was not favorable. As he pointed out, even the heartiest of terrestrial bacteria would likely be destroyed with a day of leaving the Earth's protective atmosphere by the Sun's harsh ultraviolet light.[82]

One view of panspermia is that that life has always existed in the universe. Living organisms have never formed from non-living matter; rather, life is spread from one planet to another.[83] The complexity of DNA has led some scientists to wonder if it was designed by some agency outside of the Earth. After all, the information encoded in human DNA, if written out in ordinary language, would occupy a hundred thick volumes.[84] Some biologists even regard living cells less as matter and more as a kind of supercomputer.[85] Some serious scientists, including the co-discoverer of the structure of DNA Francis Crick (1916–2004), speculated that extraterrestrial aliens may have actually "seeded" planets in the universe by sending microbe-laden probes out into space in a process called "directed panspermia" as a way in which to spread the building blocks of life.[86]

Crick and Leslie Orgel (1927–2007) found it puzzling that organisms with somewhat different genetic codes—DNA—do not coexist on Earth. They believed that the protein synthesis machinery in terrestrial life was so complex that it was unlikely that it could have been created by the slow, stepwise progression of evolution. As Crick opined, "The origin of life appears…to be almost a miracle, so many are the conditions which would have had to be satisfied to get it going."[87] This quandary was so serious, they argued, that we must consider even the wildest alternative hypotheses. One theory they proposed would be that terrestrial life, with its seemingly impossibly complex biochemistry, may have been artificially designed by intelligent beings from another planet in our galaxy. The aliens then purposefully sent it to Earth by "space mail." These alien engineers may have originally been derived from a much simpler biochemistry, and unlike ours, could have developed by chance. Crick and Orgel presented the theory of directed panspermia as a viable scientific alternative to biochemical theories on the origins of life on Earth. As evidence, they noted that all terrestrial life shares the same genetic code, which suggested that either life encountered an early evolutionary bottleneck, or that life evolved from organisms that originated on another planet.[88] To Crick, the double-helix of DNA seemed so intricate that it suggested it was designed. Adding a humorous spin on the directed panspermia thesis, the Cornell astronomer Thomas Gold, once speculated that eons ago an alien spacecraft might have landed on Earth and left behind garbage; from this bacterial waste we descended.[89]

The directed panspermia theory presupposed an advanced extraterrestrial civilization preceded ours by eons. Crick and Orgel speculated that spaceships could carry large samples of a number of microorganisms, each having different, but simple nutritional requirements. As they explained, because of their extremely small size, a vast number of these microorganisms could be carried on small spacecraft. The ability of the microorganisms to survive without special equipment for long periods of time would also be a special advantage. The conditions on many exoplanets are likely to favor microorganisms rather than higher organisms, which would present a much larger number of prospective habitats to seed. If we someday conclude that we are alone in the universe, Crick and Orgel speculated that we might decide to infect other planets in order to propagate our DNA in the cosmos.[90] Orgel once admitted that he wrote the paper in a rather tongue-in-cheek tone, but noted that Crick took it more seriously.[91] Recently, this idea was captured in the popular culture in the 2012 film *Prometheus*. In the opening scene with a hovering spacecraft overhead, a humanoid alien drinks a liquid which causes him to disintegrate. As his remains cascade into a waterfall, his DNA triggers a biogenetic reaction, thus seeding complex life on Earth.

Even if the seeds of life were planted by some extraterrestrial agency, it was far from certain that life on our planet would evolve as it did. Ernst Mayr (1904–2005) argued not only that it was unlikely that a human form would emerge on another planet, but that it was also nearly impossible for intelligence to emerge.[92] Even if life is abundant in the galaxy, an extraordinary conjunction of favorable conditions would be necessary to give rise to intelligent life. In

1964, George Gaylord Simpson (1902–1984) criticized scientists who envisioned an evolutionary path that culminated in the creation of intelligent beings, such as humans. As he noted, evolution is an opportunistic and unpredictable process that does not deterministically move toward some pre-established goal. Instead, evolution makes do with what is available at a particular period of time and under a given set of circumstances. Consequently, it would be highly unlikely that the chain of evolution would similarly unfold on another planet unless the planet had a history identical in every detail.[93] On the other hand, some scientists have argued that periodic catastrophes that engender mass extinctions provide a "pump for evolution," which leads to more and more complex life forms.[94] Over time, life becomes more complex because it randomly explores the range of possibilities, most of which are more complex than the starting state.[95] Perhaps there is an ecological niche for intelligence. Challenging environments, for instance ice ages, tend to selectively dispense with the less adaptable individuals in a given species. At a minimum, intelligent life would require: 1) eyesight to explore the environment; 2) an opposable thumb that could be used for grabbing objects; and 3) some form of communication, such as speech.[96]

Social environments might be even more important insofar as they favor the more intelligent members of a species. In the 5 million years since the *Homo sapiens* line broke off from the human-chimpanzee ancestor, the size of the human brain has tripled. Why did the brain grow so much and so rapidly? Some evolutionary biologists speculate that the brain grew so that human beings could both cooperate and compete with other human beings. The psychologist Nicolas Humphrey and the biologist Richard Alexander have separately suggested that human beings entered into an arms race with one another. The winners were those groups that could create more complex forms of social organizations based on new cognitive abilities to interpret each other's behavior.[97] With the capability of anticipating the behavior of other group members, an individual would have an advantage in obtaining food and finding a mate.[98]

Alan J. Penny discussed the connection between the genetic evolution of the brain and the expected lifetimes of scientific civilizations. The framework of his analysis is the Drake equation, a formula SETI scientists use to estimate the number of communicative civilizations in the galaxy, which will be discussed in the next chapter. The "L" variable in the equation is the length of time that an extraterrestrial civilization remains detectable. He points out that certain traits inherent in *Homo sapiens* were the result of genetic events. According to Penny, "conceptual" thought arose a mere 3,000 years ago, citing Hellenic mathematics and early Greek and Chinese literature as the first clear evidence of modern abstract thinking. He proposes that the gene mutation that triggered conceptual thinking occurred only in a limited number of persons in the population. These gifted individuals, however, initiated a culture of conceptual thinking. Children without this inherent ability were raised to acquire it through nurture rather than nature. In fact, it now appears that under certain environmental pressures, the expansion of the brain can be reversed. There is archaeological evidence that suggests that the brain size of *Homo sapiens* has fallen by 10 percent in the last 30,000 years. Subsequent genetic events might reverse the capacity for scientific thinking, and in doing so, limit a civilization's technological longevity.[99]

Marvin Minsky once noted that intelligence is experimental. Most organisms survive in a hostile environment by following instinct. Relying on experiments would be dangerous insofar as most intellectual experiments are almost always fatal.[100] Intelligence is not automatically progressive; members of a species may not automatically get smarter as the generations go by.[101] As intelligence is an unstable adaptation, it is far from certain that it will develop in higher organisms.[102] Based on the experience on Earth, intelligence emerged in only a few species, including *Homo sapiens*, Cetaceans, and Cephalopods. Ernst Mayr speculated that extraterrestrial civilizations could be rare because high levels of intelligence do not always benefit organisms. As he noted, intelligence came to humans with a high price: It required a large brain and a complex nervous system in addition to the metabolism to maintain them. He concluded that intelligence was fluke in history that was not inevitable or a necessary consequence of the development of life on Earth.[103]

If, some scientists ask, intelligence has such good survival value, why didn't dinosaurs evolve big brains, build rockets, and fly to the Moon? After all, they had plenty of time to do so.[104] As the paleontologist Niles Eldridge once mused, "The dinosaurs had 150 million years to get smart and didn't. So what would another 65 million years have done for them?"[105] On the other hand, Bernard E. Finch speculated that if the great reptiles had survived, a species of dinosaurs could have one day evolved into a bipedal humanoid form, walking erect and with high intelligence.[106] About 65 million years ago, a species of dinosaur the same height and weight as humans appeared to be well on its way to developing intelligence. The *Saurornithoides* stood on their hind limbs and used their front appendages to catch food. Their "hands" had the equivalent of an opposable thumb, which permitted a precision grip not unlike that of humans. Their brains were not puny like other species of dinosaurs; rather, they were on the order of 100 grams, or just a little smaller than that of a human infant. Perhaps if the *Saurornithoides* survived for another 10 to 20 million years, they might have become the first intelligent creatures on Earth.[107] Dale Russell of the Canadian National Museum produced a scale model called the Stenonychosaurus—a humanoid dinosaur with an upright posture, a large head, and intelligent eyes—as an extrapolation of the *Saurornithoides* genus if it had survived.[108]

But even if intelligence life has emerged outside of Earth, its form could be radically different than humans. Isaac Asimov found it highly improbable that extraterrestrial aliens would be humanoid in form.[109] Likewise, Carl Sagan predicted that extraterrestrial beings would most likely look nothing like any organisms or machines familiar to us.[110] John Lily, who studied the communications of bottle-nosed dolphins, pointed out that their brains were larger than ours and packed with neurons; this proved that more than one intelligent species had evolved on Earth.[111] But, because dolphins have no hands or other manipulative organs, they cannot use their intelligence in conjunction with technology.[112] Although the sea is an excellent incubator of life, it is a poor incubator of intelligence. Even if dolphins were able to conceptualize fire and work out the steps necessary to tame and use it, their environment precludes them from putting any of it into practice.[113] This has implications for SETI.

Proponents of convergent evolution are more sanguine about the prospect of intelligent life on other planets assuming humanoid form. Similar environments, so the theory goes, will produce similar adaptations, even in unrelated organisms.[114] There are only a limited number of evolutionary possibilities, which over time leads to a convergence of biological adaptations. As a consequence, extraterrestrial aliens could indeed end up looking a lot like us. Frank Drake believed that convergent evolution would produce intelligent aliens not dissimilar to humans: "They won't be too much different from us. If you saw them from a distance of a hundred yards in the twilight you might think they were human."[115] As Christian de Duve noted, natural selection can act only on the variants that are offered to it.

A cornerstone of evolutionary theory posits that the variants on which natural selection acts on, arise by chance factors. Whatever the causes of these variations (mutations), they are assumed to be entirely unrelated to the anticipation of some future needs. Thus, natural selection has no foresight, as it applies only to current ecological conditions. This reasoning has been interpreted to suggest that natural selection is ruled by contingency. As Stephen Jay Gould once put it, rewind the tape of life, and you would get a very different result. But according to the Nobel Prize-winning cytologist and biochemist Christian de Duve (1917–2013), there is a theoretical flaw in this line of reasoning. As he explains, the fact that an event occurs by chance does not exclude its inevitability.[116]

Another proponent of convergent evolution, Seth Shostak, believes that alien animal life, especially those on rocky or oceanic planets similar to Earth, might have familiar anatomical designs. For instance, the heads of animals contain the sensory organs—eyes, ears, whiskers—close to the brain, thus reducing the reaction time and increasing the animal's chance of survival. As a consequence, Shostak reasons that a design that efficient would probably be common.[117] Likewise, Andrei Finkelstein believes that aliens would probably resemble us, not only having heads, but two arms and two legs as well.[118] Although the evolution of *Homo sapiens* is unlikely to occur on another planet, functionally equivalent intelligent beings may exist, albeit in a much different physical form.To illustrate this point, Shostak uses the example of the streamlining of ocean predators. The torpedo-shaped body plans of both the barracuda and dolphins reflects good engineering for two species whose survival depends on moving through the water quickly. Although these two species are only very distantly related, their similar physiologies are the result of convergent evolution.[119]

Perhaps intelligence could evolve in different species as well. A study of dolphins and toothed whales fossils from the last 50 million years indicates that some species have dramatically increased their intelligence (as measured by the ratio of brain to body mass). The fact that several species have increased their cognitive faculties leads Shostak to believe that convergent evolution is at work adapting different species to a niche market for intelligent animals.[120] He concedes that most planets in the galaxy will be unsuitable for evolution leading to intelligence, but he counters that in a galaxy in which habitable worlds could easily number in the billions, "it requires heroic pessimism to argue that *only* [italics in original] Earth has the *mise en place* for cooking up thinking beings."[121]

Paradoxically, some scientists have invoked convergent evolution to reject the extraterrestrial life hypothesis. Simon Conway Morris of the University of Cambridge argued that the outcomes of evolution are remarkably predictable. He rejects the theory that evolution was the result of a series of random events. Instead, environmental constraints limited the direction of evolution, including the nature of intelligence. As he points out, biological systems may look radically different, but are in fact similar in operation. He likens the process of evolution to a factory floor, with components scattered around. The trick, as he sees it, is to have the right set of "plans" on hand. At the molecular level, much is already available for adaptation. He rejects the preeminent role ascribed to mass extinctions in explaining the trajectories of evolution. He asks, if the massive asteroid would have sailed harmlessly past the Earth and the dinosaurs survived, would evolution have turned out much differently? According to him, probably not because by the Oligocene era, the Earth was finally leaving its greenhouse state and entering a cold world culminating in the Plio-Pleistocene ice ages. In this alternative world, "warm-blooded, socially-adept, large-brained tool makers" would presumably have an evolutionary advantage. In the grand scheme of things, the dinosaurs were doomed, and the end-Cretaceous extinction simply accelerated the inevitable. If humans were to recklessly destroy themselves, he believed that other intelligent beings would be waiting in the wings to follow the converging paths that led to intelligence, culture, and tool use.[122]

Although the evolutionary routes are many, Morris concluded that the destinations are limited: "what is possible has usually been arrived at multiple times, meaning that the emergence of the various biological properties is effectively inevitable." Therefore, a planet with similar environmental characteristics to Earth would likely produce beings that resemble humans. Although Conway maintained that extraterrestrial life and culture would reflect its terrestrial counterparts, he doubted that they existed beyond Earth.[123] How does he reach this conclusion? He reasons that solar systems depending on metal-rich stars would have begun to form billions of years ahead of ours. If life takes holds on a planet with a similar biosphere as our planet, then intelligence is an evolutionary certainty, yet he notes that the Fermi paradox still holds. The fact that we have not encountered an advanced extraterrestrial civilization leads him to conclude: "We never had any visitors, nor is it worth setting up a reception committee in the hope that one day they might turn up. They are not there, and we are alone."[124]

Despite his involvement in SETI, Paul Davies has argued that the probability that life could originate through the random arrangement and rearrangement of molecules is so low that he would not be surprised if life arose only once.[125] As he noted, there is no inherent orderliness in the chemical sequences that gave rise to life. The process was haphazard, which leads to a bleak conclusion. The arrangement of DNA is both random and highly specific, producing a unique combination of qualities that would be hard to explain by deterministic forces. Viewed in this way, he concludes that life is a "freak phenomenon that arose by an exceedingly lucky fluke, a process of such staggering improbability that we can safely say it happened only once in the observable universe." Invoking the reasoning of the weak anthropic principle, he points out that the fact that we can observe this can be explained by an inevitable selection effect, that is, observers can exist only where there is life.[126]

To know if life is truly common in the universe, we must find another example of life that had an independent origin.[127] Concerning the question of extraterrestrial life, no other planet has attracted more interest and prolonged attention as Mars. Ancient water erosion and volcanic activity suggest that Mars was once a more dynamic place.[128] Evidence suggests that 3 to 4 billion years ago Mars was a markedly warmer and wetter planet, presumably as a result of a thicker atmosphere, which led to a massive greenhouse effect.[129] Data collected from NASA's Curiosity Rover impliy that water was once plentiful on the planet and of such quality that it would have been good enough for humans to drink. Water on Mars may have evaporated entirely or disappeared beneath the surface of the soil; it is not a stretch to imagine that life once inhabited the planet as well. It is also not unreasonable to assume that the planet could have once supported microbial life.[130] Furthermore, the nutrient phosphate is plentiful on the red planet, perhaps five to ten times more abundant than on Earth. Phosphate, which serves as the backbone of DNA, is an essential part of the molecules that cells use for energy. In fact, the dominant phosphate-loaded minerals on Mars are distinctly more soluble and thus release more phosphate into water, than those most commonly found on Earth.[131] Life on Mars may have followed the water underground and adapted to changing conditions.[132] Possibly, life may still reside beneath the surface of Mars, hibernating in the frozen soil or in the form of spores.[133]

The discovery of life on the red planet, though, would not automatically answer the issue of the plurality of life. It is quite possible that life began on Mars and later arrived to Earth via a Martian rock. There is speculation that Mars might have seeded Earth with the building blocks of life. For billions of years, the two planets have traded rocks blasted into space by comet and asteroid bombardments. Over the course of geological history, trillions of tons of Martian material have rained down on our planet.[134] Conceivably, microbe-laden debris blasted from Mars' surface could have survived this journey in space and reached Earth. Matter is far more likely to travel from Mars to Earth than vice versa because escaping Earth's gravity requires more than two and a half times the energy required to leave Mars.[135]

In August 1996, then-President Bill Clinton enthusiastically announced that NASA scientists had found evidence for life on Mars in the form of microscopic features inside a meteorite recovered from Antarctica in 1984 by a meteorite-collecting team. The meteorite—Allan Hills 84001— landed in Antarctica about 13,000 years ago and was estimated to have formed about 4.5 billion years ago, in a period when Mars was warmer and had a denser atmosphere.[136] Analysis discovered organic molecules in the meteorite believed to have been blown off of Mars roughly 16 million years ago. The original finding suggested that life once existed on Mars when the planet was warmer and wetter, though subsequent analysis chipped away at that conclusion.[137]

In recent years, biologists have discovered that the majority of terrestrial species are microbes. Biologists have only scratched the surface of the microbial realm. Paul Davies speculated that it is possible that there could be billions of microbes contained in, for example, a sample of soil or seawater that are representative of some form of life of which we are unaware. This would have very important implications for the field of astrobiology because if it could be established that life originated on Earth more than once, it would imply that life would emerge readily in Earth-like conditions on Earth-like planets as well.[138]

The discovery of so-called "extremophiles"—organisms that can live in severe environments from the very cold to the very hot to the very toxic—which biologists previously believed were inhospitable for life, have strengthened the extraterrestrial life hypothesis.[139] Tullis Onsott, a geologist at Princeton, published an article that described how a previously unknown species of roundworm survived in an African gold mine nearly a mile beneath the Earth's surface, which is 100 times as deep as multi-cellular life had been previously observed. If such an organism could survive underneath the surface of the Earth, Onsott speculated that it may be possible on Mars as well.[140] What is more, solar energy is not required to sustain life. In fact, there are microorganisms that directly metabolize hydrogen sulfide and minerals in the warm water seeping through vents from the Earth's interior.[141] The microbiologist Lyle Whyte of McGill University in Montreal discovered bacteria living at subzero temperatures in a methane-rich spring on Axel Heiberg Island in the Canadian Arctic Ocean. Scientists speculate that similar life forms could be the source of recently discovered methane plumes on Mars. As Whyte explained, microorganisms in the deep subsurface of Mars could be responsible for producing the methane gas.[142]

Astrobiologists hold out hope that life could reside beneath the frozen surface of Europa, a moon of Jupiter. As Europa spins around Jupiter, the planet's huge gravitational field squeezes the moon, creating friction deep within its core. The heat generated from the friction could cause some of the ice cover to melt.[143] On planets with higher or lower average temperatures, organisms could have markedly different biochemistries than organisms on Earth, which allow them to survive.[144] Perhaps organisms on other planets could adapt to environments in which there is a higher concentration of ultraviolet light. In that vein, Carl Sagan once speculated that Martian turtles, or organisms with ultraviolet-opaque shields on their backs, could roam other planets.[145]

Conceivably, organisms could find innovative ways to survive elsewhere in the cosmos. Freeman Dyson conjectures that it might be possible for life to adapt to cold environments far from stars. To keep warm, these organisms might develop optical concentrators to focus sunlight onto their vital parts. He proposed that one way to distinguish the living from the non-living objects in space would be to analyze the back-scattered light with a spectroscope and search for spectral features that might be identified with biologically interesting molecules.[146] As Dyson notes, the experience on Earth has demonstrated two fundamental qualities of life—adaptability and invasiveness. Life on Earth has been found on the most inhospitable corners of the planet. As such, he finds it reasonable to believe that life would exhibit this quality elsewhere in the cosmos. He even conjectures that it might be possible for life to survive in a vacuum free of air. These life forms would have a distinct advantage in spreading throughout the cosmos as they could survive the lone journeys through the vacuum of space.[147] These organisms might be able to live on habitats such as comets or asteroids. Such environs, according to Dyson, would be more efficient because they have a far greater surface area to mass ratio compared to planets. This would allow life not only to efficiently collect energy but dispose of heat waste as well. Moreover, on a planet, only a tiny fraction of its mass is useful to its inhabitants.[148] Based on this line of reasoning, Dyson argued that scientists should search for life based on the premise of detectability rather than probability. As he points out, estimates of the probability of life existing in various places are grossly unreliable; however, estimates of detectability are generally quite reliable. Therefore, it behooves astrobiolgists to consider that "all kinds of unlikely things might be out there, and we have a chance to find them if and only if they are detectable."[149]

NASA has embraced astrobiology as an important goal. Since 1996, when NASA created its current astrobiology program, the agency has increased its budget from $10 million to $55 million.[150] But the origins of the program go back further. In 1959, the agency's first administrator, T. Keith Glennan, appointed a Bio-science Advisory Committee whose goals included the search for extraterrestrial life. A leading pioneer in the new science of astrobiology is Joshua Lederberg. His and Carl Sagan's belief in the existence of extraterrestrial life found favor in some NASA circles.[151] Under their influence, NASA began research on developing prototypical instruments to detect life on Mars and other planets.[152] In 2008 NASA's *Phoenix* Mars lander spent five months analyzing the soil of Mars. The probe confirmed the presence of patches of underground water ice. Moreover, it detected the mineral, calcium carbonate, which could indicate the occasional presence of thawed water on the planet.[153] As a general rule, Freeman J. Dyson noted that the space program flourishes when there is agreement on goals. Currently, scientists are divided on one side by those whose goal is to understand the physical and chemical processes going on in the universe, while another group seeks to find evidence of extraterrestrial life. To keep the space program healthy, he advises missions that satisfy both of these goals so that they will generate support from both sides.[154] Therefore, every mission searching for life should also include other exciting scientific objectives so that the mission is worth undertaking regardless if it finds life or not.[155]

Currently, the field of astrobiology concentrates on discovering simple life forms, as it is assumed that it is highly unlikely that complex life exists outside of Earth in our solar system. Some of the starlight that reaches the Earth during the transit passes through the distant planet's atmosphere which allows for the spectroscopic evaluation of its composition. At the present time, our instruments are able to analyze only the very large exoplanets. Currently, Earth-like exoplanets are too small to monitor with this approach.[156] Even within our own solar system, our technology is limited when it comes to detecting life. For instance, in the early 1990s NASA equipment aboard an orbiting space probe only 1,000 miles from the surface of the Earth failed to detect any signs of life on the planet.[157] But as exoplanets are discovered at the rate of about one a week, scientists now assume that planets are plentiful in the universe. As we examine in the next chapter, it is not unreasonable to speculate that advanced civilizations could exist beyond Earth.

CHAPTER 4

The Structure of Extraterrestrial Civilizations

THE UNIVERSALITY OF THE LAWS of chemistry suggests that life exists on planets besides Earth. But what makes the prospect of extraterrestrial life exciting is the prospect that there could be intelligent alien beings not unlike humans. As the discussion of astrobiology in the previous chapter made clear, extraterrestrial intelligence is far from a certainty, for the widespread belief in its existence "rests on logic and intuition, not observations and experiments."[1] Be that as it may, the example of life on Earth illustrates increasing complexity of life over time. As James Grier Miller explained, since life formed on Earth, it has undergone a steady progression in the direction of larger biological and social entities.[2] Single cells eventually conglomerated to form multi-cellular organisms. A biological division of labor led to the creation of more complex organisms with organs. Finally, *Homo sapiens* originated in sub-Saharan Africa around 200,000 B.C.

By their nature, humans are social animals that seek to establish communities. Enlightenment philosophers, including Thomas Hobbes, John Locke, and Jean-Jacques Rousseau, asserted that human beings in a state of nature were isolated individuals for whom society was unnatural. But as Francis Fukuyama points out in *The Origins of Political Order: From Prehuman Times to the French Revolution*, this notion of primordial individualism is a fallacy.[3] Instead, he argues that Aristotle was correct in the sense that humans are political by nature. After all, everything modern biology and anthropology tell us about human nature suggests that there was never a period in human evolution when humans existed as isolated individuals. At first, humans congregated into families and later, groups. It would seem that the human brain is hardwired to facilitate many forms of social cooperation. In fact, the primate precursors of *Homo sapiens* had already developed extensive social, and indeed, political skills. Our closest primate relatives—chimpanzees—are also very sociable and build hierarchical organizations known as troops. But because chimps do not have a language, they are unable to move to higher levels of social organization. Approximately 50,000 years ago, behaviorally modern humans emerged that were capable of communicating with language, which enabled them to build much more complex forms of social organization. Tribes were formed, followed by communities. Nations brought more people together and in the twenty-first century, supranational organizations loom larger and larger.

Tool-making enabled humans to master their environment, but other species of animals, including beavers, have used tools as well. To date, beavers have demonstrated no capacity to build a civilization. There are, however, other animals that are capable of building complex societies. For instance, social insects such as bees, ants, and termites build colonies; however, they result from instinctual behavior, the guidelines of which are built into their genes and nervous systems at birth.[4] But even if a civilization does emerge, it might not develop sophisticated technology that would enable it to communicate with other intelligent beings in the universe. To create complex inventions, science must be practiced. It follows that the genetic structure of the brains of the members of the civilization must permit the kind of thinking that makes science possible.[5] But even with intelligence, science is not inevitable. As Paul Davies noted, even the ancient Greek philosophers did not practice science in the contemporary sense by using experiment and observation; rather, they believed that answers could be deduced by pure reason alone.[6] Nevertheless, they made remarkable advances in reason and mathematics that were preserved and nurtured for centuries by Islamic scholars during the European Dark Ages. Without their stewardship, Davies believes that it is very doubtful that science and mathematics would ever have taken root in medieval times.[7] He attributes the spectacular success of science to the fertile interplay of theory and experiment.[8] To be sure, humans can arrive at knowledge simply through trial and error. However, the true power of science, Davies argues, derives from its sound theoretical basis that leads us to design novel contraptions based on understanding the principles that govern them.[9]

With science, intelligent beings are much more capable of producing sophisticated technology. Isaac Asimov once argued that the ability to tame and use fire is what distinguished man from other animals.[10] In large part, energy extraction determines the level of sophistication of a technological civilization. On that note, in 1964, Nikolai Kardashev, a Russian astrophysicist who mounted the first Soviet search for extraterrestrial intelligence, proposed a classification of alien civilization based on their methods of energy extraction. His scale has three categories. A Type I civilization can harness all available energy sources on its planet. A Type II civilization harnesses energy directly from the star in its solar system, not merely using solar power, but the mining of energy from the star. Finally, a Type III civilization is able to harness the power not of only its solar star, but also other stars in its galaxy.[11] For his part, Paul Davies added one more level—Type IV—a civilization that commandeers the entire cosmos.[12]

For Kardashev, energy consumption defines civilizational progress and could one day enable interstellar travel.[13] He once estimated that there could be civilizations with ages of technological development 6 to 8 billion years longer than that of the Earth. However, given the estimated age of the universe—13.7 billion years—this proposition seems unlikely, for it probably would not have been possible for complex life to survive and build a civilization that far back in the universe's history. Nevertheless, as Michio Kaku explained, Kardashev's system of classification is reasonable because it relies on available supplies of energy: "Any advanced civilization in space will eventually find three sources of energy at their disposal: their planet, their star, and their galaxy. There is no other choice."[14]

How would a civilization advance on this scale? A civilization might advance to Type I status by applying fusion power or by producing antimatter to be used as an energy source. Alternatively, one might be able to harness zero-point energy.[15] To advance beyond Type I status, Freeman Dyson theorized that a hypothetical megastructure could be utilized to encompass a star as a system of orbiting solar power satellites to capture most of the energy output.[16] Constructing such a device—a Dyson sphere—would be a gargantuan engineering undertaking, but theoretically possible. To complete the project, Dyson theorized that it might be possible to use solar energy to dismantle planets and rearrange their pieces as parts of a massive solar collection sphere.[17] The proposed biosphere would not be a solid shell, but rather, a "loose collection or swarm of objects" independently orbiting the star.[18]

Human civilizations, as Dyson noted, have a tendency to constantly increase their energy consumption. Based on this reasoning, eventually, civilizations would be compelled to search for more ways to harness energy.[19] He estimated that the time required to restructure the entire planetary system into a vast solar collector would be about 100,000 years. An astronomer observing from another solar system would notice a striking transition as the process was underway. The star in the solar system undergoing the transformation would become obscured, but due to the law of the conservation of energy, just as many kilowatts would emit from the dark object as used to flow from the star when it was once visible. Taking this scenario into account, Dyson proposed a SETI strategy that would focus on detecting infrared radiation instead of visible light. He reasoned that an advanced civilization using astro-engineering on a grand scale would create a distinctive infrared object in the night sky.[20]

A suitably advanced extraterrestrial civilization may find it preferable to construct habitats around a variety of stellar types.[21] A Type III Civilization might also be able to harness energy by building a device around a spinning black hole, which would release far more energy than a star can through nuclear fusion. By exploiting the law of conservation of angular momentum, the prodigious power of its rotation could be used to extract energy.[22] Even more ambitiously, an advanced extraterrestrial civilization might be able to tap into the energy released from massive black holes that reside at the center of some galaxies.[23] Joseph Baugher once suggested that a Type III civilization could control the process of star creation and even alter the structure of entire star clusters. Such a civilization would be immortal because if one star died, it could create a new one or move somewhere else in the galaxy.[24]

Currently, our civilization occupies a position that has not quite attained

Type 1 status. Carl Sagan once classified Earth as 0.7 on the Kardashev scale.[25] As he and I.S. Shklovskii noted, Earth's civilization is in its infancy; hominids have only inhabited the planet for about 0.1 percent of its history. Civilization has existed for only one-millionth of the lifetime of Earth, and a technical civilization capable of transmitting and receiving interstellar communications has been present for only just over one one-hundred millionth of geological time.[26] Building on Kardashev's framework, Sagan proposed a classification using letters to designate a civilization's acquisition of information. A Type A civilization would be analogous to a hunting and gathering society on Earth. Writing in 1973, he characterized Earth as a Type H civilization.[27]

How can we determine the number of extraterrestrial civilizations that reside in the Milky Way? Astronomers pondered this question in 1961 at the first scientific conference on Communication with Extraterrestrial Intelligence (CETI), which was sponsored by the U.S. National Academy of Sciences in Green Bank, West Virginia. Participants in the conference called themselves "The Order of the Dolphin."[28] Otto Struve, the director of the National Radio Astronomy Observatory, organized the meeting. At the conference, Frank Drake unveiled his epynonomous Drake equation as a formula designed to estimate the number of communicative civilizations in the galaxy:

$$N = R\ fp\ ne\ fl\ fi\ fc\ L$$

According to the equation, the letter (N) stands for the number of detectable communicating civilizations in space, which equals the rate of star formation (R), multiplied by the fraction of stars with planets (fp), multiplied by the number of planets that are hospitable to life (ne), multiplied by the fraction of those planets where life actually emerges (fl), multiplied by the faction of planets where life evolved into intelligent beings (fi), multiplied by the fraction of those planets with intelligent creatures that are capable of interstellar communication (fc), multiplied by the length of time that an extraterrestrial civilization remains detectable (L).[29] The bad news for SETI is that even if one of the numbers in the equation is near zero, then N will also be very close to zero regardless of how big all the other numbers are.[30]

Because the values of so many of the variables are unknown, there is no "solution" to the Drake equation at this time.[31] Much educated guess work is applied to arrive at an approximation. After discussing each variable at length, the conference participants estimated that there were somewhere between 1,000 and 100 million advanced civilizations in the Milky Way.[32] As Drake once put it, SETI scientists assume that technology-using civilizations are abundant in the universe.[33] To those critics that describe the equation as speculative, he counters that just the opposite is true insofar as each variable has already taken place at least once as evidenced by our civilization on Earth.[34] His colleague Carl Sagan maintained that habitable planets and intelligent life were common in the universe, speculating that one technological civilization was formed in the Milky Way every ten years.[35] He estimated that there were about 1 million advanced civilizations distributed more or less randomly throughout the galaxy, a proposition that in retrospect seems unlikely because of contemporary knowledge of the "galactic habitable zone" which was discussed in Chapter 2. Based on Sagan's assumption, on average, we could expect the nearest extraterrestrial civilization to Earth to be about 200 light-years away. If that is the case, our radio waves would not yet have reached the nearest extraterrestrial civilization insofar as Earth has only been emitting radio waves into space for only about 60 years.[36]

Even if a species succeeds in creating a functioning civilization, how long it will survive is an open question. As the political commentator Charles Krauthammer once noted, ultimately the survivability of extraterrestrial civilizations would come down to politics—that is, how conflict is managed within their societies.[37] Political systems evolve in a manner comparable to biological organisms. According to Darwin, evolution proceeds through the processes of variation and selection. Random genetic combinations cause variation in organisms. As a consequence, those that are better adapted to specific environments have greater reproductive success and propagate themselves at the expense of those that are less well adapted. Likewise, in a long historical perspective, political development also proceeds through variation in the sense that those forms of political organization that are more successful—for example, those that are capable of generating greater military and economic power—displace those that are less successful.[38]

It is important to understand, however, that political evolution differs in significant ways from its biological counterpart. As Francis Fukuyama noted, for one thing, variation in political institutions can be planned and deliberate, as opposed to random. Another way in which political development differs from biological evolution is that institutions, unlike genes, are transmitted culturally rather than genetically. Humans tend to invest institutions with intrinsic value, which leads to conservatism over time. By contrast, biological organisms do not reify their own genes. If the genes do not permit creatures to survive and reproduce, then nature ruthlessly eliminates them. It is certainly possible for cultural traits such as norms, customs, laws, beliefs, and values to be changed, sometimes rapidly, in the event of a major upheaval. But generally, institutions resist change. Therefore, institutional evolution can theoretically be both faster and slower than biological evolution.[39] To be sure, the conservative quality of institutions has a clear adaptive value, for if people did not have a biological proclivity to conform to rules, then they would have to be constantly renegotiated at enormous cost to the stability of the society. But societies whose institutions have become so conservative may not be able to adapt quickly enough to adjust to new circumstances when they arise.[40] To survive in the long run, a civilization must be capable of adaptation.

For many SETI scientists, the most uncertain variable in the Drake Equation is L, the temporal length of the civilization, a factor that is determined by what we call human nature. If most civilizations avoided self-destruction, then as Carl Sagan reasoned, "the sky is softly humming with messages from the stars."[41] His Russian colleague I.S. Shklovskii, an astrophysicist from Ukraine who spearheaded the Soviet interest in life in the universe, once mused that the lifetimes of extraterrestrial civilizations may involve a race between cooperation and competition. He propounded a taxonomy consisting of three types of societies. The first type solved the social organization problem before it developed high technology. The second type solved the social organization problem by accident, for example, a nuclear dictator who could wipe out all resisters. Finally, the third case would be analogous to Earth, where the social organization problem was not solved before the onset of high technology.[42] Soviet ideology conjectured that socialism—and eventually communism—were characteristic of advanced political and economic development. As a result, Soviet scientists reasoned that advanced extraterrestrial civilizations would likely be socialistic or communist, and thus would not be hostile to Earth.[43] As a citizen of the Soviet Union, Shklovskii believed that communism was the superior social system and that a capitalist world order would eventually self-destruct. Obviously influenced by his country's ideological currents, Shklovskii argued that as long as capitalism existed on Earth, a violent end to intelligent on the planet was probable. Still, he held out hope that nations could resolve their differences, opining that the forces of peace in the world were great.[44] In later years, however, he lost interest in SETI, becoming increasingly depressed with the prospect of global nuclear Armageddon.[45]

Civilizations on Earth, as George Basalla pointed out, "are unstable social entities with relatively brief lifetimes." Therefore, for the sake of SETI, civilizations should not be viewed as enduring political entities; rather, they ought to be characterized by science and technology, specifically, those that emit electromagnetic radiation.[46] Even when civilizations fall, they seldom collapse into total chaos. Instead, civilizations usually tend to fall back into a state of lower complexity.[47] Ironically, there is even the possibility that human intelligence could regress despite advances in technology. In a highly technological society, where intelligent machines carry out so many important tasks, there could be very little mental challenge for most humans. Intellectual burdens could lighten, as professional and technical services are taken over by computers. Lacking pressures for increased intelligence, advanced civilization could backslide. Paradoxically, as Albert A. Harrison mused, there could be civilizations that have become technologically advanced, but whose citizens could have an appalling low level of intelligence.[48]

Using cultural anthropology as her framework, Kathryn Denning also examined the L variable in the Drake equation. As she explains, anthropology demonstrates that the notion of unilinear social evolution is inaccurate and overly simplistic. Nevertheless, the idea of unilinear progress has taken root in Western thinking and been assumed by many SETI scientists. For instance, Carl Sagan counseled that to survive, humans would have to make it through a "technological adolescence" where the threat of nuclear annihilation, global warming, environmental despoliation, and overpopulation loomed large. Once humans passed through this perilous transition, they would be prepared to colonize the cosmos. However, archaeological studies indicate some societies have merely down-shifted to a less energy-intensive form of social organizations and survived. Rather than taking unilinear progress for granted, she counsels that we should consider the range of possibilities on other planets.[49] On that note, Isaac Asimov once mused that there might be second, third, perhaps up to ten generations of civilizations in a planet's history before its star leaves its main sequence.[50]

The L variable in the Drake equation remains a contentious issue for SETI scientists. Peter Ulmschneider once calculated that 99.8 percent of all intelligent civilizations that ever existed in the Milky Way are now extinct. As a consequence, the most active interstellar travelers might be archeologists studying the remains of departed civilizations.[51] For his part, Paul Davies sees no reason why once an advanced extraterrestrial civilization is established why it should not endure for an extraordinary length of time, perhaps millions or tens of millions of years or more.[52] Optimistically, Ray Norris once estimated that the average extraterrestrial civilization is 2 billion years older than our own.[53] But to survive for so long would require tremendous advances in science and technology.

As Christian de Duve once noted, humans are not the ultimate achievement of evolution; rather, they are only at a transient stage.[54] In that vein, Ray Kurzweil, a noted inventor, entrepreneur, and futurist, predicted that by the year 2042, we will reach the "Singularity," which he defines as "a future period during which the pace of technological change will be so rapid, its impact so deep, that human life will be irreversibly transformed."[55] Specifically, he refers to an era in which humans will transcend their biological bodies and meld with computers. He predicts that by the end of the twenty-first century, the non-biological portion of our intelligence will be trillions of trillions of times more powerful than unaided human intelligence. Technological advances, he explains, tend to occur not in linear, but rather, exponential trends.

In his book, The *Singularity is Near: When Humans Transcend Biology*, Kurzweil identified six epochs of human evolution. The first epoch centered on physics and chemistry. We can trace our origins to a state that represents information in its basic structures, that is, patterns of matter and energy. A few hundred thousand years after the Big Bang, atoms came together to form molecules. Chemistry was born a few million years later. Of all the elements, carbon proved to be the most versatile. In the second epoch, which started several billion years ago, carbon-based compounds became more intricate until complex aggregations of molecules became self-replicating mechanisms, thus ushering in biological life. The third epoch began when animals began recognizing patterns, which still accounts for most of the activity in our brains. Our endowment for abstract thought combined with our opposable thumbs enabled human-created technology thus ushering in the fourth epoch. The singularity, when human technology merges with human technology marks the fifth epoch. Finally, the sixth epoch entails the propagation of human machine-based intelligence throughout the universe. In this final stage of evolution, intelligence will begin to saturate all of the matter and energy in its midst.[56]

The transition to the singularity will be the natural next step of evolution. Ultimately, he predicts, most of the intelligence in our civilization, will be non-biological. Rather than fear the singularity as a scenario in the style of The *Terminator* films, Kurzweil exclaims that it will be an era when humans realize their highest potential: "Our civilization will remain human—indeed, in many ways it will be more exemplary of what we regard as human than it is today, although our understanding of the term will move beyond its biological origins."[57] He dismisses fears that radical life extensions will lead to overpopulation and the exhaustion of limited natural resources. Such dire predictions, he counters, ignores the wealth of creation from nanotechnology and artificial intelligence.[58] Nanotechnology—which involves ultra-small-scale engineering—will have a spectacular impact on information storage.

As intelligence saturates the matter and energy available to it, Kurzweil predicts that it will turn "dumb matter" into "smart matter." By doing so, this smart matter, although nominally following the laws of physics, will be able to harness the most subtle aspects of these laws in such a way as to manipulate matter and energy to its will. When that point is reached, intelligence would become more powerful than physics, and even cosmology. A civilization than can harness this intelligence will be able to maneuver and control gravity (through exquisite technology) and other cosmological forces to engineer the universe it desires.[59] It would stand to reason that if this is the case, then the "L" variable in the Drake equation would be large because presumably, machine-based life would be more durable and adaptable than biological life.

Once this intelligence is achieved, Kurzweil reasons that it would require only a modest number of centuries for it to vastly expand throughout the universe. As he explains, it is unreasonable to assume that every single extraterrestrial civilization that is more advanced than ours would only be a few decades ahead. Instead, most would he ahead of use by millions, if not billions, of years.[60] Once the singularity is achieved, he predicts that it will begin to rapidly expand throughout the universe at the speed of light, if not faster. Yet, if it takes only a few centuries at most to go from the advent of computation to the singularity, at which point an advanced extraterrestrial civilization begins to expand outward at the speed of light, why have we yet to detect any such galactic empire? Inasmuch as we have not encountered any expansion of extraterrestrial intelligence, Kurzweil suspects that we are actually the most advanced civilization in the universe.[61] In recent years, several leading SETI scientists have echoed Kurzweil's thesis as it applies to extraterrestrial intelligence. Even as far back as 1964, Shklovskii and Sagan noted that in the future, the division of life into two categories—natural and artificial—could become meaningless. Biological organisms could synthesize with cybernetic devices. A gradual transition could take place concomitant with substantially extended lifetimes for these organisms. Such long-living entities would be better positioned for interstellar dialogue given the extended length of time necessary to communicating between extraterrestrial civilizations.[62]

Perhaps a day may come when biological intelligence will be viewed as merely a midwife of the "real" intelligence of the machine realm. Far more powerful, scalable, and adaptable, machine intelligence could be immortal.[63] In the future, Paul Davies predicts that the longstanding relationship between the biological and non-biological realms with be inverted. Instead of humans making specialized machines, machines will design and make specialized life forms.[64] When the day comes that humans determine genetic modification, then pure natural selection will break down. Although there are currently legal prohibitions against certain types of human genetic engineering, Davies notes that this taboo is rooted in the norms and mores of a particular era. These circumstances will likely change over time.[65] As a consequence, he predicts that the distinction between living and non-living, organism and machine, natural and artificial, is set to evaporate. Eventually, these "auto-teleological super-systems" will be self-designed and redesigned. They would grow, improve, and adapt, not by the long arduous process of Darwinian evolution, but through their own intellectual creativity.[66] Unlike humans who have a unique sense of self, artificial intelligence might not possess a personal identity of the type that we have come to know. The power of computers is that they can be linked together to pool resources. This quality would give a machine-based intelligence an enormous advantage over human intelligence because it could redesign itself, merge with other systems, and grow. As Davies explains, "feeling personal" would be a distinct impediment to progress.[67] Once an "extraterrestrial quantum computer" has secured its safety, stability, and an extreme sense of isolation, Davies believes that its own future would be guaranteed for trillions of years barring unforeseen accidents that could not be dealt with by automatic repairs.[68]

According to this line of reasoning, we can expect advanced extraterrestrial biological intelligence to have long since transitioned to machine form. Based on the history of Earth, the evolutionary period frame from the emergence of simple hunting and gathering bands to complex societies is relatively short on geological and astronomical time scales. Consistent with Kurzweil's thesis, Seth Shostak points out that the relative timescale to go from a technological society to artificial-based intelligence is very short. It would seem to follow then that the majority of intelligence in the cosmos would be artificial, not biological.[69]

If intelligent beings emerge elsewhere in the cosmos, it is likely that they have contemplated the issues discussed in this chapter and would be interested in communication with intelligence outside of their planets. With that assumption, the SETI (Search for Extraterrestrial Intelligence) enterprise looks for evidence of advance life in the cosmos which will be explored in the next chapter.

CHAPTER 5

The Search for Extraterrestrial Intelligence

HUMANS HAVE LONG MULLED OVER the prospect of aliens, but it was not until around the middle of the last century that a viable methodology of detecting extraterrestrial intelligence was put forward. The origins of the SETI (Search for Extraterrestrial Intelligence) can be traced back to the so-called "new astronomy," which recognized that celestial bodies radiated energy in many regions of the electromagnetic spectrum other than just the optical that we are familiar with in everyday experience. While studying radio interference in 1931, Karl Jansky of Bell Telephone Laboratories detected static, or radio emissions, by accident. This discovery led to the creation of the radio telescope, which is in essence a large antenna used to detect, amplify, and analyze radio emissions from celestial sources.[1]

Travelling at the speed of light, radio waves seemed to be the most viable known method of interstellar communication.[2] Taking this into account, in 1959 Giuseppe Cocconi and Philip Morrison published an article that called for searching for radio signals as a method to detect extraterrestrial intelligence.[3] Titled "Searching for Interstellar Communications," the article marked the first time in print that scientists took notice that radio telescopes were sensitive enough to detect radio signals from distant stars.[4] Morrison and Cocconi knew that space was a kind of radio, at least with respect to the short-wavelength microwave variety. Although stellar gas and dust block, or distort, visible light, microwaves are given a free pass. Furthermore, radio waves travel through Earth's atmosphere with ease. Thus they reasoned, radio seemed the logical medium that extraterrestrial aliens would use to attempt to make contact with other civilizations.[5]

The next year, Frank Drake initiated *Project Ozma*, which used a radio telescope at the National Radio Astronomy Observatory in Green Bank, West Virginia, to search for signs of extraterrestrial intelligence. The initial project took two months to complete.[6] Drake took the name from the princess in L. Frank Baum's children story *Ozma of Oz*, which was a book in a series called *The Wonderful Wizard of Oz*.[7] Perhaps no other figure than Drake has done more to advance the SETI cause. According to his autobiography, he developed a fascination with extraterrestrial intelligence at a very early age, despite his family's Christian fundamentalist background; he sought the truth through science.[8] While an undergraduate attending Cornell University, he was intrigued by a guest lecture presented by Otto Sturve, the famed astronomer. Drake was excited when Sturve opined that life could exist on planets outside of the our solar system.[9] Coincidentally, years later, Sturve took over as the director of the observatory in Green Bank, where Drake would work.[10]

Drake divided the history of the search for extraterrestrial intelligence into four eras. The first dates back at least 3,000 years ago when people started contemplating the nature of the universe, without any hard scientific data or use of the scientific method. Instead, they applied the principles of pure philosophy, most notably—logic, to deduce the nature—origins, and history of the cosmos. The second era commenced with the Copernican Revolution in the sixteenth century. Using new telescopes, astronomers such as Kepler and Galileo recognized other objects in the solar system to be planets similar to the Earth. It was in the early seventeenth century that the first hands-on proposals for signaling our planetary neighbors emerged. In 1959–60, the third era began, when scientists first employed quantitative measures to measure the strength of possible extraterrestrial signals traveling across interstellar space. It was during this era that SETI first embodied philosophical, qualitative, and quantitative elements. Initially, the effort was designated as CETI, or Communication with Extraterrestrial Intelligence; by the mid-1970s, however, the word *communication* was considered too presumptuous because no alien civilizations were discovered with which to communicate. Instead, the term *search* was considered more appropriate. NASA was the first to adopt the new term to christen its nascent search activities; SETI was soon accepted by the international community as well.[11] Finally, the fourth era, which began in the early 1990s, is not only quantitative, but thorough in the sense that an expanded array of methodologies is used to look for signs of extraterrestrial intelligence.[12]

SETI uses numerous scientific methods to scan the galaxy for electromagnetic transmissions from civilizations on other planets. At the present time, SETI concentrates on detecting radio signals for reasons of detectability and the assumption that it is the most viable means of interstellar communication. Although there is a chance that we may one day detect unintentional leakage radiation from an extraterrestrial civilization, deliberate signals that are transmitted with the intention of being detectable are the most likely to be found. Radio waves with certain properties—for instance, those that are persistent in time, frequency, and position in their area in the sky—are good candidates for re-observation.[13] In the main, the American SETI strategy hunts for continuous-wave signals with narrow band receivers.[14] The reason for this approach is because any signal that has a very narrow line width would likely be artificially rendered. The so-called "cosmic waterhole" is the transmission frequency where intelligent civilizations might gather to communicate with one another.[15] Currently, SETI assumes a wavelength of 18 cm (centimeters).[16] But not all SETI programs agree with this assumption. For instance, Soviet astronomers used a SETI strategy that assumed that extraterrestrials would realize the inherent difficulty in selecting a particular radio frequency, and thus would be more inclined to transmit a signal that could be picked up on any frequency. A short burst or pulse-type signal would fit this bill because it would cause radio power to appear over a very broad range of frequencies.[17]

In 1971, a band of 24 scientists and engineers created a blueprint for a vast array of telescopes known as Project Cyclops, which was proposed for detecting extraterrestrial life under the joint direction of Bernard M. Oliver (1916–1995) and John Billingham (1930–2013). Rather than build one large antenna, many small ones could be built, which would combine the signals collected from this multi-element array. By doing so, it would be possible to mimic the resolution (the ability to see detail) of an antenna as large as the distance between the farthest two members of the array.[18] A major reason for this approach is cost effectiveness. It is much less expensive to build many small antennas rather than one big one, even when the total collecting area is the same.[19] The Cyclops report estimated that unintentional leakage radiation from sources such as television and radio transmissions could be detected approximately 30 light years away with a high-quality radio receiver tuned to the right frequency, as long as the receiver was connected to an antenna three kilometers in diameter and pointed in Earth's direction. Unfortunately, Project Cyclops never got beyond the planning stages.[20] With an estimated $10 billion in 1971 dollars, the heavy price tag discouraged many people.[21]

On February 2, 1995, a new search effort, Project Phoenix, was christened. The venture continued for nine years and was serially conducted on three primary instruments: the Parkes dish in Australia, the 140-foot radio telescope in Green Bank, and a 1,000-foot Arecibo reflector in Puerto Rico.[22] Even more ambitious is the Allen Telescope Array (funded by Paul Allen, a co-founder of Microsoft), which relies on computation that can extract highly accurate signals from many low-cost dishes.[23] A joint program managed by the SETI Institute and the University of California at Berkeley, the Allen Telescope Array can perform many research tasks faster and more effectively than competitive radio telescopes. It became operational in 2007 and is located at the Hat Creek Radio Observatory in Northern California.[24] The array is a collection of 42 radio dishes—each 20 feet wide—that are connected together through computer technology to simulate the performance and sensitivity of a very large aperture telescope. As a result, it was built at much lower cost than would have been for constructing a larger telescope.[25] By the year 2030, the array could be able to check for signals in the direction of a million or more star systems.[26]

Humans have been broadcasting signals into space for only a little over 60 years. Furthermore, our radio and television transmission are exceedingly weak and could escape detection by extraterrestrial intelligence, even those equipped with enormous radio antennas.[27] Although radio waves travel at the speed of light, they also fade rapidly with distance.[28] At the present time, the power necessary to transmit detectable messages over long interstellar distances is beyond our capabilities.[29] Unless an extraterrestrial civilization resides close to our solar system, it is unlikely to detect our inadvertent radio transmissions.[30] It is conceivable, however, that an extraterrestrial civilization could operate technology that would enable it to detect even our relatively weak transmissions. If that is the case, then as Neil de Grasse Tyson wryly noted, our first messages to be detected by aliens could be television shows like *I Love Lucy* and *The Honeymooners*, which might lead them to question our intelligence.[31]

SETI scientists have been patiently waiting to hear a call from an extraterrestrial interlocutor. In 1968, a group of Cambridge radio astronomers discovered what they called "rapidly pulsating radio sources." At first, SETI scientists wondered if they could have been artificial signals sent by an extraterrestrial civilization.[32] It later transpired, though, that what they detected was a pulsar. First named by Frank Drake, it was later determined that pulsars are rapidly rotating neutron stars that emit massive amounts of radiation and produce highly regular radio transmissions.[33] Not to be dissuaded, the contrarian scientist Paul LaViolette once went so far as to argue that some of the pulsars that have been discovered may have actually been modified by an extraterrestrial intelligence. Although our radio telescopes pick them up, he laments that so far, we refuse to accept them for what they truly are.[34]

Over time, our civilization's electromagnetic footprint has been fading. Indeed, the use of radio communications could be a fleeting practice for advanced civilizations. In the early days of SETI, scientists believed that a relentless rise is radio traffic would develop as wealth and technology advanced. However, just the reverse actually occurred, as we have come to rely on methods of communications that minimize radio frequency power leakage, such as cable television and submarine telephone cables. First, low-power satellites that direct their signals earthward came to dominate point-to-point, which have been engineered so that almost all of the radio power is carefully aimed at the Earth with little radiation leakage into space. Second, the bulk of telecommunications shifted away from radio to fiber-optics buried beneath the ground.[35] Digital signals tend to be weaker and do not contain any strong narrow bands; as such, they resemble noise and would be much more difficult to detect. As

Frank Drake noted, the power released into space from these systems is a 100,000 or more times fainter than the power from the traditional television stations. As a corollary, "the bulk of electromagnetic transmissions of human civilization are fast fading to black!"[36] If current trends continue, by the end of this century, there will be no substantial radio output from Earth. From a SETI perspective, these trends are disheartening.[37] Based on the example of Earth, as extraterrestrial societies become more technologically sophisticated, they will probably reduce their wastage of power, and by doing so, become undetectable.[38] Consequently, it is possible that the galaxy could be bustling with advanced civilizations, yet they leave no detectable artificial radio signature.[39]

Aside from radio transmissions, some SETI scientists conjecture that extraterrestrial aliens would use lasers as a communications medium. Taking this into consideration, another search option employed is the Optical SETI, which was first proposed in 1961 by R.N. Schwartz and C.H. Townes, not long after the development of the laser.[40] Pulsed lasers could outshine our Sun briefly by at least four orders of magnitude in brightness, and could offer the best means of interstellar communication. An advantage to using lasers is that they could be narrowly beamed, thus providing a high amount of information per unit of energy.[41] One drawback, though, is that the beam is very narrow and could easily go undetected.[42] Although no evidence of interstellar message transmission via laser has been detected, the proposition in now seriously considered by the SETI community. Working on this assumption, Harvard University's Advanced All-Sky Optical SETI aims to scan the entire sky of the northern hemisphere for pulsed laser beacons.[43]

The optical SEIT actually has a long pedigree. The German mathematician, Karl Gauss (1777–1855), once considered using lanterns and mirrors to send signals to inhabitants on Mars. In a similar vein, the astronomer Joseph von Littrow (1781–1840) proposed in 1840 that a 24-mile wide ditch be dug in the Sahara Desert, filled with a mix of water and kerosene, and set on fire to serve as a beacon. Even more extreme, the French poet Charles Cros (1842–1888) suggested that a huge mirror be used to burn numbers onto the deserts of Mars as a surefire way to get the attention of our Martian neighbors.[44] In retrospect, these schemes may seem ludicrous, but if an alien civilization had the proper astronomical equipment, they may have been viable. In fact, such thinking continues today in the SETI community. Although it is currently infeasible to search for the lights of cities on exoplanets at night, Frank Drake mused that sometime this century, it might be possible.[45]

As a civilization advances, it develops innovative ways in which to harness energy. As mentioned in the previous chapter, Nikolai Kardashev proposed a classification of extraterrestrial civilizations based on their methods of energy extraction. Presumably, an advanced extraterrestrial civilization exploiting vast amounts of energy would produce a signature that could be detected. Kardashev reasoned that the handiwork of an advanced alien civilization would radiate enormous amounts of waste heat. Based on this premise, he argued that an appropriate SETI strategy should search the infrared region of the electromagnetic spectrum.[46] Although no telltale signs of extraterrestrial life have yet to be detected, Kardashev countered that there have not been proper searches.[47]

Energy from a planet's home star would provide a very large and renewable source of energy; it stands to reason that an extraterrestrial civilization would seek to harness it. With the proper instruments, Louis K. Scheffer contends that the large-scale use of solar power could be visible across vast interstellar distances, thus introducing another potential SETI strategy. For this to be detected, however, there must be methods sensitive enough to see the reflection of the solar rays against the glare of the star. Although an extraterrestrial civilization may not deliberately send signals out into space, there is a growing consensus on Earth that utilizing renewable energy is a good idea; it would seem to follow that an advanced extraterrestrial civilization would come to the same conclusion. Scheffer speculates that if only 1 percent of extraterrestrial civilizations will attempt explicit communication, but 50 percent go on to develop solar power, then the number of civilizations we might observe would increase by a factor of 50 if we concentrate our SETI efforts on this technique.[48]

Assumedly, a Type II or Type III Civilization would leave a significant footprint in the galaxy, such as radio transmissions or infrared emission, which would be discernible. According to the Second Law of Thermodynamics, an advanced civilization would generate large quantities of waste heat that could be detected by our instruments.[49] Almost certainly, a device such as a Dyson sphere would dramatically alter the light spectrum of the enclosed star, and in doing so, create a noticeable infrared glow that could be identified by peering astronomers, even on the far side of the galaxy.[50] Freeman Dyson urged SETI scientists to focus on technology and its effects when conducting searches for signs of extraterrestrial civilizations. A highly developed technological society, he argued, would emit intense infrared radiation that might be detectable to Earth's instruments.[51] In his estimation, too much attention has been given to planets. In fact, it is quite possible that a Dyson's sphere could outlive its creators and remain an artifact not unlike the Pyramids of Egypt.[52] In the summer of 2013, the Templeton Foundation—a philanthropic organization that funds interdisciplinary research relating to "big questions" of human purpose and reality—awarded a $200,000 grant to Geoff Marcy, a researcher with NASA, to examine data collected from the Kepler telescope to identify possible alien technologies in space. A distinguished scientist and SETI chair at the University of California at Berkeley, Marcy has identified over 100 exoplanets.[53]

In upcoming years, improvements in astronomy could make an optical SETI strategy more feasible. Donald E. Tarter advocates the placement of radio telescopes on the far side of the Moon as that would be an excellent location not only for astronomy, but as a SETI listening post as well. Because the dark side of the Moon is insulated from most, or all of the background of Earth's microwave noise, it would be an excellent site for a SETI satellite array.[54] Even more ambitious would be the creation of a "gravitational lens" that would use the Sun to provide a detailed view of the entire universe. The concept stems from Einstein's theory of relativity, which posits that the presence of mass bends space. In our solar system, the massiveness of the Sun significantly warps the space it occupies and could thus serve as a gigantic lens through which to observe the cosmos. If a receiving device could be placed in a far region of space—say 300,000 astronomical units[55] from the Sun—we could have a window on the universe.[56]

Inasmuch as the Milky Way is so vast, SETI scientists must limit their search efforts to certain areas in space. Presumably, the best place would be to look at nearby stars, but according to Ray Kurzweil's thesis, intelligent life will one day evolve into a post-biological entity. Perhaps, this artificial intelligence would no longer require a habitable planet, but could survive anywhere in the cosmos where computers could operate. This presumption could have implications for SETI, for as Michael Michaud noted, if biological extraterrestrial intelligence is unlikely to sustain beacons for millennia, machine intelligences might be even less inclined to do so. As a consequence, an extraterrestrial intelligence "whose presence is in the voids of interstellar space, would be extremely difficult to detect."[57]

Machines would not require a habitable planet in the traditional biological sense. Rather, they could roam the cosmos in spaceships. Taking into account the prospect of extraterrestrial quantum computers, Paul Davies recommends a SETI strategy that would seek to identify signatures of intelligence through the impact that alien technology would make on the astronomical environment.[58] Conceivably, these machine-based extraterrestrials could be voracious consumers of energy. If that is the case, they could be easier to detect in SETI experiments. Based on this assumption, SETI scientists might look for areas near the galaxy's hot spots where aliens might decide to saddle up. On that note, Clement Vidal, a researcher at the Free University of Brussels, mused that advanced alien civilizations might migrate toward black holes to harness their massive energy.[59] The galactic center at which massive black holes reside, could be yet another location for aliens seeking a prodigious stream of energy.[60]

On the other hand, as a technologically advanced civilization would presumably consume energy more efficiently, its demand for energy resources might actually decrease over time. The computer scientist Marvin Minsky once questioned the assumption that an advanced technological civilization would radiate more infrared emission. Because radiation at any temperature above the cosmic background level is wasteful, he reasoned that the more advanced the civilization, the lower would be its infrared emission. His assumption implies that the more advanced technological civilizations could actually be less detectable.[61] As a society experiences a natural progression toward efficiency by reducing its leakage into space, it could belie its presence in the cosmos. Just as we no longer use smoke signals, advanced extraterrestrial civilizations could have found superior means of communications to radio transmissions.[62]

Still another possible mode of communication would be to send physical artifacts throughout the universe. Stanley Kubrick's film *2001: A Space Odyssey* was based on this theme, as a vertical rectangular "monolith" is discovered by astronauts on the Moon. Scot Lloyd Stride of the Jet Propulsion Laboratory and Richard Burke-Ward, a television journalist and novelist, respectively, have recommended broadening SETI to encompass a Search for Extraterrestrial Artifacts.[63] Likewise, two computers scientists at Rutgers University, Christopher Rose and Gregory Wright, argue that aliens are most likely to communicate in a manner not unlike messages in bottles: Aliens would record their messages on floppy disks, thumb drives, flash cards, or their equivalents and pack them into interstellar rockets that are aimed at their extraterrestrial interlocutors.[64] In fact, a similar method has already been practiced by humans. In 1972, the Pioneer 10 spacecraft, which was directed at a celestial sphere near the boundary of the constellations Taurus and Orion, contained a plaque designed by Carl Sagan that detailed a diagram of a hydrogen atom, a pulsar map with the Sun at the center, figures of a nude man and woman set in front of a to-scale silhouette of Pioneer, and a sketch of the solar system. The next year, an identical plaque was launched on the Pioneer 11 spacecraft. Still another artifact was sent out into space in 1977: On board was a collection of Voyager Golden Records, which included 116 images detailing life on Earth and the methods of finding the planet, as well as the recorded voices of U.S. then-President Jimmy Carter and the Secretary General of the United Nations, Kurt Waldheim.[65] The expected erosion rate is so low that the messages should remain intact for millions of years, which could make them the longest-living man-made artifacts.[66]

Taking into account the tremendous cost of sending astronauts to other solar systems, Donald E. Tarter speculated that a Type III civilization might instead choose an option based on miniaturization using nanotechnology and quantum engineering. Though this method, the civilization would not make itself known through gargantuan projects or beacons or ferocious energy consumption; instead, they would opt to send miniature probes and forms of communication that might not be obvious to us.[67]

At the present time, microwaves seem to be the most efficient and least expensive medium to transmit messages into space, but in the future, more exotic methods could be used.[68] Some scientists propose that someday neutrinos—nearly massless particles that travel at the speed of light—could be used as a medium for interstellar communication. For reasons that are still unknown, neutrinos effortlessly pass through matter.[69] But only rarely will a neutrino hit a nucleus and bring about a detectable transmutation.[70] As their interaction with matter is extremely weak, they could travel unimpeded throughout the galaxy and even beyond. As such, they could be an effective medium of interstellar communication.

All of the transmission and detection methods discussed so far assume that the speed of light limit is inviolable. But as Frank Drake opined, the solution to the SETI question may lie in laws of physics yet undiscovered.[71] He and Martin Harwit once speculated that tachyons—hypothetical particles that travel faster than the speed of light—could be used as a medium of interstellar communication.[72] Even more promising, would be the use of quantum entanglement to transmit information. When particles such as photons, electrons, or even molecules the size of buckyballs,[73] interact and then separated, they are said to be entangled. Once entangled, the observation of one particle will produce instant effects for the observer of the twin particle irrespective of the distances (even light years) separating them. Entangled particles are mysteriously linked together and remain intertwined forever. Such a notion seemingly violates Einstein's dictum of locality, which postulates that what happens in one place cannot affect something in a faraway location unless a signal is sent to the other location at a speed equal to or less than the speed of light.[74] Be that as it may, this effect has been demonstrated on numerous occasions. Nicholas Gisin, among other physicists, confirmed entanglement with numerous experiments.[75] Likewise, the experiments of John Bell proved that under certain conditions, Einstein's "spooky action at a distance" does indeed occur.[76] As of yet, however, it does not appear possible to send readable messages faster than the speed of light.[77]

Could quantum entanglement someday be used effectively as a means of interstellar communication? Donald E. Tarter speculates that someday we might be able to do so, which would allow us to communicate instantaneously across great distances, thus avoiding the speed of light limitation as outlined in Einstein's Special Theory of Relativity. For his plan, Tarter calls for sophisticated "photon traps" that would grab and store the photons for polarization measurement. The information transmitted would be carried in the changing polarization of the photon stream. Once two civilizations have entangled photons in the position, instant communication would then be possible. However, the initial contact could not exceed the classical limit of the speed of light. Tarter imagines that advanced civilizations might use massive and distant cosmic phenomena such as gamma-ray bursts to carry their quantum keys throughout the universe. In that sense, the gamma-ray bursts could operate as an intergalactic network of information transfer not unlike the human brain, in which the interstellar signals approximate the firing of neural synapses.[78]

In 1992, Frank Drake boldly predicted that by the year 2000, we would detect a signal from an extraterrestrial civilization.[79] He assumed that in all likelihood, any civilization that we discovered would be more advanced than ours, thus in a sense providing us with a glimpse into the history of the future.[80] To date, there have been some exciting signals detected, but alas, they later transpired to be natural and not artificial in origin. For example, in 1965 Nikolai Kardashev and Evgeny Sholomitsky discovered a suspicious signal in space. The signal was so powerful that they thought it could have been transmitted by a Type III civilization. The Sternberg Institute quickly convened a press conference, where it was announced that an alien signal had been detected. It later transpired, however, that the signal emanated from a quasar. Quasars, or quasi-stellar objects, are turbulent galaxies from whose cores tremendous amounts of energy are released.[81] Another probable false positive—the "Wow" signal—occurred on August 15, 1977, when Jerry Ehman, a SETI volunteer at The Ohio State University, observed a startling strong signal received by telescope. After circling the indication on a printout, he scribbled "Wow!" in the margin. The radio telescope indicated that the signal was 30 times as loud as normal deep-space radiation.[82] Some observers consider the message to be the most likely candidate from an extraterrestrial source ever discovered. Subsequent searches, however, have failed to detect it again. Most likely, the powerful narrowband spike emanated from a man-made satellite.[83]

To create an institutional basis for the search for alien intelligence, the SETI Institute was founded on November 20, 1984, to seek private funding for a targeted search.[84] As the mission statement of SETI proclaims, "We believe we are conducting the most profound search in human history."[85] In his autobiography, Seth Shostak recounted that over the course of his career, he met no more than a handful of astronomers who doubted that extraterrestrials were out there. Nearly all astronomers agree that both life and intelligent life are abundant in the universe, but not all agree with the effort to search.[86] Thus funding for the endeavor has at times been contentious.

At one time, SETI succeeded in acquiring federal funds, but the project has not without its critics. On February, 16, 1978, Senator William Proxmire (D-WI) bestowed his infamous "Golden Fleece of the Month Award" to SETI. The dubious distinction was given to those programs that Proxmire deigned to be wasteful. Not to be deterred, Carl Sagan was able to secure a private meeting with Proxmire and convince him of the efficacy of the SETI program. In his meeting with Proxmire, Sagan discussed the L variable—that is, a civilization's longevity in the Drake equation and its relevance to Earth. As Sagan explained, the way for a civilization to attain great age would be to work through its period of technological adolescence by avoiding self-destruction after the advent of nuclear technology. Upon hearing this, Proxmire was incredulous, asking Sagan, "So you mean that if we find some evidence of extraterrestrial intelligence, somebody elsewhere has avoided self-destruction?" Sagan replied, "That's very much what many of us believe." Proxmire then followed, "Well, if that's the case, if other worlds might be able to teach us how to survive, maybe SETI's worth the investment."[87] As a result, Proxmire ended his campaign against SETI.[88] But in September 1993, a freshman Democratic senator from Nevada, Richard Bryan, led an effort to end federal funding for NASA's SETI program. By a vote of 77 to 23, the Senate voted to allow his amendment to stand, which defunded the program.[89] After the SETI program was canceled, Senator Bryan remarked, "This hopefully will be the end of the Martian hunting season at the taxpayer's expense."[90] NASA had planned on spending $100 million on the SETI mission during the decade of the 1990s.[91]

NASA's decision to cut funding for SETI is somewhat surprising given the large public interest in extraterrestrials. As the noted astrophysicist Neil de Grasse Tyson once intimated, whenever a person seated next to him on a plane discovers that he is an astrophysicist, "nine out of ten times she'll query [me] about life in the universe." Tyson found that no other topic triggers so much interest from the public.[92] Although NASA cancelled funding for SETI, the agency continues to support astrobiology.[93]Over the long haul, it could be difficult to sustain political support for a highly speculative program like SETI, which requires a long-term commitment for success. A science fiction writer and advisor to SETI, Paul Anderson explained, "If SETI succeeds, CETI will then stretch over a period of years, decades, or possibly centuries. An organization dedicated to it may become something like a church, outliving nations as it carries on its magnificent mission."[94] Albert A. Harrison once mused that the culture of North America, with its emphasis on individualism, may not be suitable for a long-term project such as SETI; by contrast, Asian societies, which are more group-oriented and favor long-term planning, might be more conducive to the nature of SETI.[95]

Despite the loss of federal funding, the SETI Institute enlisted the help of fundraising experts who were able to acquire substantial support from a number of private donors, including William Hewlett and David Packard (from the Hewlett Packard Company), Mitchell Kapor (of Lotus), Gordon Moore (co-founder of Intel Corporation) and Paul Allen (co-founder of Microsoft). As a result, SETI was able to rebound as a private enterprise.[96] Still, sporadic and uncertain funding hampers the SETI project. For instance, in April 2011, SETI temporarily shut down its Allen Telescope Array due to lack of funding.[97] But within a few months, the SETI Institute was able to raise enough money from over 2,000 donors, including Jodie Foster, who played the fictional SETI scientists Ellie Arroway in the 1997 film *Contact*, to resume the search for alien civilizations.[98] As Michael Michaud noted, SETI is "dependent on the personal decisions of very wealthy people. To loosen purse strings, we may need a motivating event such as finding evidence of alien biology on an extrasolar planet."[99]

With its limited resources, the SETI Institute has developed innovative ways to carry on its efforts. In 1999 a downloadable program—SETI@home—was introduced to computer users who want to contribute their computer power to SETI. In a quintessential exercise of crowdsourcing,[100] data from the Arecibo radio telescope are sent via the Internet to the personal computers of volunteers who have donated their computers' idle time toward the search. By 2009, more than 8 million persons had downloaded SETI@home.[101] The idle time by such a vast number of computers provides an enormous computational resource that can be exploited.[102] Yet another crowdsourcing project—the nonprofit SETI League—was founded in 1994 to organize search efforts by amateurs using small dishes with the goal of networking these capabilities into a global system of 5,000 observing stations.[103]

So far, SETI has detected no evidence of extraterrestrial radio transmissions. Is anyone really out there? Or perhaps, is SETI using the wrong methodology? After all, even communications traveling at the speed of light would not make for a speedy dialogue by interstellar standards: A radio message to our nearest star—Alpha Centauri—would take 4.2 years to arrive at its destination. Could there be more effective means of communication employed by extraterrestrial civilizations? Some possible alternatives were discussed earlier, but at the present time, they are beyond the reach of our current science and technology. As it now stands, searching for radio signals seems to be the most viable SETI strategy, but the community should be open to other methodologies as they become available. Thus the SETI institution should not become too committed to a particular method lest other avenues show promise in the future.

Although we have not detected aliens, might they have detected us? If so, why then do they not make an effort to communicate with us? Some scientists have wondered if extraterrestrial civilizations might fear aggression from galactic neighbors and are thus loath to broadcast signals. Perhaps some alien civilizations are xenophobic and seek to hide their communications signals with the hopes of not being detected by potential competitors.[104]

Or perhaps extraterrestrial aliens are aware of our presence, but are not interested in us. Michio Kaku uses the analogy of human contact with ants. When we come upon an anthill, he explains, we do not request to see their leader or bring trinkets to them and offer unparalleled prosperity through the fruits of our technology.[105] Because of astronomical time scales, a civilization capable of visiting Earth could be millions of years ahead of us and thus might find us uninteresting. As he points out, the main danger ants face is not that human want to invade them or eradicate them. Rather, we might simply "pave them over because they are in the way."[106] As Kaku, pointed out, the real danger would be if Earth got in the way of the aliens' highway.

By 2009, SETI had examined only 0.0000005 percent of the Milky Way.[107] Not to be discouraged, Jill Tarter points out that so far SETI efforts are analogous to scooping a glass of water and examining it to see if there are any fish in the ocean. The vast length of the Milky Way—about 100,000–120,000 light years in diameter—will require much time for a thorough scan for radio signals. However, as Tarter explains, Moore's Law (the axiom that computing power tends to double every 18 months) will enable SETI programs to scan greater areas in the cosmos and search for more complex signals in the near future.[108] If someday SETI's efforts are indeed successful and an extraterrestrial signal is detected, what effect would it have on our civilization? How should we respond? The next chapter addresses these questions.

Chapter 6

Preparing for Indirect Contact

If the SETI enterprise is someday successful, it could have serious implications for our institutions and our conception of our place in the cosmos, for we would know for certain that we are not alone. The detection of a message from an extraterrestrial civilization would be the culmination of the Copernican principle, which posits that the Earth occupies no central or favored position in the universe. And just as Copernicus's heliocentric model of the solar system bedeviled the church during the Renaissance, a detection of extraterrestrial intelligence could prove to be a conundrum for some religious traditions—to wit, Christianity—insofar as its adherents assume a special relationship between *Homo sapiens* and a divine creator. A discovery of such profundity could affect other institutions as well. Perhaps governments will fear that their authority could be challenged if a more advanced civilization is discovered. On a psychological plane, detecting an alien message might lower our collective self-esteem if the sender of the message exhibits scientific and technological superiority.

With the receipt of a message, we will have to decide if we should remain silent or respond. Some scientists counsel that we should listen for, but not respond to, extraterrestrial radio transmissions. The so-called "passive SETI" entails the use of general purpose astronomical instruments for the purpose of detecting signals only. By contrast, "active SETI," sometimes referred to as

"CETI" (Communications with Extraterrestrial Intelligence), is the intentional transmission of signals with significant power toward selected stars and targets.

The first international conference on the prospect of contact with extraterrestrial intelligence was held in September 1971, at the Byurakan Astrophysical Observatory of the Armenian Academy of Sciences in the Soviet Union. The acronym CETI was chosen for the meeting for a variety of reasons. First, "Communication" as indicated in the acronym, was one of the stated goals of the conference participants. Second, *ceti* is the Latin genitive for whale. As cetaceans are an intelligent species, they might hold out the best hope for inter-species communication on Earth. If we could communicate with dolphins, for example, then perhaps someday we will be able to communicate with extraterrestrial civilizations as well. Finally, one of the two stars, which was first examined by Frank Drake in Project Ozma (the first SETI effort inaugurated in 1960), was Tau Ceti.[1] Carl Sagan and I.S. Shklovskii were the main organizers of the conference. Together, they would go on to author a book—*Intelligent Life in the Universe* (1966)—which became the bible of the SETI enterprise. Like Frank Drake, Sagan rejected the religion of his youth—Judaism—but seems to have sublimated the spiritual impulse in the search for extraterrestrials. As Sagan's biographer Keay Davidson explained, the famed astronomer developed a "quasi-religious belief" in extraterrestrial intelligence, which from a psychological perspective suggested that these alien beings "were secular versions of the gods and angels he had long since abandoned."[2]

To date, SETI has never committed a sustained transmitting effort in the sense of using enough power and large enough antennas to broadcast effectively across interstellar distances.[3] Nevertheless, there have been a few symbolic efforts to reach out to extraterrestrials and notify them of our existence. Sagan was involved in these projects. For instance, in 1974, he and Frank Drake created the Arecibo Message, which was beamed in the direction of the M13 star cluster. As the first message intentional message sent into space, it consisted of 1,679 binary digits that collectively formed an image of the Earth. The message was controversial, as some commentators feared that it was folly to announce our position in the cosmos to extraterrestrial aliens who might harbor malicious intent toward us. For instance, the British astronomer and Nobel Laureate

Sir Martin Ryle opposed the Arecibo Message and other active CETI efforts for fear that humans could be attacked.[4] The "Great Transmission Debate" centers on the appropriateness of sending intentional signals into space. The main concern is the risk that contact, whether direct or indirect, could be detrimental. This presents a conundrum because presumably, it will be very difficult to detect an unintentionally transmitted radio signal; thus, communication between civilizations is unlikely to occur until one side initiates contact. The science fiction author and futurist David Brin has counseled strongly that there should be a moratorium on messaging extraterrestrial intelligence until an informed debate on the topic takes place.[5] Even Sagan conceded that Earth's civilization was still too young to try to contact extraterrestrial civilizations.[6] The Arecibo Message, he explained, was not really intended as a serious attempt at interstellar communication; rather it was an indication of the remarkable advances in terrestrial technology.[7]

Even a seemingly benign proposition as receiving an alien message might prove to be dangerous. For instance, a malicious extraterrestrial broadcaster could send a message analogous to a computer virus that damages human technology or coerces humans into a seemingly benign, yet ultimately destructive course of action.[8] In some science fiction stories, humans receive instructions to construct a device, which they dutifully build once completed, the machine takes over the Earth.[9] In this same vein, Allen Tough once speculated that extraterrestrials might offer us a Trojan horse, that is, something that appears good but is actually malevolent and could lead to our destruction. For example, a computer-controlled biochemical or genetic experiment purporting to prolong human life might introduce a disease or create new dangerous life-forms on our planet. If we do not understand the full implications of the step-by-step instructions, we would have to rely on good faith.[10] Similarly, in Fred Hoyle's British television drama—*A for Andromeda*—an alien civilization beams biological information to trusting Earth scientists and persuades them to incubate copies of the extraterrestrials in a sort of long-range version of *Jurassic Park*.

Considering the potential pitfalls, Mark C. Langston warned that sending an electronic message with an embedded virus might be an attractive measure for a paranoid extraterrestrial civilization seeking to neutralize potential competitors in the galaxy. For his putative "Centaurian virus" to do its damage, however, it would have to be able to detect and use modern networking hardware on Earth. Thankfully, as Langston points out, the virus would likely to be stopped in its tracks because the odds of an alien computer sharing the same configuration of computers on Earth are essentially zero.[11]

Currently, there is no legal framework that regulates the transmission of messages into space. As Kathryn Dennig explains, a "tragedy of the commons" problem inheres in this dilemma in the sense that an individual could take it upon himself to transmit signals without concern for the welfare of his planet. Soon the technology will be available to many who would choose to do so, and it will be well-nigh impossible for governments to prevent these transmissions.[12] In a controversial project announced in 2013, a group of scientists and entrepreneurs created the world's first continuous message beacon in an attempt to communicate with extraterrestrial civilizations. Jacob Haqq-Misra and a group of investors took over the Jamesburg Earth Station radio dish in Carmel, California. From this facility, they plan on sending messages to various star systems. For a fee, people can use the device to transmit their own messages into space.[13]

Other observers believe that these fears are overwrought. Douglas A. Vakoch calls for integrating active and passive SETI programs, that is, the transmission of intentional signals before detecting direct evidence of extraterrestrial intelligence. Some would argue against this idea, noting that our civilization is too young to attempt to communicate with extraterrestrial civilizations with which Earth would be at a comparative technological disadvantage. Statistically, it follows that any civilization detecting our signals would be many years ahead of ours. However, as Vakoch explains, even a million-year-old civilization could find reasons for not sending intentional messages because there could still be others that are more advanced that might have hostile intent toward it. Based on this assumption, all civilizations would thus settle for a passive, rather than an active search strategy and as a result, no interstellar dialogue would ever commence.[14]

This seeming impasse to interstellar dialogue notwithstanding, James Benford et al. note that it is part of the human condition to communicate our existence to both our contemporaries and posterity. The drive to communicate with other civilizations could take several forms. For instance, a civilization near the end of its life could feel compelled to announce its existence or plead for help. Or, perhaps a civilization could send out a beacon based purely on pride. Religion could be common in the cosmos, in which case, aliens might send beacons to spread the good news and look for converts.[15]

If we decide to transmit messages into space, what should we say? And how should interstellar messages be formulated so that extraterrestrial alien civilizations could decipher them? A number of factors have to be considered when transmitting messages. Astronomers must decide what frequency to use and where to point the message. The timing of the message is important as well. Finally, the senders have to settle on a language to use and determine how the message is to be encoded (i.e., technical details such as bandwidth and the form of modulation—AM, FM, or other).[16] Pointing out the limits of interstellar messages, Seth Shostak, the senior astronomer at the SETI Institute, notes that because of the long distances that signals must traverse to reach a recipient, senders might assume that they are engaged in one-way communication. Even using Steven Dick's optimistic estimate that there are between 1,000 to 100,000 civilizations throughout the Milky Way, the average separation between them would be in the range of hundreds to a few thousand light years.[17]

Based on our current knowledge of physics, a two-way conversation is unlikely to be lively because even messages traveling at the speed of light would take many years to be received. Consequently, senders might be inclined to transmit the content of long messages all at once, since an actual dialogue between interstellar civilizations may never occur. But if the messages are too long, it would be difficult to pick up the signal just as the message begins. As Ronald N. Bracewell once noted, a message containing a long story runs the risk that the listener will tune in near the end of the broadcast; conversely, a short message repeated again and again could bore the listeners to tears while they try to acknowledge it. Taking this into account, Bracewell recommends the transmission of messages that are nested within other messages. That is, messages containing short items that are frequently repeated and sandwiched between episodes of a longer story that is repeated less frequently, all of which would be contained in an even larger communication.[18] To be effective, repetition would be necessary and should occur at intervals. But because of cost constraints, Shostak believes that unless the transmitting society is so advanced that it can afford both the instruments and the energy to broadcast simultaneously to a vast number of stars, it will most likely adopt a communications strategy that focuses on sending short, frequently repeated messages.[19] Concerning the response, Carl DeVito recommends that our reply should consist of a series of messages—perhaps a week or a few days apart—that contain more information than the last, but also some redundancy as a check on interpretation.[20]

The cognitive foundations of interstellar communication could be quite complicated. As David Dunér observes, communication is not something that is pre-given; rather, it is a process that evolved over millions of years that involved interplay between humans and their environment. Furthermore, it would be anthropocentric to assume that extraterrestrial aliens would have senses exactly like humans insofar as their different physiques and brains would lead to different patterns of cognition. To maximize the chances that our communications will be received and understood, Dunér recommends that we aim our CETI efforts at exoplanets with similar characteristics to Earth. He bases this recommendation on the theory of convergent evolution. As evolution is to a great extent a process of adaptation to specific environments (as well as mutations and accidental occurrences), there would be a greater probability that life forms emerged similar to ours on these planets and would thus be more likely to understand us. Rather than sending abstract messages based on some assumed universal scientific or mathematical knowledge, Dunér argues that a CETI message should be very specific and concrete in content so that it would force the receiver to interpret it in just one way: "it must be direct, immediately understandable, nonambiguous, and tied to the situation and locality."[21] To be sure, achieving this intersubjectivity would be challenging; however, he posits that it would be the most effective method to establish viable communications between two civilizations. By establishing a mutual referential behavior, this intersubjectivity is more likely to be achieved.[22]

Inasmuch as anthropologists are trained in deciphering and translating other cultures, their cross-cultural analysis could be applied to interpreting extraterrestrial messages.[23] Drawing upon cognitive archaeology, Paul K. Wason notes the inherent difficulties in communication between even closely related species. As he explains, modern humans and Neanderthals coexisted in regions for quite some time; it is likely that they shared the Levant for some 60,000 years, although evidence of them living side by side is unclear. Modern humans entered Europe roughly 40,000 years ago. Neanderthals did not die out until about 30,000 or 32,000 years ago, so there was considerable time for the two species to overlap while there. Although Neanderthals made tools, they were far less sophisticated than those created by Upper Paleolithic hunters. Neanderthals did indeed make limited use of symbolism, but there is virtually no accepted evidence that they engaged in artistic activities, such as painting or music. Wason argues that the interrelationships between social intelligence and technical intelligence found in modern humans would be opaque to Neanderthals. In short, Neanderthals would have much difficulty interpreting the lives of modern humans. The implications for SETI are sobering, for if two such closely related species as Neanderthals and *Homo sapiens* would have such a difficult time understanding each other, meaningful dialogue with an independently evolved species could be even more challenging. Rather that expecting to fully understand another species, he suggests that a more reasonable goal is to identify the intentions in their signals.[24]

Even within our own species, communications between different cultures can be challenging. Jason T. Kuznicki disputes the SETI assumption that extraterrestrial aliens, by virtue of their communicative abilities, will share with humans a certain set of concepts such as mathematics, logic, and spirituality. Examining the historical case of contact in the seventeenth century between the French Jesuits and the Montagnais, Huron, and Iroquois tribes in what is now Eastern Canada, Kuznicki demonstrates that there were significant differences between the Jesuits and their Indian interlocutors over the nature of metaphysical concepts such as the soul. From this case, he infers that humans, at least initially, may find the reasoning of extraterrestrial aliens to be woefully defective. Likewise, they may find ours equally perplexing.[25]

Guillermo A. Lemarchand and Jon Lomberg argue that we must find "cognitive universals" that could be understood by alien civilizations. Specifically, they recommend that this strategy should focus on four types of cognitive universals: physical-technological, aesthetic, ethical, and spiritual. An advanced civilization presupposes a certain mastery of science and mathematics, which might lead one to believe that the principles underlying these fields are universal. Based on this assumption, Paul Davies, the Chair of the SETI Post-Detection Task Group, recommends that if presented with the opportunity to reply to an extraterrestrial intelligence, he would choose "Maxwell's equations, the field equations of general relativity, Dirac's equation of relativistic quantum mechanics, and a selection of mathematical theorems."[26] As humans and extraterrestrials share a common physical reality, their independent attempts to understand the universe through science will tend to converge over time. After all, the laws of physics and chemistry are universal. Furthermore, because they resist interpretations as having moral or religious overtones, messages based on science and mathematics are less likely to offend anyone and would thus make good starting points for an interstellar conversation.[27] All the same, these scientific disciplines could vary considerably, as there is sure to be a degree of both methodological and semantic incommensurability between civilizations on these matters.

For our messages to be deciphered, some SETI scientists have suggested the creation of a Rosetta Stone that would enable the recipients to interpret our communications. The original Rosetta Stone was discovered in 1799 by a French soldier in the Nile Delta town of Rashid. In 1824, the French orientalist Jean Francois Champollion deciphered the Rosetta Stone, which enabled him to read the Egyptian hieroglyphics.[28] Stéphane Dumas puts forward a proposal based on mathematics and physics. Information about the properties of the hydrogen atom, for example, is universal and stands a good chance of decipherment. By contrast, human languages are culturally dependent, imprecise, and difficult to use. To effectively communicate, Dumas suggests that a synthetic language should be created.[29] Likewise, Hans Freudenthal once proposed the development of a lingua cosmica for the purpose of communicating with alien civilizations.[30] To that end, he created a language called LINCOS, based on mathematical and logical precepts. Alexander Ollongren later refined his system and christened it NEW LINCOS.[31]

Besides considering decoding, when formulating messages, the sender should also make the messages entertaining so that they hold the recipient's interest. Ian O'Neill, an astronomer and science producer for *Discovery News*, recommends that we send messages that are as interesting and attractive as possible; otherwise, the aliens may find us boring and not respond. He argues that so-called universals, such as mathematic formulas, might be ignored by intelligent extraterrestrials who probably would have mastered these disciplines eons ago. In their stead, he suggests taking a lesson from Twitter and transmitting short messages that are full of fascinating information.[32] Similarly, Seth Shostak proposed that if we want to send a message that is clearly understood, then we ought to feed Google servers into the transmitter and effectively send the World Wide Web to aliens. He estimates that this method would take less than a year to transmit using microwave; using infrared lasers would shorten the transmission time to no more than two days.[33] Rather than concerning ourselves with the content, we could send them everything—"the good, the bad, and the unattractive."

Perhaps a SETI message would be more interesting if composed in a journalistic fashion. As Morris Jones points out, historically, SETI scientists have maintained good relationships with journalists; however, these relationships have been one-sided in the sense that the SETI scientists provide material for the journalists' stories. Jones proposes that the SETI community could benefit by encouraging input in the opposite direction when contemplating how to compose interstellar messages. As an extraterrestrial civilization receiving a message from Earth is likely to be far more technologically advanced than we, the former is likely to find nothing novel in our communications if it is based on scientific or mathematical principles. To make our SETI messages more interesting, Jones counsels that it might be useful to spice them up with entertaining hooks to attract their attention. These hooks could be in the form of art, music, or anything else that is likely to be distinctive to our species.[34]

Likewise, Harry Letaw Jr. recommends that if we should decide to send messages in space to extraterrestrial civilizations, they should include a form of "cosmic storytelling." Human beings are a species of storytellers; hence, a narrative embedded in a message could impart a reasonably accurate portrait of the human condition. Images of human activities could be used to convey stories that would provide a context for interstellar communication rather than messages containing only abstract symbols such as an alphabet and numbers.[35]

Species on other planets are sure to have taken different evolutionary trajectories than ours which will certainly affect their outlook and aspirations. As history on Earth has demonstrated, increased technological capacity carries with it the potential for self-annihilation of our species in the form of global nuclear war, pandemics, environmental degradation, and overpopulation. For a civilization to survive over the long haul, it must produce an ethical breakthrough among members of its community to achieve comity on its planet. An extraterrestrial civilization that has achieved such harmony could conceivably impart an ethical universal that would be applicable to Earth.[36]

What would be the implications if we someday receive an alien message? According to the optimistic view as expressed by Carl Sagan and Frank Drake, we might gain access to an "Encyclopedia Galactica," which would answer all of the important questions in science, engineering, and social sciences. By contrast, pessimists warn that any extraterrestrial civilization with which we come into contact will almost certainly be technologically far more advanced than we are. Consequently, it would be prudent to proceed with the utmost caution. In fact, it would probably be best for us to remain silent.[37] In 1960, a committee under the leadership of the psychologist Donald N. Michael completed a report on the peaceful uses of space for the US Congress. What became known as the "Brookings Report" discussed the prospect of the discovery of life beyond Earth. The committee expressed concern about how this discovery might affect people, noting that it could have potentially devastating effects on religion. The discovery of life elsewhere would contradict the religious parables that depend heavily on what occurred on Earth, such as Christ's birth and redemption.[38] Moreover, the report noted that historically whenever a more advanced civilization encountered a less advanced one, the results have often been disadvantageous for the weaker party.[39] By the 1970s, SETI scientists began to assume that the establishment of communication with aliens would have profound effects on science, technology, and philosophy on Earth. In 2002, The SETI Committee of the International Academy of Astronautics accepted the Rio Scale—a scale ranging from zero to ten—for use in evaluating the impact on society of any announcement regarding the discovery of evidence of extraterrestrial intelligence.[40]

How does the public feel about the prospect of discovering alien intelligence? Examining attitudes in America about this issue, George Pettinico drew upon a 2005 telephone survey conducted by the National Geographic Channel, the SETI Institute, and the University of Connecticut which found that a majority of Americans—60 percent—expressed a belief that "there is life on other planets in the universe besides Earth."[41] His study identified two key demographic factors that influence people's attitudes about extraterrestrial life—that is, level of education and religiosity. Americans who are more religious (as measured by the frequency of religious service attendance) are twice as likely to reject the notion of life on other planets. This finding is consistent with Pettinico's expectations insofar as conservative Christians are more likely to view the universe—despite its size and scope—as existing for the benefit of humanity alone. Although most religious Americans accept the notion that the Earth is not at the physical center of the universe, in their minds our planet remains the single spot of divine focus where God created life. Still, nearly half (45 percent) of religious Americans believe that life exists on other planets besides Earth. Those people who are open to the possibility of the existence of extraterrestrial life are also more positive about eventually making contact someday. College-educated respondents were less fearful of the discovery of extraterrestrial life than their less educated counterparts.[42]

Hollywood's portrayal of hostile alien invasions notwithstanding, historic anecdotes suggest that the general public is sanguine about the existence of extraterrestrial civilizations. Rather than being fearful, they are more likely to be excited about the proposition. Albert A. Harrison noted several precedents when a sizable portion of the population believed that the existence of extraterrestrials had been verified. For example, in the 1830s many scientists actually believed that there was life on the Moon, claiming to have observed roads and artificial features on its surface. In 1835, *The New York Sun* printed a series of articles purportedly based on those in the *Edinburgh Courant* and the "highly respected" Edinburgh *Journal of Science*. These sources, the *Sun* claimed, stated that the well-known discoverer of Uranus, Sir John Herschel, had developed a powerful telescope through which he was able to observe a species—"vespertilio-homo," or "bat man"—on the Moon. Standing about four feet in height, these creatures flew as bats but walked as bipeds. They were described as intelligent, expressive and communicative, and capable of fine arts and architecture. It later transpired that the author of these sensational articles—which came to be known as "The Great Moon Hoax"—was Richard Adams Locke (1800–1871), who had no connection to Herschel. Although the articles were fiction, the readers were generally delighted with the discovery of life on the Moon and excited by the richly detailed descriptions.[43] Likewise, many people were intrigued by the accounts of canals on Mars as detailed by Giovanni Schiaparelli, Camille Flammarion, and Percival Lowell. To be sure, some people were understandably upset in 1938 when they listened to Orson Welles' broadcast of the *War of the Worlds*, which depicted a hostile Martian invasion of New Jersey. Broadcasts of the story in Chile in 1944 and Ecuador in 1949 occasioned distress as well.[44]

On April 13, 1965, an article in *The New York Times* reported that the Russian astronomer Evgeny Sholomitsky had attributed flickering radio waves emitted by a stellar object known as CTA 102 to a super civilization. His colleague, Nikolai Kardashev supported his assertion, opining that CTA 102 was definitely an artificial radio source.[45] Further analysis, however, revealed that the signal emanated from a quasar and was not artificial in origin. The public reaction was not distress, but rather, subdued excitement. These examples lead Harrison to believe that the public would not panic when informed of an extraterrestrial civilization unless it demonstrated hostile intent, as in the *War of the Worlds* scenario.[46]

Traditionally, mass media images of extraterrestrials have concentrated on three general themes. The first image is that of a benign alien who represents some higher force that seeks to save our planet, as portrayed in *ET* and *Close Encounters of the Third Kind*. The second is menacing, as depicted in films such as *The Thing*, *Independence Day*, and *Alien*. Finally, the third image is the cyborg—part living being and part machine that can have either menacing or protective qualities as depited in *The Transormers* films.[47]

According to the "celestial savior model," the sentiment is that extraterrestrial alien visitors would be magnanimous in the style of Klaatu in the film *The Day the Earth Stood Still.* In this theme, visiting aliens capable of interstellar travel have progressed much further than humans, not only in science in technology, but in morality and ethics as well. Not unlike on Earth, where advanced economies provide technical assistance to Third World nations, Ted Peters once speculated that extraterrestrial aliens might provide Earthlings with quick technological fixes for our serious social problems.[48]

Some optimists believe that extraterrestrials might try to guide human development in a progressive trajectory. This was the theme expressed by Arthur C. Clarke, an esteemed scientist who worked as a radar officer in the Royal Air Force during World War II. After the war, he served as president of the British Interplanetary Society. His non-fiction books *Interplanetary Flight: An Introduction to Astronautics* (1950) and *The Exploration of Space* (1951) established him as a noted advocate of space travel.[49] But he is best known for his work in science fiction. He and Stanley Kubrick produced the popular science fiction film *2001: A Space Odyssey.* Based on his short story, "The Sentinel," the film centers on a monolith of extraterrestrial origin, which appears at critical junctures in humankind's development.[50] In the opening scene, a tribe of proto-humans is driven from a waterhole from a competing tribe. The next day, the proto-humans awaken to an upright vertical monolith (shaped as a rectangle), which emits an eerie sound. Soon thereafter, the tribe's leader has an epiphany of using bones left over from an animal carcass as a crude weapon. With this newfound knowledge, he and his tribe confront the rival tribe that earlier drove them from the waterhole. Using the new weapons, they pummel the leader of the opposing tribe, and in exultation, one of the proto-humans lofts a bone in the air at which point, the scene moves forward to the year 2001 with a space cylinder floating in space. A mission to the moon has unearthed a monolith, which the astronauts surmise is extraterrestrial in origin. Soon thereafter, the space agency sends a mission to Jupiter. Unbeknownst to the astronauts, the impetus for the mission was to find out more about the nature of the monolith, another of which has been detected near Jupiter. A seemingly personable computer on board the spacecraft—HAL—turns on the crew, turning off the life support systems of those astronauts in suspended animation and severing the lifeline of another astronaut while he floats in space on a repair operation. The lone surviving astronaut manages to disable HAL and later encounters the monolith in space, at which point he is rushed through a tunnel at great speed, which could be construed as a wormhole. At the end of his journey, he observes a human figure that resembles himself eating a meal. Next, his attention is drawn to an elderly dying man on a bed who appears to be a much older version of himself. The film concludes with a large incandescent human fetus in the womb floating in cosmos.

Concurring with the premise of *2001*, Frank Drake once suggested that an advanced extraterrestrial civilization might station an automatic technology monitor, which would function as an alarm with beacons across interstellar space. When a local level of technology advances to a certain point—for example, a monitor might analyze the content of radioactive elements in the atmosphere—a message would be sent to alert the civilization that had planted the device.[51] Interestingly, in April 2012 NASA detected an anomalous-looking object jutting out from the surface of Mars. The seemingly perfectly rectangular structure bears a striking resemblance to the monolith depicted in the film *2001: A Space Odyssey*.[52]

A less optimistic view of encounter is the "alien enemy model," as exemplified by H.G. Wells' *War of the Worlds*. This model posits that any close engagement with an alien civilization which has overwhelming scientific and technological superiority will lead to conquest and subjugation, if not extermination.[53] As societies advance, they face a myriad of problems relating to population growth and resource depletion. As Albert A. Harrison mused, some will die off, in which case, we will never hear from them. Others will survive through moderation either by rationing resources, or prohibiting certain technologies, or both. As they probably will not be able to afford interstellar communication, we probably will not hear from them either. Finally, the remaining societies will have worked through their problems without depleting their resources or inhibiting their economic growth or development of their technologies. It is these successful societies with which we are most likely to establish contact. With their large resource base, they are unlikely to wage war in search of plunder. On Earth in the twenty-first century, war is a less likely undertaking for economically strong nations. If this pattern holds elsewhere in the cosmos, Harrison believes that our "newfound neighbors will be peaceful, and this should affect our decision on how to respond to them." If, however, alien civilizations have high capabilities and evil intentions, he points out that "their remote location should buy us time to find constructive ways to respond."[54] The hostile alien scenario will be examined in greater detail in Chapter 12.

What effect would the receipt of an alien message have on organized religions? Ted Peters and Julie Louise Froehlig devised "the Peters ETI Religious Survey" to gauge the impact of the discovery of extraterrestrial life on religion. They received more than 1,300 responses worldwide from persons in multiple religious traditions. According to their findings, religious adherents overwhelmingly registered confidence that neither they as individuals nor their religious tradition would suffer anything like a collapse in the event of contact with extraterrestrials. Conversely, those respondents who self-identified as non-religious were far more fearful (or perhaps gleeful) of a religious crisis than were the believers.[55] According to Peters, it was the agnostics and atheists, not the people affiliated with religious groups, that felt the discovery of extraterrestrial life would be most catastrophic for religion.[56] As a generalization, theologians with strong academic credentials are more open to the prospect of extraterrestrial life than those who interpret the Bible literally.[57] Father Theodore M. Hesburgh, who once wrote a foreword to a NASA report on interstellar communication, reasoned that an infinite God would probably not confine his creation of intelligent life to one planet, Earth. In fact, these same arguments were used by Christian philosophers to challenge the Aristotelian doctrine in the fourteenth century.[58]

It is assumed that Christianity would be the religion most challenged by the concept of extraterrestrial beings. After all, the Christian doctrine teaches that God came down to Earth in human form—Jesus—to save one species on one planet.[59] Arguably, the confirmation of extraterrestrial intelligence would deal a blow to any theology based on human uniqueness. Nevertheless, the Vatican has funded an observatory, and the cleric in charge, Father Coyne, announced that there is no reason why extraterrestrial aliens, if they do indeed exist, should not be baptized and welcomed into the Catholic Church.[60] Some Christian theologians have speculated that perhaps Jesus Christ died for all sentient beings in the universe, in which case it is the destiny of mankind to spread the word throughout the cosmos. For example, Emmanuel Swedenborg (1688–1772), a Swedish scientist turned religious prophet, asserted that Christ had come only to the Earth and not to any other planet. His chief purpose was to communicate the scriptures so that they could be disseminated from the Earth to the rest of the universe.[61] In that same vein, the fifteenth-century theologian William Vorilong claimed that the crucifixion of Jesus simultaneously redeemed all intelligent creatures in the cosmos. Building on this concept, E.A. Milne (1896–1959) conjectured that an interstellar radio system operated 2,000 years ago and broadcasted news of Christ's sacrifice to the entire universe. As a result of this radio transmission, it is unnecessary to recreate the crucifixion throughout the cosmos.[62] Alternatively, there could be a multitude of incarnations in which one savior dies for each deserving species.[63]

These theological adaptations could ultimately undercut the credibility of some religious doctrines. Some observers opine that if contact with an extraterrestrial civilization is ever established, terrestrial religions will no longer be able to go back to business as usual. Be that as it may, even Paul Davies conceded that theology, like science, is thirsty for knowledge and may be able to adapt. Nonetheless, the discovery of extraterrestrial intelligence would diminish the sense of terrestrial human uniqueness.[64] Enamored of an advanced extraterrestrial presence, perhaps some people on Earth might create a cargo cult in which they deify the aliens.[65] David Wilkinson, a Methodist theologian who also holds a Ph.D. in physics, argued that humans would not necessarily come to doubt their special relationship with God; instead, he developed the idea of non-exclusivity to explain that humans, along with intelligent beings elsewhere, are special in God's eyes.[66] Likewise, Steven J. Dick's "cosmotheology" and Diarmuid O'Murchu's "quantum theology" broaden the concept of religion for a universe full of life.[67] Contact with a more advanced alien civilization would not necessarily disenfranchise humans from the *Imago Dei*—that is, image of God.[68] Based on this philosophy, all intelligent beings are special in God's eyes.

The issue of contact raises an important question: Who would speak for

Earth, and what should be said? In 2001 the SETI Institute established the SETI Post-Detection Task Group, whose mission is to prepare for the day when contact with extraterrestrial intelligence is confirmed.[69] To that end, an international group of SETI scientists established a protocol—The Declaration of Principles Following the Detection of a Signal from Extraterrestrial Intelligence—for how to respond to this event.[70] According to this protocol, those who discover the signal should first determine if it is indeed authentic and not man-made interference or natural cosmic static. After that, the appropriate national authorities should be informed. Finally, the detection is supposed to be disseminated promptly and openly through scientific channels and the public media.[71] The protocols have no legal standing. Likewise, the Task Group has no official writ to impose its policy recommendations on anybody.[72] For his part, Seth Shostak has argued that any attempts to proscribe METI (Messaging to Extraterrestrial Intelligence) were rooted in paranoia and would amount to anti-science. Based on his reasoning, there is no need for a consultation process as it is not really clear who would have the right to decide the issue. Moreover, he questioned how such a moratorium could possibly be enforced.[73]

In preparation for an interstellar dialogue if contact should ever be made, Douglas Vakoch organized SETI's Earth Speaks project to involve ordinary people in the decision-making process on how we should respond to an alien's message. The project's website invites people from all over the world to submit ideas for messages that they would like to send to extraterrestrials.[74] A professor of clinical psychology at the California Institute of Integral Studies, Vakoch serves as the Director of Interstellar Message Composition at the SETI Institute.

If contact is ever made, will the Earth establish diplomatic protocols with extraterrestrials not unlike the United Nations on Earth? In 2010, rumors circulated that a Malaysian astrophysicist and director of the U.N. Office for Outer Space Affairs, Mazlan Othman, had been appointed as the "alien ambassador" in the event that contact with an alien civilization is established. Othman denied the story, but conceded that her office, and the U.N. Committee on the Peaceful Uses of Outer Space, would be appropriate forums for managing the response to the discovery of an extraterrestrial civilization.[75]

Considering the tremendous distances between solar systems, diplomacy between humans and extraterrestrial aliens could be challenging. Some SETI scientists have speculated on the possibility of establishing direct contact with extraterrestrial aliens by way of computer emulation. According to William Sims Bainbridge, cognitive science and information technology are progressing in a direction that will one day make it possible to transmit realistic avatars of individual human beings to extraterrestrial civilizations.[76] Not unlike Ray Kurzweil, who predicts a "singularity" in the year 2042 when humans will transcend biology, Bainbridge notes that the convergence of artificial intelligence and cognitive neuroscience has prepared the basis for the prospect of personality capture and emulation. In the not-so-distant future, people might be able to capture their personalities in computers. This scanned information could then be transmitted at the speed of light to other planets. Bainbridge predicts that the first interstellar expeditions will not be carried out by biological-based humans; rather, "infonauts" could travel at the speed of light as bits in a communication path. Bainbridge envisages a virtual reality "Cosmopolis," where multiple civilizations will send avatar ambassadors who emulate the original biological beings from which they were derived. There, they will engage in varieties of social interaction including trade and cultural innovation. Stretching the bounds of physics, Bainbridge explains that because these avatars would reside locally in the Cosmopolis within information systems equipped with sensors and robots (perhaps quantumly entangled with instruments elsewhere), communications between civilizations could be instantaneous and would not be bound by the speed of light limitations on interstellar radio transmissions. As a result, the avatars could serve as real-time ambassadors not unlike their counterparts on contemporary Earth.[77]

To some scientists, the receipt of an alien message will be the most important event in human history, for it will prove that we are not alone in the cosmos.[78] What is more, it would demonstrate that an advanced civilization was able to work through the myriad of problems associated with modernity and survive. If a civilization can surmount these problems, presumably its technology will continue to advance, in which case interstellar travel could someday be possible. Working off of that assumption, some researchers assert that extraterrestrial aliens have already visited our planet, as detailed in the next chapter.

CHAPTER 7

The Ancient Alien Hypothesis

For over 50 years, SETI scientists have labored hard to detect extraterrestrial radio transmissions, but alas, no tangible evidence has been acquired. Perhaps, we are alone after all. The so-called Fermi paradox suggests that advanced civilizations are not common in our galaxy. Enrico Fermi, who won the Nobel Prize in physics in 1938, was a principal scientist in the Manhattan Project—the U.S. effort to develop the atom bomb during World War II. In 1950, while having lunch with his colleagues at Los Alamos, they discussed the prospect of extraterrestrial life. Initially, he and his colleagues considered the probability of the existence of alien civilizations to be quite high. About a half hour later, when the group moved on to discuss other topics, Fermi blurted out, "Where is everybody?" As he reasoned, if technologically advanced civilizations were common in the universe, and assuming many of them had preceded Earth by many, many years, then it followed that they should be detectable in one way or the other by now. In fact, based on his rapid computations, he determined that we should have been visited very long ago many times over.[1] Even if its top speed was limited to but a small fraction of the speed of light, Fermi calculated that a civilization with even a modest amount of rocket technology could colonize the Milky Way within a few tens of millions of years—not a terribly long period of time by astronomical standards. Even if it took each new colony half a million years to set up two colonies of its own, this exponential growth would lead to more colonies than there are stars in the Milky Way in just 20 million years.[2] Those who advance the extraterrestrial hypothesis must address this enigma.

Perhaps extraterrestrial aliens have already visited our planet. A poll conducted by *Life* magazine in the spring of 2000 indicated that 30 percent of the American population believed that aliens had landed on Earth at one time or another.[3] Evidence of ancient extraterrestrial visitation would answer the Fermi paradox: There is no paradox after all, because aliens have already been here. In fact, anomalous sightings in the sky date back to the beginning of recorded history. For example, in 1450 B.C. during the reign of Pharaoh Thutmose, Egyptian scribes recorded an incident involving "circles of fire" that appeared for several days before finally ascending back into the sky. Some chroniclers of Alexander the Great claim that he and his army were harassed by a pair of flying objects in 329 B.C.[4] In 91 B.C., the Roman author Julius Obsequens wrote about a round object that traveled in the sky. In the year 1235 General Yoritsume and his army reported seeing strange globes of light dancing in the sky near Kyoto, Japan. And at dawn on April 4, 1561, hundreds of residents observed a large number of objects appearing to be engaged in aerial combat over Nuremberg, Germany.[5]

Historically, witnesses describe Unidentified Flying Objects (UFOs) in a manner that is reflective of their eras. For instance, in the early fourth century Emperor Constantine claimed to have seen a cross in the sky with the message, "In this sign, conquer." After the apparition, he kept his pledge to convert to Christianity when he won a battle in 312 A.D.[6] According to interpretations of a Sanskrit text—the *Drona Parva*—ancient flying machines,—or vimana—,were built by an alien race, the Rbhus. Supposedly, these crafts could travel not only through the air, but to other worlds as well.[7] Some scholars even wonder if these ancient aliens used what we would call a nuclear weapon to destroy an ancient civilization located at Mohenjo Daro in India.[8] Interestingly, J. Robert Oppenheimer, the father of the American atomic bomb, invoked a passage from the ancient Hindu text, Bhagavad Gita, when witnessing the first atomic test at the Trinity site in New Mexico on July 16, 1945: "I have unleashed the power of the Universe, now I become Death, the destroyer of worlds."[9]

According to the ancient alien hypothesis, if we replace transcendental sky gods with advanced technocrats, the ancient texts appear more plausible. Purportedly, these extraterrestrial interlopers began meddling in human evolution and culture in the distant past.[10] Proponents of the theory frequently cite the case of Ezekiel in the Old Testament. In the year 593 B.C., the prophet described an extraordinary vision that included four winged creatures who moved on a four-wheeled vehicle, from the center of which came what appeared to be burning coals of fire. The story intrigued Josef E. Blumrich, a native of Austria and NASA engineer who played a leading role in the design of the *Saturn V* rocket, which took astronauts to the moon. His curiosity piqued, he decided to work out a design for the craft described by Ezekiel. In his book, *The Spaceships of Ezekiel*, Blumrich went so far as to assert that the description as explained in the Bible could be adapted into a practical design for a landing module launched from a mother spaceship.[11]

Numerous scholars have advanced the ancient alien hypothesis. As mentioned in Chapter 1, as far back as the seventeenth century, Bernard le Bovier de Fontentell speculated that inhabitants of the Moon might have once visited the Earth.[12] But the ancient alien thesis really came into fruition in the twentieth century. In a three-volume book—*Interplanetary Contacts* (1928–1932)—a Russian scientist, Nicholas Rynin discussed theories of rocket technology and space flight, along with an analysis of ancient legends regarding air and space ships, from the Greek legend of Icarus to the Hindu Epic of the Mahabharata. In 1959, the Soviet ethnologist M.M. Agrest postulated that extraterrestrial beings had visited the Earth. He conjectured that the events described in the Bible were in reality based on actual visits of extraterrestrial astronauts. Extant monuments of ancient civilizations and cultures were cited as evidence of the remnants of construction projects on which humans and extraterrestrials collaborated.[13] In the Hebrew Bible, the word "Nephilim" has been deciphered as "giants" or "titans," but Agrest averred that the correct meaning of the term was "beings fallen" (from the sky).[14]

In his 1954 book—*Flying Saucers on the Moon*—the British author Harold T. Wilkins combed through ancient literature and legends inferring that our ancestors had witnessed strange objects in the sky.[15] In 1964, another British writer—W. Raymond Drake—published *Gods or Spacemen?*, which examined the ancient astronaut issue.[16] Still another early advocate of the ancient alien hypothesis, Morris K. Jessup wrote a book in 1956 titled *UFOs and the Bible* in which he argued that numerous references to extraterrestrial visitations could be found in the scriptures. For his fieldwork, he claimed to have explored Incan ruins in Peru. He concluded that the Incan stonework had been "erected by the levitating power of space ships in antediluvian times."[17] As Jessup argued, there was no evidence to suggest that there was a gradual development of advanced civilizations in Baalbek, Easter Island, Peru, or Egypt; rather, the megalithic stone work in those areas appeared to have been ready-made as though colonies were set up directly by some outside agency.[18] Citing the research of Colonel James Churchward, Jessup speculated that India was originally settled by the Nagas from Burma, who in turn were the descendants of the original Mayans.[19] Prefiguring Gerard Kitchen O'Neill's "generation ships," Jessup speculated that alien colonies could exist naturally and easily in space. As such, they did not necessarily have to come from other stellar systems, or even other planets.[20]

Even Carl Sagan and I.S. Shklovskii in their book *Intelligent Life in the Universe* (1966) counseled that scientists and historian should seriously consider the possibility that ancient aliens established contact at different periods in Earth's history. According to Sagan, once a civilization realized that interstellar flight was feasible, it immediately developed the technology no matter how difficult or costly the undertaking. He reasoned that the scientific advantages gained by a civilization's contact with other space communities would justify the huge investment. Based on his calculations, he concluded that each star in the galaxy received a random visit from another galactic civilization at least once every 100,000 years.[21] Foreshadowing Zecharia Sitchin, Sagan once mused that the ancient Sumerians might have owed their civilization to creatures from another planet that conveyed knowledge to them.[22] He even conceded that Sumerian legends implied contact between humans and non-human civilization of immense powers on the shores of the Persian Gulf in the fourth millennium B.C. or earlier. According to certain passages in the ancient scribes, strange beings appeared over the course of several generations whose main purpose was to instruct mankind. These beings were described as gods in a variety of forms, not all of which were human. Their home was believed to be the stars, which was governed by a representative and democratic assembly.[23] Sagan speculated that contact might have previously occurred, but the chronicles of these visitations could have been modified and distorted in legends and theological narratives.

Arguably no other person has done more to popularize the ancient alien theory than the Swiss author Erich von Däniken, whose book, *Chariots of the Gods?*, became an international best-seller in the 1970s. Two years after the book was released, 600,000 copies had been sold. In 1970 a television film with the same title was released in the United States.[24] According to his thesis, extraterrestrial aliens have had a longstanding relationship with humans, as the former guided the latter. In antiquity, the people on Earth received visits from extraterrestrial aliens. These aliens annihilated part of the human species, but chose to keep a segment of the population alive. The humans the aliens initially encountered were not the *Homo sapiens* that we are today. Rather, modern man developed through a type of ancient eugenics program carried out by extraterrestrial scientists, which involved the artificial fertilization of select female members of the human species. Their offspring represented a new race that had skipped a stage in natural selection.[25] Periodically, the aliens would return to Earth to evaluate the progress of their genetic experiments until they produced creatures intelligent enough to live under the rules of society that their visitors imparted to them.[26] The most intelligent of the humans were chosen to be kings. As a visible sign of their power, they received radio sets through which they could contact and address the gods.[27]

According to this narrative, ancient aliens influenced a number of ancient cultures, including the Sumerian, Mayan, Hebrew, Egyptian, Tibetan, and Scandinavian. The ancient Sumerians connected their symbols of gods with specific stars, which von Däniken infers as an extraterrestrial connection.[28] He speculates that the destruction of Sodom and Gomorrah, as depicted in the book of Genesis in the Old Testament, could be interpreted as a deliberate nuclear explosion.[29] The Flood described in the parable of Noah's Ark was a deliberate plan to exterminate "the human race except for a few noble exceptions."[30]

The evidence von Däniken adduces for his thesis is circumstantial, consisting in large part of the surviving edifices of ancient civilizations. Amazingly large geoglyphs—figures drawn on the ground—can be found in Peru. Some observers believe that it would have been well-nigh impossible for primitive cultures to design them from the ground without surveying instruments.[31] To von Däniken, this suggested that they were airfields designed by extraterrestrial travelers. To buttress his thesis, he cites the legends the Incan legends imparted to the Spanish conquistadors that told of a lost race who came from the sky that would one day return to Earth.[32] Even with modern equipment, von Däniken believed that it would be impossible for contemporary architects and engineers to build ancient structures, such as the pyramid of Cheops. According to his reading of history, the engineering and manpower requirements for such a massive enterprise would not have been autonomously available to seemingly primitive cultures.[33] For instance, the Great Pyramid of Cheops in Egypt is a massive structure that contains vastly more material than a modern skyscraper.[34] Based on his historiography, he claims that the ancient Egyptians did not have the logistical capabilities to carry out a project of such a grand scale. According to his estimates, there was not enough food in all of Egypt to feed the workers. Nor were there accommodations for them. Mummification, he concludes, was a procedure, not for preparing ancient Egyptians for some nebulous afterlife; rather, it was intended to be a form of suspended animation whereby those mummified could be later reanimated.[35]

To support his case, von Däniken invokes numerous ancient artifacts. For example, in 1929 scholars discovered an old map dated from 1513 and signed by an admiral of the Ottoman Empire named Piri Reis, which proved to be amazingly accurate, notwithstanding the limited extent of sixteenth-century geography. The bottom of the map appeared to show a land mass that could be interpreted as Antarctica. Presumably, it would have been impossible for Reis to know of Antarctica insofar as the frozen continent was not discovered until 1823 by James Weddell. Furthermore, the coast of Antarctica was not mapped until 1840 after the expedition of U.S. Navy Lieutenant Charles Wilkes. Some authorities have come forward to validate Reis' map. In a letter dated July 6, 1960, U.S. Air Force Colonel Harold Z. Ohlmeyer of the 8th Reconnaissance Tactical Squadron of the Strategic Air Command stated that the geographic detail of the map agreed remarkably with the results of the seismic profile made across the top of the ice-cap by the Swedish-British Antarctic expedition of 1949. To him, this suggested that the coastline had been mapped before it was covered by the ice-cap.[36] Based on these analyses, von Däniken concluded that Piri Reis' source documents must have been drawn from aerial photographs.[37] Inasmuch as humans did not have such technology at the time, it followed that extraterrestrials were the likely agents.

Von Däniken sees himself following in the tradition of Heinrich Schliemann, who accepted the legend of Troy as depicted in Homer's *Odyssey* an historical fact. As a result of his forbearance and labor, in 1876 he discovered the ancient city in what is now Turkey.[38] Still, som-e of von Däniken's more fantastic claims stretch credulity. Interestingly enough, on the copyright page of *Chariots of the Gods?*, it states that "This is a work of fiction. Names, characters, places, and incidents are either the product of the author's imagination or are used fictitiously, and any resemblance to actual persons, living or dead, business establishments, events, or locales is entirely coincidental."[39]

Building on von Däniken's thesis, Philip Coppens examined the histories of numerous archaic civilizations to make the case for the ancient alien hypothesis. In his book, *The Ancient Alien Question: A New Inquiry Into the Existence, Evidence, and the Influence of Ancient Visitors*, he reviews the literature on this controversial hypothesis. Coppens asks, how were Stone Age peoples able to perform massive engineering projects which involved the movement of stones weighing many tons?[40] He notes that the three pyramids of the Giza Complex in Egypt were laid out in the formation of Orion's Belt, implying some connection to the stars. Furthermore, he points out that not a single pyramid has been found which contained a mummy. Citing Christopher Dunn's book, *The Giza Powerplant*, he speculates that they might have served as power plants, rather than mammoth mausoleums.

Examining Mayan mythology, Coppens concluded that the ancient Amerindians were in contact with extraterrestrial aliens. The lid of Lord Pacal's tomb in Palenque is often cited as among the most convincing evidence for the ancient alien hypothesis. The carving on the lid depicts a human sitting in what could be inferred to a space capsule. Coppens notes the similarity in the layouts of the three pyramids at Giza with the three main structures of Teotihuacán and their representation of the Belt of Orion. For Coppens, these arrangements are no coincidence; rather, they suggest that the same extraterrestrial agents were involved in the construction of ancient edifices on different continents.[41]

Still, Coppens concedes that much of what has been written on the ancient alien theory is unreliable, noting that under greater scrutiny, several of the claims later transpired to be spurious. He cites the example of Robert Temple, who in his book—*The Sirius Mystery* (1975)—wrote about the Dogon, a tribe in Mali, West Africa, that purportedly possessed extraordinary knowledge of the star system Sirius. As the brightest star in the sky, it was the most important marker for the ancient Egyptians, who based their calendar on it. According to Lynn Picknett and Clive Prince in their book *The Stargate Conspiracy*, Temple was influenced by his mentor Arthur M. Young, an American inventor, helicopter pioneer, cosmologist, and philosopher. Young gave Temple an article written by two French anthropologists, Marcel Griaule and Germaine Dieterlen, on the star lore of the Dogon tribe—after which Temple began his research for *The Sirius Mystery*. Subsequent research, however, cast doubts on the credibility of Temple. The scholar Walter Van led a team of anthropologists to Mali and declared that they could find absolutely no trace of the detailed star lore as reported by the two French anthropologists who wrote about the Dagon tribe in Africa.[42]

Surprisingly, as Coppens notes, some academics reject the ancient alien hypothesis as racist—as it attributes an otherworldly agency to the architectural achievements of ancient civilizations, some of which were non-white. As Monty Dobson explained in an essay entitled "History Undoctored": "There is an underlying tone of cultural superiority, which implies that 'ancients,' who were likely brown, were incapable of independently developing the sophisticated technology and culture we know they had without help."[43] Nevertheless, Coppens is reluctant to throw the baby out with the bathwater. Until his untimely death at the end of 2012, he remained an indefatigable advocate of the ancient alien hypothesis appearing on numerous television documentaries on the topic.

The most prolific author of the ancient astronaut theory was Zecharia Sitchin. Born into a Jewish family in Soviet Azerbaijan in 1920, in 1952 he settled in New York and later moved to Israel. He was one of fewer than 200 persons in the entire world who could read the Sumerian language.[44] The provenance of the Sumerian language is unknown, for it has no cognates with any Indo-European, Semitic, or other language.[45] Sumer was the earliest recorded civilization, appearing suddenly about 6,000 years ago in the delta of the Tigris and Euphrates Rivers. It preceded the Egyptian civilization by several hundred years. Sitchin proposed that both were derived from a common antecedent civilization.[46] In a highly controversial thesis, Sitchin asserts that the flowering of Sumerian civilization circa 3,800 B.C. was occasioned by some outside agency.

In his first book, *The Twelfth Planet* (1976), Sitchin advanced the theory that various Sumerian texts told of the existence of a planet in our solar system whose inhabitants colonized Earth more than 400,000 years ago. Sitchin proclaims that the Nibirians originated from the "twelfth planet" in our solar system. According to his framework, for humans, our planetary system contains eleven known major celestial bodies—the Sun, the Earth, the Moon, Mercury, Venus, Mars, Jupiter, Saturn, Uranus, Neptune, and Pluto—plus one more body located on the outskirts of the solar system that has yet to be detected by astronomers.[47] From the perspective of the Nibirians, who approached our planet from the periphery of the solar system, Earth was the seventh planet that they encountered after Pluto, Neptune, Uranus, Saturn, Jupiter, and Mars.[48] One Nibiru orbit around the sun took 3,600 Earth years to complete.[49] According to his very detailed yet bizarre story, life on their planet—Nibiru (the Sumerian name for the twelfth planet)—faced a slow extinction 450,000 years ago as the atmosphere eroded.

In an effort to stave off this crisis, a Nibirian mission was sent to Earth, which uncovered rich deposits of gold. It was discovered that gold—one of the best conductors of heat and electricity—could be used to replenish the atmosphere of Nibiru. Besides gold, supposedly, the Nibirians coveted the Earth's fuel and energy resources as well, which seems odd for an advanced civilization able to traverse the long distance to the planet.[50] Presumably, they would have learned of a more sophisticated way to harness energy in accordance with the Kardashev scale. Be that as it may, according to Sitchin, the Anunnaki—as they were known by their human subjects—founded the ancient civilizations. The

Nibirian supreme leader—Anu—supervised the colonization of Earth from the home planet. The Nibirians established a thriving colony on Earth initially under the leadership of Anu's two sons, Enil and Enki. Enil served as the mission commander, while Enki served as the science officer. Most essential were the settlements in the Persian Gulf region on Earth established for the purpose of mining gold. Eventually, the number of Anunnaki on Earth reached 600. According to Sitchin, about half—the Igigi (those who observe and see)—remained skyborne as they performed an interplanetary liaison-type function.[51]

But all was not perfect in the new world. There was a longstanding antagonism between the two brothers, Enil and Enki, due to Nibirian protocol.[52] Originally, Nibirians mined the gold in the Middle East and Africa, but a revolt among the alien laborers prompted the leaders to create modern man as a race of pliant slave workers. At first, the Nibirians attempted to domesticate humans so that they could serve their masters. But later, Enki came up with an idea to speed-up the Anunnaki's mission to harvest the gold supply on Earth. He proposed the creation of a breed of primitive workers who would toil in the gold mines. Sitchin conceded that Man is the product of evolution; however, he maintains that *Homo sapiens* is the creation of the "gods."[53] Enki was said to have had an altruistic view of humanity and sought to improve the species. Through genetic manipulation, he imprinted the image of the gods onto his subjects.[54] Invoking scripture, Sitchin notes that the Old Testament makes reference to "Elohim" creating humans in their own image. Some researchers have interpreted the biblical term "Elohim" to signify many gods instead of a single deity, which would lend credence to Sitchen's thesis that the gods were actually alien overlords.[55] According to Sitchin, this first breed of intelligent worker was derived from a hominid, or proto-*Homo sapiens*, in southeast Africa. This feat was achieved by mixing the blood of a young Anunnaki male with the egg of a female hominid. From there, the fertilized egg was inserted into the womb of a female Anunnaki. Thus was born the first man, Adamu.[56] Many of the early prototypes were defective, often sexless beings. Sitchin's interprets the ancient Sumerian texts to imply that the Anunnaki geneticists produced a variety of mutated creatures—or chimeras—including animals such as bulls with lion heads, winged animals, and apes and humanoids with the head and feet of goats.[57] But eventually, a genetic type was produced that was capable of procreation, which according to Sitchin, is implied in the Biblical vernacular of "knowing," denoting sexual relations.[58]

Roughly 300,000 years ago, a slave race was produced. Over the centuries, this race was the object of continued experimentation and eventually resulted in changing the Neanderthal to the Cro-Magnon.[59] An elite subspecies—the Nephilim, or "giants"—were the hybrid offspring of the gods and humans. Supposedly, they were created as a royal race for the purpose of keeping the upgraded hominids in slavery and ensuring that they remained civilized. Approximately 50,000 years ago, the humans fathered by Anunnaki were permitted to rule in select cities.[60]

Sitchin found it odd that living organisms on Earth contained too little of the chemical compounds that abound on the planet. As he points out, *Homo sapiens* ("Thinking Man") seemed to appear out of nowhere, sweeping the Neanderthal man off the face of the Earth. These modern men—Cro-Magnon—were no longer "naked apes": unlike their forbears, they made specialized tools and weapons of wood and bone. Their societies were organized, as they lived in clans with a patriarchal system. They buried their dead, which suggests that they had philosophies concerning the afterlife. Sitchin rejects the conventional wisdom that modern Man appeared a mere 700,000 years after *Homo erectus* and some 200,000 years before the Neanderthal Man. For him, the emergence of *Homo sapiens* indicates an extreme departure from the slow evolutionary process, suggesting that they were totally unrelated to earlier primates.[61]

To Sitchin, the stories told in the ancient texts should be read as records of real, historical events. As regards legends of a great flood, Sitchin claims the gods did not cause the cataclysm; rather, they connived to withhold information of the event from the Earthlings. The Nibirian colonizers were ordered to abandon Earth. However, some of the Anunnaki had grown fond of their human subjects and warned them of the impending cataclysm. A small number of humans built arks to ride out the deluge.[62] The biblical account of the Tower of Babel is interpreted by Sitchin as a failed revolt by humans who attempted to construct their own launch tower in the hope of propelling a flying vehicle so that they could beseech Anu not to break up humanity. This plan alarmed Enlil, who decided to disperse the various elements of the human community. This was part of a "divide and rule" imperial policy.[63] According to Sitchin's reading of the Bible and the Sumerian texts, the descendants of Sem—the Semites—populated the area encompassing Mesopotamia. Ham and his descendants were taken to Africa and parts of Arabia. Japeth's people were transported to the Indus Valley, thus explaining the origins of the ancient Aryans.

The noted conspiracy theorist and author Jim Marrs built upon Sitchin's framework. He claimed that in a military action against unruly humans, Enlil along with his two sons, Ninurta and Adad, destroyed the Sodom and Gomorrah with a nuclear blast in the year 2024 B.C. Furthermore, the fallout resulting from the detonation obliterated all life in the surrounding areas, which ended the Sumerian civilization. According to Marrs, evidence of this ancient example of nuclear warfare can be found in the spring water near the Dead Sea, which supposedly still contains harmful amounts of radioactivity.[64]

Studying mythology, Jim Marrs found striking similarities in legends from various cultures around the world that suggested an extraterrestrial connection. For instance, ancient Hindu texts describe gods who flew in fiery crafts. Likewise, the ancient Teutons point to ancestors in flying "Wanen." Mayan myths claimed that their progenitors came from Pleiades (an open star cluster in the constellation Taurus), while the Incans professed that they were the "sons of the sun." Similarly, some Native American traditions proclaim that they are the sons and daughters of the great "Thunderbird." And Chinese texts detail tales of long-lived rulers from the heavens who sailed through the sky in "fire-breathing dragons." The common denominator Marrs found in all of these gods was their ability to fly through the air.[65]

Marrs argues that the best evidence points to the existence of the major pyramids prior to the Egyptian empires. The ancient Greek historian Herodotus attributed the construction of the major pyramids to the Egyptian ruler Cheops (the second pharaoh of the fourth dynasty), his son, and his brother. Marrs cites the opinion of Graham Hancock who chided contemporary historians for uncritically following in Herodotus' footsteps.[66] Marrs claims that there is no indisputable evidence to support Herodotus' claim. He suspects that there is a connection between the pyramids and the stars, concluding that whoever built the Great Pyramid was in possession of astronomical and geophysical knowledge unsurpassed even to this day.[67] As further evidence of alien intervention, Marrs noted that in 1900, Greek sponge divers discovered the remnants of an ancient ship in waters near Antikythera, a small island off the coast of Crete. The divers managed to retrieve several artifacts from the ship that were later analyzed. The date of their manufacture was estimated to be about 80 B.C. One artifact—the Antikythera Mechanism—was described by the British scientist Dr. Derek J. de Soll Price as a predecessor to the modern computer.[68]

Some researchers have found evidence outside of Earth to support the ancient alien hypothesis. For instance, in the 1950s astronomers noticed a strange marking on Mars that looked like a giant M that was hundreds of miles across.[69] Even the distinguished Soviet astronomer I.S. Shklovskii once made the extraordinary claim that the Martian moons were not natural. As he saw them, their eccentric orbits suggested that the moons were actually gigantic hollow satellites placed in orbit around their planet.[70] Likewise, in July 1970 two Russian scientists—Michael Vasin and Alexander Shcherbakov—published an article in which they propounded the theory that the moon was actually a planetoid that had been hollowed out eons ago in the far reaches of space by some alien intelligence.[71] Later observations, however, have not supported this hypothesis, as astronomers have concluded that the moons are solid bodies, probably asteroids, captured by the gravitational pull of Mars.[72] Although not taken seriously by mainstream scientists, the theory has caught on with some UFO researchers, including Jim Marrs.[73]

When the *Viking* spacecraft took pictures of the surface of Mars during its orbit in 1976, some observers thought that an ancient alien monument in the image of a face had been discovered. In his 1987 book *Monuments of Mars: A City on the Edge of Forever*, Richard Hoagland argued for the artificiality of the face on Mars, claiming that it was part of a city that contained a pyramid, a fortress, and other municipal constructions.[74] Astronomers, however, have not been impressed. For example, Carl Sagan wrote that the face on Mars was probably natural and not artificial; nevertheless, he conceded that it deserved further study because the hypothesis posed by its supporters belonged in the scientific arena.[75] Furthermore, because of the public's curiosity into the matter, he recommended that NASA give it more scrutiny. Subsequent probes by the Mars Global Surveyor in 1998 and 2001, and the Mars Odyssey in 2002, photographed with much higher resolution the area where the face was spotted in 1976. Alas, the new pictures revealed natural formations and not an artificial structure. Evidently, a combination of lower resolution and different lighting produced the image of the face in 1976.[76] Not to be deterred, Hoagland countered that the face was still authentic, perhaps the remnants of a ruined structure not unlike ancient edifices on Earth, such as the pyramids.[77] But in 2006, the European Space Agency, using the Mars Express orbiter, took images which once again showed mountains and not evidence for a face. Finally, a detailed photo made in early 2007 taken by the HIRISE (High Resolution Imaging Science Experiment) camera showed only mountains in the area.[78]

The scientific community has not been supportive of the ancient alien hypothesis. Isaac Asimov once dismissed the evidence presented in von Däniken's thesis as "utterly worthless."[79] Sitchin's interpretations are derived from his understanding of the Sumerian language. However, no scholars of the Sumerian language have ever corroborated his claims.[80] Moreover, his astronomical assertions seem spurious as well, as his "twelfth" planet supposedly occupies a highly elliptical orbit in our solar system and has gone undetected. As astronomers have pointed out, it is highly unlikely that a planet with a major elliptical orbit to be life-sustaining because there would be dramatic fluctuations in temperature. What is more, new, high-powered telescopes have not uncovered the mysterious twelfth planet.[81] Finally, inasmuch as the orbit of Nibiru would be in the outer reaches of the solar system, it would fall far outside the habitable zone.

In his book *Broca's Brain*, Carl Sagan expressed his disapproval of von Däniken's thesis.[82] He remained unconvinced, noting the complete absence of any corroborating artifacts that would support the hypothesis. Furthermore, he found the theory that aliens erected airfields, employed rockets, and exploded nuclear weapons on Earth to be implausible in the extreme as these technologies had only recently been developed on Earth. If extraterrestrials did indeed visit Earth, he reasoned that they would be far more technologically advanced than we are, and would not employ such comparatively primitive technology.

Analogously, it would be similar to someone framing the ancient alien idea in 1870 and concluding that the extraterrestrials used hot-air balloons.[83] Furthermore, Sagan pointed out that even preliterate cultures were able to preserve their histories for generations by way of oral narratives and would be articulate enough to explain the concept of ancient astronauts for posterity. Based on this assumption, he argued that if extraterrestrial aliens had visited a preliterate culture, we could still expect to have some recognizable record of the encounter preserved. He could not find a single case in which a legend reliably dated from an earlier pre-technological era could be unambiguously understood in terms of contact with an extraterrestrial civilization.[84] Be that as it may, although he rejected von Däniken's thesis, Sagan still believed that it was possible that alien visitors had journeyed to Earth in ancient times.[85]

Likewise, Ronald N. Bracewell was not convinced by von Däniken's thesis, as he pointed out erroneous claims in his research. For instance, von Däniken once detailed his journey to subterranean galleries hidden beneath the tropic jungle of the Ecuadorian province of Moreno-Santiago. While there, he purportedly saw metal tablets not unlike library books suggesting some kind of extraterrestrial agency was responsible for placing them there. Investigating the claim, the German magazine *Der Spiegel* sent an expedition to follow von Däniken's footsteps and interview the personages described in his book, but the findings repudiated many of von Däniken's assertions.[86] Although the architectural and engineering projects of some of the ancient civilizations were impressive, Bracewell found alternative mundane explanations rather than the extraterrestrial hypothesis. For one thing, ancient murals depict slaves laboring under the supervision of guards with swords and batons, not alien overlords. Furthermore, the ancient Egyptians had the capability of transporting massive cargoes across great distances. Finally, the ropes they used were durable and capable of hoisting the massive objects used in the construction of the ancient edifices.[87]

In addition to the scientific community, von Däniken has been attacked by religious conservatives as well. Like Sitchin, his thesis rejects the notion of God intervening in man's temporal affairs. Rather, extraterrestrial aliens take the place of the gods in ancient Holy Scriptures, a position that amounts to blasphemy.[88]

Nevertheless, the ancient alien hypothesis has gained currency in popular culture. It is featured in films such as *Alien versus Predator* and is the topic of a television series produced by the History Channel called *Ancient Aliens*. But to date, the ancient alien hypothesis has not found acceptance in the scientific community. Generally speaking, skeptics find claims of ancient alien interventions to be highly spurious. The research of von Däniken, Sitchin, and Marrs is dismissed as pseudo-scholarship that appeals to credulous people who are wont to believe sensational stories. Likewise, skeptics reject claims of extraterrestrial origins for contemporary UFOs as well. But as the next chapter makes clear, a number of seemingly credible eyewitnesses have reported aerial anomalies suggesting an extraterrestrial origin.

Chapter 8

The UFO Controversy

The study of Unidentified Flying Objects (UFOs) has long been consigned to the fringe of the research community. Although the SETI project is regarded as a serious scientific effort, so-called Ufology suffers from a huge credibility gap despite the fact that many people in America take the extraterrestrial hypothesis seriously. According to a 2002 poll conducted by the Roper organization, 56 percent of those Americans surveyed said that they believed that UFOs were extraterrestrial in origin.[1] And according to a 2005 survey, many Americans—72 percent—believe that it is at least somewhat likely that there are extraterrestrial aliens that have the technology to travel through space and visit different solar systems.[2] Moreover, many Americans believe that they have seen UFOs. A 1996 Gallup poll reported that 12 percent of the respondents to a survey claimed to have seen what they described as UFO, which when extrapolated, translated to about 30 million Americans.[3] This does not automatically imply what they claimed to have seen were extraterrestrial spacecraft. After all, by definition, UFO merely stands for *unidentified* flying object. The failure of UFOs to be taken seriously in academia can be ascribed in large part to a dearth of physical evidence and the questionable reliability of UFO claimants. Nevertheless, numerous sightings of anomalous aerial objects over the years have fired the imagination of people who wonder if extraterrestrials might be visiting our planet.

From the earliest days of recorded history in America, UFO sightings have been documented. According to one legend, an unidentified submerged object was observed by the watch officer on the deck of the *Santa Maria*, the flagship of Christopher Columbus' fleet.[4] In another incident, the captain's log on the *Santa Maria* recorded "a marvelous branch of fire" fall from the sky and land into the sea.[5] Another early recorded UFO sighting took place on the Charles River in Boston in 1639. James Everell, described as a "sober, discreet man," and two others, reported being carried by a great light at the river.[6]

UFO sightings are certainly not confined to the United States. In the spring of 1909 people all over the British Isles, particularly along England's east coast, reported seeing what was dubbed "scareships."[7] Some observers thought that they might have been German zeppelins; however, very few military zeppelins existed at that time. Moreover, none were able to navigate the 700-mile round trip to and from the British Isles. In the 1930s, people in Norway and Sweden claimed to have seen "ghost aircraft" in the sky. Some researchers speculate that the aircraft were experimental prototypes which were later developed into the VI and V2 rockets used by Germany in 1944 and 1945.[8] The ghost aircraft, however, were often seen flying in formation, which suggests that they were not rockets. At the time, the Swedish military, as well as the militaries of Norway and Finland, took the reports quite seriously.[9]

Not surprisingly, there is usually a spike in UFO sightings during times of war and human conflict.[10] In the early morning hours of February 25, 1942, there were numerous sightings of UFOs over the city of Los Angeles. Their appearance triggered a city-wide panic, as residents feared that a Japanese strike might have been launched on the west coast. Roughly 1,400 anti-aircraft rounds were fired during the ordeal. It does not seem plausible, though, that the Japanese were responsible because their military did not have the capability for such an attack, as they had no aircraft carriers capable of sailing that close to the United States—nor were any of their planes able to reach California from Japan. These facts suggest to some researchers that the anomalous objects may have been extraterrestrial in origin.[11] Reportedly, at least six civilians died as a result of automobile accidents and heart attacks attributable to the gunfire.[12]

Also during World War II, Allied pilots reported strange objects tailing their planes, which they thought were some type of unconventional German aircraft. Encounters with so-called "foo-fighters" occurred throughout the war, but most sightings covered a period between September 1944 and April 1945.[13] After the war, it was discovered that German pilots had observed similar objects tailing their planes as well. Puzzled by reports, according to some accounts, the Luftwaffe even established a unit—Sonderburo 13—to investigate the matter.[14] In 1992, the UFO researchers Barry Greenwood and Lawrence Fawcett uncovered the first official military records on the topic, which consist mainly of very brief excerpts from mission reports that flight commanders were required to complete upon returning to base. The U.S. government was concerned that these encounters suggested evidence of advanced German or Japanese technology; however, postwar investigations of German documents and research facilities failed to reveal that the Germans had any aircraft that remotely matched the foo fighters.[15] David Tressel Griggs, a scientist who worked on radar technology at the MIT Radiation Laboratory during the war, concluded that the foo fighters were real and constituted some unexplained military technology. Be that as it may, there is no record of any aircraft—Allies or Axis—having ever been damaged by them.[16]

Mysterious flying objects—ghost rockets—were spotted again over the skies of Scandinavia between February and October 1946. At the time, they were assumed to be Soviet missiles.[17] Usually sighted in the daytime, they were described as missile-shaped, without wings, and flying horizontally. Some were even reported as flying in formation. On July 6, 1946, a special committee was formed with Colonel Bengt Jacobsson of the Royal Air Management serving as chairman.[18] Concerned about the sightings, the Swedish government requested that Great Britain send modern radar equipment to track the purported objects. Reportedly, 200 of the rockets were eventually tracked on radar. The Swedish government questioned representatives of AB Bofors—a major Swedish manufacturer of weapons—but they denied making any launches during the times of the sightings.[19]

Some representatives of the Swedish suspected that the rockets were launched from Peenemünde, where the Nazi rocket base was located and later captured by the Soviets. According to the theory, the Soviets may have been experimenting with captured German V-weapons and were launching them from the coast near the border. On several occasions during the autumn of that year, Swedish reconnaissance aircraft were sent to pick up signals near the Baltic border, where they were immediately confronted by Soviet aircraft.[20] However, when the Soviet Red Army took control over the base, they found the rocket facility in ruins, the result of both intensive Allied bombing and retreating Germans that did not want to leave anything of value behind. Furthermore, even at the height of production, the Nazis were unable to launch the vast number of rockets as were reported sighted over Sweden. Finally, there is the puzzling question as to why the war-ravaged Soviet Union would fire hundreds of rockets over a neutral and seemingly innocuous Sweden. On October 10, 1946, the Swedish Defense Staff released a statement on the subject, noting that in some of the reported incidents, "clear unambiguous observations have been made that cannot be explained as natural phenomena, Swedish aircraft or imagination on the part of the observer."[21] Although not expressed in public, the Swedish position was that the rockets were launched by the Soviets as a political move to intimidate the Swedish government. Likewise, the recently appointed director of the newly created CIA, General Hoyt Vandenberg, opined to President Truman that the missiles were likely Soviet in origin and launched from the Peenemünde site. The final Swedish report was more skeptical. It conceded that many of observations could be attributed to fantasy and misinterpretations; still others seemed authentic, but defied a conventional explanation.[22]

The U.S. government even took an interest in the reports of the ghost rockets over Sweden. On August 1, 1946, Colonel Edwin K. Wright, the Assistant Director of the Office of Strategic Services (OSS) sent a memorandum to President Truman on the topic.[23] General James Doolittle in his capacity as Vice President of Shell Oil Company visited Sweden that same month. An article in *The New York Times* reported that the general had gone to Sweden to examine Swedish radar systems used to locate the ghost rockets.[24] A rumor soon spread that Doolittle was there as part of a mission to assist the Swedish government in its investigation, an allegation that the Swedish military denied.[25]

In the late 1940s, another sudden influx of sightings was observed in the

United States and around the world. The contemporary UFO era commenced on June 24, 1947, when a pilot, Kenneth Arnold, reported seeing nine shiny objects flying past Mount Rainer Plateau in Washington State while searching for a missing C-46 marine transport plane. An experienced search and rescue pilot, Arnold had logged more than 4,000 hours in the air.[26] He estimated that the flying speed of the UFOs was more than 930 miles per hour. For three minutes, he tracked the crafts as they coasted from Mount Rainer to Mount Baker and beyond. He described the UFOs as crescent-shaped craft that emitted blue-white flashes from their wing tips. Observing their erratic flying patterns, he recounted that they looked "like a saucer would if you skipped it across the water."[27] Afterwards, the term "flying saucers" stuck in the popular vernacular. Another witness, Fred Johnson, corroborated Arnold's story and claimed to have witnessed six of the craft while he was on Mount Adams.[28] Arnold went on to write a book *The Coming of the Saucers* (co-authored with Ray Palmer), which promoted the extraterrestrial hypothesis.[29]

For its part, the military would not say whether it did or did not have any planes aloft near the Cascade Mountains at the time of Arnold's reported incident, and dismissed the sighting as an optical illusion in which the tips of the mountains appeared to float above the Earth as a consequence of a layer of warm air.[30] Nevertheless, immediately after the incident, the Army Air Force began to take an interest in UFOs. In late June of 1947, Brigadier General Roger Ramey, chief of the 8th Air Force in Fort Worth Texas, and Colonel Alfred Kalberer held a press conference. Ramey stated that he thought people had "been seeing heat waves," mistaking them for anomalous aircraft. Dismissively, Kalberer referred to the reports as "Buck Rogers stuff," opining that Arnold just saw a few ordinary planes.[31]

Not long thereafter, the U.S. military encountered UFOs. Shortly after noon on January 7, 1948, a number of people in western Kentucky reported seeing a strange object racing through the sky at a high speed. The craft was described as huge—between 250 and 300 feet in diameter—which one observer said looked like "an ice cream cone topped with red."[32] The strange craft was observed near Godman Air Force Base in Kentucky. Four Air National Guard F-51 Mustangs that were about to land were radioed to take a closer look. The Godman tower alerted Captain Thomas Mantell that a strange object was flying over the air field and asked him to help identify it. Led by Mantell, three other Air Force pilots pursued what appeared to be a metallic object high in the sky. As they passed through an altitude of 22,000 feet, the oxygen equipment of one of the three fighter planes that accompanied Mantell failed; consequently, two planes turned back to the field. Soon thereafter, the third pilot radioed that he was abandoning the intercept as well, but Mantell did not acknowledge the message and pushed on.[33] An official Air Force board of inquiry concluded that Captain Mantell lost consciousness at approximately 25,000 feet. Having never regained consciousness, his aircraft hit the ground at great speed near Franklin, Kentucky. The investigation determined that he blacked out because of lack of oxygen, which caused him to lose control of his aircraft. The fact that he did not attempt to eject from his plane suggested that he did not regain consciousness during his descent.

As Mantell reached 13,000 feet in altitude during his pursuit of the UFO, he stated that it appeared to be a "metallic object of tremendous size." The comment caused a bit of a furor with the government.[34] Initially, the Air Force decided that the UFO sighting was actually the planet Venus; however, because the planet would have been nearly impossible to see at that time due to the sky conditions of that day, it was later decided that the object was probably a weather balloon. Once again, the Air Force later changed its explanation and decided that the sighting had been a combination of both the planet Venus and a high-altitude Skyhook weather balloon, which was part of a secret project unknown to local commanders at the time.[35] Others, however, believe that he encountered an alien craft that shot him from the sky. In death, Mantell became Ufology's first known martyr.[36]

Yet another wave of UFO sightings began in 1964 and lasted for about ten years. On April 24, 1964, a police officer, Lonnie Zamora, observed an oval object that appeared to make a crash landing in a ravine near Socorro, New Mexico. Two figures described as the size of boys and wearing white overalls, stood nearby. After he radioed the incident to headquarters, the figures were gone, and the craft began ascending into the air. The publicity surrounding the event ushered in a new era of more incredible UFO sightings.[37]

Compellingly, a significant proportion of UFO sightings have reported by military pilots and their crews. A notable example involves Parviz Jafari, a major in the Iranian Air Force, who in 1976 approached a luminous UFO that was observed over Tehran. Jafari described the object as brilliant and rectangular/cylindrical in shape. The object emitted flashing strobe lights sequencing blue, red, and orange. At one point, the large object (estimated to be the size of a Boeing tanker) ejected a smaller object, which flew directly to Jafari's jet. The ejected object later rejoined the larger one, but another small craft was released and flew downward.[38] Just as Jafari prepared to arm and fire a Sidewinder missile at the flying object, his control panel and communications went dead. His equipment returned to normal only after his jet moved away from the object. Apparently, the U.S. government took an interested in the case, as a once-classified memo from the Defense Intelligence Agency on the incident was sent to the National Security Agency (NSA), the White House, and the CIA.[39] At a UFO conference held at the National Press Club in 2007, Jafari stated that he made visual contact with the object but was so blinded by its brilliance that he could not make out its exact size.[40] He went on to become a member of the Iranian general staff and supervised the defenses against the Iraqi invasion, which forced an eight-year stalemate that saved the Iranian government. For his strategy, he became a national hero and his story about the dogfight with the UFO has become popular lore in Iran.[41]

Another purported military encounter with a UFO occurred on April 11, 1980, when Lieutenant Oscar Santa Maria Huertas of the Peruvian Air Force was ordered to intercept what was initially believed to be an aerial spying balloon. When Huertas got close to his target, he realized that it was not a balloon after all, but appeared to be an object with a shiny dome on top. According to his description, the object lacked the typical features of an aircraft, as it had no wings, propulsion jets, exhausts, or windows. He fired a burst of sixty-four 30 mm shells that would normally obliterate a target, but the barrage had no effect. The bullets seemed to be absorbed by the object, which ascended very rapidly away from the air base.[42]

According to several accounts, the UFOs involved in military encounters seem to have demonstrated an uncanny capability to repeatedly evade an attack at the very last moment when the pilots "locked on" to their targets and were ready to fire. It was as if they somehow knew that the pilots were ready to push the fire button.[43] Inasmuch as these last-minute evasions were so perfectly timed, the pilots concluded that they could not be coincidental. Thankfully, only a small fraction of UFOs have demonstrated even a remote semblance of hostility, and those rare occasions occurred only after severe provocation, such as an attack by military aircraft. After many years of study, the U.S. military concluded that UFOs posed no threat to national security.

Nevertheless, there have been numerous reported incidents of UFOs flying over or near Strategic Air Command (SAC) and other military bases in the United States. Some UFO researchers speculate that there is some kind of extraterrestrial monitoring going on that manifests when there is a nuclear crisis situation on the planet.[44] On October 24, 1966, the Minuteman missile system at Minot Air Force Base in North Dakota was adversely affected on an afternoon during which UFOs were sighted by multiple witnesses on the ground at three separate sites for over three hours. Two of the objects were tracked on radar.[45] As a UFO passed over one particular site, the missile's indicators flashed "Launch in Progress." A security team was immediately dispatched to investigate the situation. On arriving at the silo, the security team observed a metallic disk-shaped object surrounded by bright flashing lights moving above their heads. In order to avert an unauthorized launch, the control room personnel activated the missile's "Inhibit" switch. Fortunately, the procedure worked and eventually the missile's indicators returned to normal.[46]

A similar incident occurred the next year at Malmstrom Air Force Base in Montana. On the morning of March 24, 1967, Air Force First Lieutenant Robert Salas, a missile control launch officer, received a call from a security guard reporting a glowing red, oval-shaped UFO hovering directly over the air base. Salas immediately woke up his crew commander, First Lieutenant Fred Meiwald, who was napping during his break. Within one minute of the phone call, the missiles started shutting down, one by one. According to Salas' account, the missiles were inoperative while the UFO was overhead. Located from five to ten miles from where the UFO hovered, the missiles were about a mile apart from one another with independent backup power systems. A formerly classified Air Force telex stated that "all ten missiles in Echo Flight at Malmstrom lost strat alert [strategic alert] within ten seconds of each other...The fact that no apparent reason for the loss of ten missiles can be readily identified is a cause for grave concern to this headquarters."[47]

It is unheard of for ten missiles to shut down in short order, as the anticipated failure rate for the ICBMs (intercontinental ballistic missiles) was one missile in the entire squadron every ten months. As a result, the probability of what happened at Echo Flight was off the charts.[48] After the incident, technicians checked every possible cause of the missile failures, but were unable to find a definitive explanation for what happened. At the time, it was suggested that the most likely cause was some kind of electromagnetic pulse that was directly injected into the equipment. If that was the case, the force involved had to penetrate 60 feet underground to disable the missiles. Although SAC termed the Malmstrom incident a "cause for great concern," the role of the putative UFOs remains undocumented. The only reference to UFOs in the report described them as a rumor with no foundation. Most of the information about the UFO connection to the incident comes from Salas, who warned U.S. officials that UFOs hovering over military installations could be responsible for deactivating weapons systems.[49]

A similar incident occurred in the Soviet Union on October 4, 1982, when a huge UFO was sighed near the missile base at Byelokoroviche. In this case, the launch control officers were stunned when they observed an unknown force begin to prepare the missiles for a strike against the United States. Fortunately, however, the system suddenly shut itself down without any seeming human intervention. A subsequent investigation from Moscow under the direction of Colonel Boris Sokolov totally dissembled the site, but found no malfunctions to explain what occurred.[50]

Robert Salas suspected that UFOs could be responsible for a more recent incident that occurred in 2010, when 50 missiles went down at the F.E. Warren Air Force Base in Wyoming.[51] A squadron of ICBMs dropped down into what is known as "LF Down," which means that the missile control launch officers could no longer communicate with the missiles.[52] Despite the seriousness of the incident, there seemed to be no repercussions insofar as there is no record of any reprimand or other disciplinary action taken against anyone in the chain of command. This is in stark contrast to the response to two previous incidents just a few years before. First, in August 2006, four nuclear cone assemblies were accidentally shipped from F.E. Warren Air Force Base in Wyoming to Taiwan. The other, and more serious, incident occurred in August 2007, when there was an unauthorized movement of nuclear warheads from Minot Air Force Base in North Dakota to Barksdale Air Force Base in Louisiana. Specifically, nuclear warheads that were mated on six cruise missiles were inadvertently transported. The crew and the personnel responsible for handling the weapons were not aware that they contained nuclear warheads. As a consequence of the two aforementioned incidents, eventually the Secretary of the Air Force and the Chief of Staff were both forced to resign.[53] By contrast, nothing comparable happened after the 2010 malfunction at the F.E. Warren Air Force Base, which suggests that no human error was responsible for the incident. But if that is the case, why did the malfunction occur?

Perhaps the most notable reported military case involving a UFO occurred on December 26, 1980, at the Rendlesham Forest, near Ipswich in England. On that night, a blip appeared on the radar screens at Royal Air Force (RAF) Watton in nearby Norfolk. The blip fell off in the area of the Rendlesham Forest. Two U.S. Air Force security police officers who manned the east gate at RAF Woodbridge at the time, John Burroughs and Budd Parker, saw an object fall into the nearby forest.[54] They described the object as shiny and shaped like a cone about five to six feet in height.[55] Likewise, a three-man patrol from the U.S. Air Force's 81st Security Police Squadron reported seeing a triangular-shaped metallic craft moving through the trees, which eventually landed in a small clearing. Upon arrival at the landing site, a member of the patrol, Sergeant James Penniston, reported seeing a brightly-lit, oval-shaped craft that was sitting on a triangular landing structure. According to his description, the craft was approximately nine feet by nine feet at the base and standing six to seven feet in height.[56] He took out a notebook and sketched the shape of the object along with strange markings that he saw on its body, which he likened to Egyptian hieroglyphs. He then touched the craft and took photographs of it.[57] The rest of the security detail described being overwhelmed by a feeling of disorientation and confusion when they approached the craft, as if it had been surrounded by a dense, but invisible force field.[58] The security team watched the object for about 15 minutes, after which it levitated and began to move silently away from them through the trees.[59] Unfortunately, the Air Force later told Penniston that his photos of the craft were overexposed and did not come out.

The next day, the local police went to the forest to investigate the incident, but found no traces of the craft. Interviews with the local residents, however, revealed that on the previous night, farm animals were highly agitated and behaving very strangely. Returning to the site the day after the incident, Burroughs and Penniston discovered indentations that formed a perfect triangle in the clearing where they believe that the object had landed. Other physical signs they identified included trees in the area that had their tops broken off and strange serrations on some of the tree trunks.[60] A team from the British Ministry of Defense that specialized in scientific intelligence examined the spot where the craft allegedly landed with a Geiger counter and determined that the radiation readings seemed significantly higher than the average background—around seven or eight times above what one would normally expect in that area.[61] Moreover, a U.S. Air Force aircraft that flew over the area in the morning reported that infrared radiation was "pouring" out of the forest.[62] Years later, it was reported that the area where the object had been seen was free of microbial growth.[63]

Two nights after the sighting, the deputy base commander, Lieutenant Colonel Charles I. Halt, observed an anomalous craft as well and became one of the highest-ranking military officers ever to go on the record about a UFO sighting. As he and his team came closer to the site, their radio communications were subjected to interference and powerful mobile generators that Halt brought with him to illuminate the forest began to cut out.[64] Halt described the UFO as a red-orange oval object with a black center in the forest. The object zigzagged through the sky, when suddenly it exploded into five white lights and then disappeared[65] Halt recalled hearing farm animals making loud calls, suggesting they were affected by the object. The local constabulary had also received reports about strange lights on that evening.[66] Halt documented his encounter on a hand-held cassette recorder. The 17-minute tape is widely available on Internet sites such as YouTube.[67]

The British UFO skeptic Ian Ridpath offered an explanation for the incident, arguing that a lighthouse was responsible for the unusual sightings.[68] He advanced a theory that the flashing, pulsating lights in the trees were actually the rotating beams from the Orford Ness Lighthouse located six miles away. For his part, Lieutenant Colonel Halt claimed that he was aware of the location of the lighthouse.[69] Taking umbrage with Ridpath's explanation, he commented, "A lighthouse doesn't move through a forest, doesn't explode, doesn't change shape, [and] doesn't send down beams of light."[70]

The enlisted personnel in the two security details were instructed not to reveal completely the details of the incident, and to refer to the object only as a light they could not identify.[71] Although Sergeant Penniston indicated that he was told that the subject was classified, neither he nor anyone else has been admonished for speaking publicly about the incident.[72] In 1996, an inquiry in the British Parliament ended with the Ministry of Defense concluding that the Rendlesham incident had no defense significance.[73] The incident is sometimes referred to as "England's Roswell" and has attained considerable notoriety. In fact, word of the incident reached Buckingham Palace, where Prince Phillip made inquiries about it. At a reception after she returned from China, when questioned about the incident by Lady Georgina, Margaret Thatcher replied, "My dear, there are things you just can't tell the people."[74]

Numerous civilian accounts of UFO sightings have surfaced over the years as well. In some rare instances, harm resulted from close encounters on the ground. A notable case—the "Cash-Landrum incident"—occurred in Texas on December 29, 1980, when Betty Cash, Vickie Landrum, and her grandson, Colby, were driving along Highway FM1485 near Lake Houston. They witnessed a swarm of helicopters—they counted 23 in a few minutes—seemingly fly out of nowhere and then move off again along with a glowing craft, which pulsed intense heat and emitted a beeping noise. Viewed from a different angle, the craft became cigar-shaped, a bright, oblong cylinder of light. Soon thereafter, the craft, along with the helicopters, vanished.[75] Corroborating their accounts, several witnesses in nearby Dayton reported seeing a UFO around the same time. Over the next few hours, Betty, Vickie, and Colby showed symptoms of illness, not unlike those that occur due to exposure to radiation. Betty fared the worst, as she suffered blinding headaches, blisters on her skin, and later developed breast cancer. Betty and Vickie sued the federal government, believing that the U.S. military was responsible insofar as some of the helicopters appeared to be the Army's CH-47 Chinook model. However, their case was dismissed after representatives of the Army, Navy, and Air Force testified that they did not operate the crafts at that time and location. The U.S. government disclaimed any responsibility for the incident.[76]

More sightings followed. For instance, the Hudson Valley wave, which began in upstate New York and parts of Connecticut in 1982 and involved repeated sightings of large silent objects hovering at low altitudes with extremely bright spotlights. The residents reported mainly delta- or V-shaped objects.[77] One of the more notable UFO sightings occurred on November 7, 2006, at Chicago's O'Hare Airport. Based on the collection of eyewitness testimony, the UFO was estimated to have ranged in size from approximately 22 to 88 feet in diameter. The craft was said to have hovered at about 1,500 feet over Gate C17 at the United Airlines terminal. The events were reported to have lasted from about five to fifteen minutes. At first, the Federal Aviation Administration and United Airlines denied having any information about the incident. A team of experts from Dr. Richard Haine's group, the National Aviation Reporting Center on Anomalous Phenomena (NARCAP), spent five months investigating the incident and its safety implications. Their 154-page report concluded that the disc-shaped UFO observed at O'Hare was a solid object that behaved in ways that could not be explained in conventional terms.[78] However, there was no corroborative radar evidence at the airport suggesting that a solid object was in the sky.[79] The witnesses all agreed that the object made no noise and kept a fixed position in the sky before it ascended. Curiously, airline officials insisted that witnesses remain silent on the issue. It took two months for the news media to pick up the story, but when they did, there was tremendous public interest. The Chicago *Tribune* received a million hits on its website for the story, which broke a record.[80]

Even more intriguing was the case of the "Phoenix Lights," which included a number of spectacular sightings that were reported in Arizona on March 13, 1997. Scores of witnesses claimed to have seen a massive aircraft measuring the size of "multiple football fields." As is often the case, the government did not appear interested in the episode and no official investigation was launched despite the public outcry from the scores of eyewitnesses. A few months after the sightings, then-Governor Fife Symington announced at a press conference that he would reveal the source of the Phoenix Lights—whereupon his chief of staff, Jay Helier, was escorted to the podium in handcuffs and wearing an alien mask over his head.[81]

Four months later, the officials at Luke Air Force Base, which is located next to Phoenix, put out a news release that attributed the incident to flares that were dropped from A-10 Warthogs of the Maryland Air National Guard, which were practicing in an area southwest to the city. However, the timing was off, as the flares were dropped after 10 P.M., which was after the reported sightings that took place between 7 and 9 P.M.[82] Although the flares could have been responsible for a second set of observations, witnesses to the first set insist that the "Phoenix Lights" belonged to a very large object that blotted out the night stars.[83] In 2010, Symington provided an essay to the investigative journalist Leslie Kean's book *UFOs: Generals, Pilots, and Government Officials Go on the Record*, in which he revealed that he too had seen the UFO, but explained that his levity at the press conference was as an effort to allay the consternation of his constituents.[84] In addition to eyewitness accounts, trace evidence has been linked to UFOs.

Cattle mutilations have often been conflated with UFOs. Sometimes UFOs are spotted near the scenes of these incidents. The first recorded cattle mutilation occurred in September 1967, when a mare named Lady was found dead in a field on a Colorado ranch near Mont Blanca in the San Luis Valley. The horse's head and part of her neck had been stripped clean of flesh; however, the rest of the body appeared to be untouched.[85] An autopsy revealed that the cuts on the horse's body had been caused by something that generated huge amounts of heat, which some researchers believe must have been a laser.[86] The pathologist who examined the horse, Dr. John Henry Altshuler, who at the time worked at the Rose Medical Center in Denver, described the wound as a "vertical, clean incision." Interestingly, as Dr. Altshuler noted years later, "There was no surgical laser technology like that in 1967."[87] Some veterinarians, however, have dissented from the opinion that the incisions have the character of laser burns.[88] Several interesting traits are consistently reported in mutilation cases, which suggest that some extraterrestrial agency might be responsible.

Despite the carnage inflicted on the animals, there is often an absence of blood. In some instances, holes punched through the animals' jugular veins can be found, suggesting that they were drained of blood. Incisions on the mutilated animals often display sophisticated anatomical knowledge as well as the use of surgical instruments, but without coagulation of the blood around the wounds. Frequently, specific parts of the animal are removed, including the tongue, the coring of the rectum, the eyes, and sexual organs. In some mutilations, residual amounts of radiation or tranquilizing chemicals were detected. The carcasses are often reported to be found in an area where there are no human footprints or hoof prints. This would seem to indicate that the animal did not walk to where it died, because it would have left tracks. Furthermore, some of the mutilated animals were found with broken bones, implying that they were dropped to the ground from a significant height.[89] There is a lack of predatory indications on the animal. In fact, predators and scavengers appear to avoid the carcasses. These findings militate against the theory that terrestrial predators are responsible for the mutilations.

Although some investigators have conjectured that nefarious cults might be behind the mutilations, there is a complete absence of evidence around the carcass suggesting any kind of ritualistic observances.[90] Retired U.S. Army Lieutenant Colonel Philip J. Corso claimed to have reviewed cases on cattle mutilations while working at the Pentagon from 1961 to 1963. He described the wounds inflicted on the cattle as being made with surgical precision, perhaps by a laser.[91] Corso's startling claims are discussed in greater detail in Chapter 11.

Some researchers suspect that crop circles could be linked to UFOs as well. Crop circles are large patterns or designs, usually imposed on a field of grain such as wheat or barley. The stalks are flattened in such a way to create a pattern. The crop circle phenomenon began in force in the early 1980s, but there were antecedents. For instance, the "Mowing Devil" appeared in Hertfordshire in August 1678.[92] According to the claims of Arthur Shuttlewood and Bryce Bond, in 1972 in a field just outside of Warminster, England, they saw a large circle forming by itself in the crops on the meadow below them while they sat on a local incline.[93] Interestingly, crop circle reports are generally confined to England.

Many of the crop circles are said to be created very quickly—sometimes in less than a minute—which might suggest that some extraterrestrial agency was responsible for them. But in September 1991, two Englishmen, Doug Bower and Dave Chorley, came forward and claimed that they were responsible for making crop circles using improvised plank shoes or by stomping down on the grain in a circular manner.[94] They said that they started their antics as far back as 1978.[95] Some researchers, though, remain unconvinced, citing that although some hoaxers' designs are quite intricate, they do not have the disturbingly precise geometric and mathematical relationships that "real" crop circles display upon analysis.[96] In the hoaxes, the stalks are said to be broken, but in a real circle, they are swirled flat. Furthermore, genuine circles are alleged to show distinct cellular changes in the stalks when analyzed, while hoaxed circles do not.[97] In the real cases, the stems are bent parallel with the ground, usually about an inch up, but not broken. The plants appear to have been subjected to brief bursts of intense heat.[98]

Some researchers conjecture that the authentic crop circles might be formed from some type of high-energy plasma. One unorthodox theory advanced by Dr. Terence Meaden, the editor of Britain's *Journal of Meteorology* and founder of Britain's Tornado and Storm Research Organization, is that crop circles are formed by an atmospheric anomaly he called a "plasma vortex."[99] According to Meaden, wind currents are broken up as they pass hills and other obstructions. When they rejoin, a vortex of air is created that can dip down and form a circle in a crop field. Consistent with this theory is the claim that at a microscopic level, there are often changes to the plants' cells after a crop circle is discovered.[100] Furthermore, the seeds of the affected plants will not germinate.[101] An alternative and more natural theory posits that the circles are caused by electromagnetic meteorological effects related to the northern lights.[102] Still other explanations implicate piezoelectric energy (an electric charge that accumulates in certain solid materials) and subterranean gas or energy pressure.[103]

The SETI scientist Seth Shostak dismisses the notion that crop circles are some sort of extraterrestrial messages, noting that wheat fields are poor memory storage devices, as they would carry very limited information. Alternatively, the idea that aliens would travel light-years to carve temporary graffiti seems unlikely to him as well.[104] Be that as it may, a hard core segment of researchers remains adamant that some intelligent source is behind the crop circles. Gerald S. Hawkins, a retired chairman of Boston University's Astronomy Department and a science adviser to the director of the U.S. Information Agency, once claimed to have discovered five separate geometric theorems concealed in crop circle geometry.[105] Various patterns were found to align with the sun and the moon over a long period of time. To some researchers, this militates against the theory that hoaxers are responsible for all of the crop circles.[106] But who then is responsible?

The consensus among researchers is that 95 percent of UFO sightings can be explained by earthly sources. The other 5 percent, however, cannot be attributed to secret military exercises or natural phenomena. Witnesses maintain that the UFOs appear to make exceptional performance maneuvers that are beyond the capabilities of known aircraft. Moreover, they insist that they are guided by some intelligence.

Marcello Truzzi once wrote "extraordinary claims require extraordinary proof."[107] Later popularized by Carl Sagan, this aphorism is often invoked by skeptics concerning the veracity of UFO claims.[108] Several explanations have been proffered to explain UFO sightings. Naturally occurring phenomena can sometimes create the appearance of unusual flying objects. Atmospheric anomalies, such as lightning storms can illuminate the sky in strange ways, thus creating an illusion of a UFO. During a temperature inversion over a swampy area, gas can hover above the area and become slightly incandescent. When smaller pockets of gas separate from the larger packet, it might even create the illusion that scout ships are leaving the mother ship.[109] Venus, the brightest object in the night sky after the moon, sometimes appears to follow people as they travel in a car. And meteors striking across the night sky can give the illusion of a piloted spacecraft. Numerous man-made phenomena can generate UFO sightings as well. The most convincing UFO cases involve multiple sightings by credible eyewitnesses that are backed up by radar.[110] However, radar is not foolproof. Radar echoes bounding off mountains create echoes producing waves that zigzag and fly at enormous velocities on radar screens.[111]

In his book, *Flying Saucers: A Modern Myth of Things Seen in the Sky*, the eminent psychologist Carl G. Jung (1875–1961) proposed that UFOs were symbolic projections of deep human yearnings for wholeness and transformation which reside deep in the human psyche.[112] Consequently, UFO reports tend to occur in waves and reflect the zeitgeist. Donald H. Menzel suggested that these sightings occur in waves because when UFOs are in the news, people are more apt to look for them and see them. Once the publicity subsides, so does the reporting.[113]

All the same, psychological tests suggest that UFO witnesses have normal backgrounds and come from all walks of life.[114] Kenneth Ring found that people who reported a UFO experience did not score appreciably higher on a measure of fantasy proneness.[115] However, Troy Zimmer found that people who believe in UFOs were more likely to be open to beliefs of alternate realities, including astrology and the occult.[116] But a selection bias could account for this finding. As James E. McDonald noted, credible witnesses are often among the most reluctant to come forward with a witness report on UFOs.[117] Taking this into account, Robert L. Hall noted that when reasonable persons report events that receive no social support from their friends and peers and do not fit their own prior beliefs—such as UFO sightings—their accounts should be taken seriously. Often the witnesses did not take seriously the topic of UFOs prior to their experiences. Their observations were often jarring to their own beliefs; nevertheless, they insisted that the accounts of what they had seen were real. For these reasons, Hall counseled that we should take such witnesses seriously.[118]

Some skeptics doubt that interstellar travel is physically possible; hence, UFOs cannot be of extraterrestrial origin. Considering the vast expanses between solar systems, it would seem that attaining superluminal or faster-than-light speeds would be necessary for biological entities to practice interstellar travel because of their limited lifespans. In an article published in *Science Magazine* titled "The Physics and Metaphysics of Unidentified Flying Objects," William Markowitz argued that insofar as the reported UFOs seemed to violate the laws of physics and engineering as we know them, UFOs could not be extraterrestrial in origin.[119] Be that as it may, some UFO researchers have deduced that these various anomalous objects must originate from some place beyond Earth. Flying saucers have been reported to have traveled at high velocity with little or no sound and are uncharacteristic of any known terrestrial aircraft.[120]

Even more arcane explanations than the extraterrestrial hypothesis have been advanced to explain UFOs. For instance, Marc Davenport conjectured that UFOs may actually be manned by time travelers from the Earth's future.[121] To buttress his case, he cites the putative discoveries of human footprints and other seemingly human traces that have been unearthed in prehistoric fossils.[122] As he points out, aliens are sometimes described as having Asiatic features. This makes sense according to his theory, because the demographic preponderance of Asians in the world today suggests that future humans are likely to have such phonotypical features.[123] But as the skeptic Paul Davies pointed out, if there were alien civilizations with super-technology such time machines, their descendants could visit us now from the future. In that case, it would open up Earth to visits not only from alien contemporaries, but also their (and possibly our) descendants. Since we have no evidence of such temporal visitors, it only compounds the Fermi paradox.[124]

An even more bizarre theory UFO theory was propounded by Ivan T. Sanderson (1911–1973) in his book *Invisible Residents: The Reality of Underwater UFOs* (1970).[125] He catalogued numerous incidents involving sightings of so-called Unidentified Submarine Objects (USOs). Anderson noted that many UFOs seemed to appear or disappear over water. His bizarre theory posited that an ancient advanced civilization inhabited the deep oceans of Earth. Although this might seem preposterous, some credible scientists now believe that most life might actually reside beneath the surface of the Earth; however, these organisms would likely be very simple life forms, for example, bacteria. Defending this theory, Sanderson argued that it might be more practical for organisms to go down rather than go up to the Earth's surface, where they would encounter changes in temperature, earthquakes, and air pollution. In the same vein, the science fiction writer, Isaac Asimov noted that there were indeed some advantages to living underground. For example, weather would not be an issue. Temperatures would hold at a fairly constant level between 55 and 60 degrees Fahrenheit. As a result, a lot of energy used for heating and cooling could be saved.[126] Alternatively, Sanderson speculated that extraterrestrial aliens could have established bases underwater long ago.[127] Some variants of this theory speculate that the USOs are controlled by an ancient humanoid race that antedates *Homo sapiens*. According to this contrarian history, Atlantis was an actual prehistoric world that was destroyed in a civil war. The survivors sought refuge from the residual radioactivity by retreating under the Earth's crust or seas.[128]

After the release of the Condon Report in 1968 (discussed in the next chapter), the extraterrestrial hypothesis for UFOs entered a period of decline, at least from the standpoint of official government attention. As Steven J. Dick noted, at least three circumstances joined to work against the extraterrestrial explanation for UFOs. First, despite years of study and official interest, no incontrovertible evidence was produced in favor of the hypothesis. Second, the New Age movement offered explanations that UFOs were metaphysical, supernatural, or interdimensional in origin, thus undercutting the credibility of Ufology. Finally, much of the new evidence adduced for the extraterrestrial hypothesis was based on subjective evidence, such as statements under hypnosis, which scientists were reluctant to accept.[129]

Although there remains considerable public interest in UFOs, academics remain highly skeptical of the topic. The occurrence of occasional blatant hoaxes has inclined many scholars to dismiss extraterrestrial claims about UFOs as pseudoscience.[130] It is suggested that UFO reports, if investigated properly, have conventional explanations. In his book, *This Demon Haunted World* (1996), Carl Sagan adamantly rejected the extraterrestrial hypothesis for UFOs. He was a strident critic of ufology, arguing that it was a powerful distraction from serious efforts to detect extraterrestrial intelligence.[131] Likewise, Stephen Hawking impugned the extraterrestrial hypothesis for UFOs as well, noting that if aliens "are going to reveal themselves at all, why do so only to those who are not regarded as reliable witnesses? If they are trying to warn us of some great danger, they are not being very effective."[132] SETI scientists are particularly scornful of contemporary Ufology.[133] For example, Seth Shostak is unimpressed with UFO claims to date, pointing out an obvious asymmetry in the logic underpinning eyewitness reports. Whereas the skeptics who seek to disprove the extraterrestrial hypothesis must demonstrate that every UFO claim is false, proponents of the extraterrestrial hypothesis need only prove that one case is authentic. Based on this reasoning, Shostak argues that the plethora of UFO reports actually undermine the extraterrestrial hypothesis because to date, no unambiguous evidence has been established. As he explains, if the evidence were convincing enough, he and his SETI colleagues would abandon their antennas "and begin crawling the countryside."[134] Although Shostak concedes that interstellar travel is theoretically possible according to our known laws of physics, he counters that it is nonetheless impractical when considering that the closest extraterrestrial civilization is probably hundreds of light years away, if not more.[135]

SETI scientists feel a need to distance themselves from Ufology because the conflation of the two activities could make it more difficult to raise much-needed research funds from the government and serious donors for whom credibility is so important.[136] As Albert A. Harrison explained, avoiding the UFO taint is important for SETI scientists so that they can gain standing with peers, granting agencies, and the public. Early SETI pioneers, such as Philip Morrison, Frank Drake, and Carl Sagan, rapidly gained the interest and support of powerful, influential scientists. Professional scientists use boundaries to define what is and what is not science, in their respective fields. Boundary work seeks to monopolize the professional authority of expertise in a particular field. By doing so, scientists create a system of checks and balances that help themselves converge upon the truth as they discover it. For instance, SETI scientists have developed elaborate protocols for confirming possible extraterrestrial messages. The basic tactic of this "border work" is to make positive attributions about one's research while denigrating the work of pseudo-scientists who ignore or misapply the rules of science. Harrison identifies various strategies that SETI scientists use to create a boundary between their field and Ufology. Bashing one's opponents is one method of border work, but as Harrison notes, can sometimes backfire. In some cases, aggressive debunking of UFO claims—for instance ascribing them to swamp gas or fireflies—can seem less plausible than the initial claim. Silence is another method of border work, but it can allow Ufologists to claim that mainstream scientists have taken neither the time nor the effort to examine the evidence. Finally, education can be applied to dissuade the public from taking UFOs seriously. However, inasmuch as much of the public in not particularly well versed in science, this approach of border work can be an "uphill battle." In an age of declining scientific literacy, vast numbers of the populace are not inclined to think like scientists or skeptics. Yet, Harrison maintains that education is the most promising strategy for minimizing confusion between SETI and Ufology in the public's mind.[137]

UFOs are difficult to study insofar as they do not submit well to laboratory experiments or allow observations under controlled conditions. What is more, the extravagance of certain UFO claims often outstrips the supporting evidence, thus straining the credulity of academics. And those researchers who do investigate UFOs often have dubious credentials and backgrounds. Finally, critics of Ufology note that descriptions of UFOs and the purported beings that operate them often appear too earthbound for space visitors, as they are said to be very humanoid in appearance and attired in human-like space outfits.[138]

Fantastic claims backed by dubious evidence have undercut the credibility of Ufology. In *UFOs: Generals, Pilots, and Government Officials Go on the Record*, Leslie Kean, an investigative journalist, breaks with this pattern by relying on accounts from numerous credible eyewitnesses to UFO encounters and authoritative sources on the topic who contributed chapters to her book.[139] None of them claimed to have experienced any repercussions from the government or men-in-black visitations. Alas, despite the impressive case she makes, smoking gun evidence remains elusive. Kean calls for a "militant agnosticism" in pursuit of UFO investigations. By agnostic, she means not rushing to ascribe UFOs to extraterrestrial sources. She advises that UFO cases should be investigated utilizing scientific techniques that evaluate physical evidence rather than relying exclusively on subjective eye witness accounts. To that end, she calls for the creation of a small agency within the U.S. government to handle appropriate UFO investigations and coordinate with other countries and the scientific community. In 2002 Kean cofounded the Coalition for Freedom of Information to obtain documents and UFOs through the Freedom of Information Act. Her effort quickly won support from John Podesta, President Bill Clinton's former chief of staff.[140]

Perhaps skepticism of the extraterrestrial hypothesis among scientists has gone too far. As Thomas E. Bullard noted, "history has come full circle as scientists have traded in their heroic status of explorers and rebels for the role of priests, too confident in their authority, too prejudiced in the rightness of their knowledge, to look before they condemn. Such loaded treatment acts as a self-fulfilling prophecy when people withhold their testimony because they fear rejection and ridicule."[141] New developments in technology could lead to a breakthrough in Ufology. For instance, the Internet has democratized Ufology, allowing more people to submit evidence and participate in the discussion on UFOs.[142] With the ubiquity of digital cameras and cell phones, there are now more opportunities for witnesses to capture UFOs in pictures that could be exciting, for as it still stands, the burden of proof is still on the UFO claimants. Although many people remain skeptical about the extraterrestrial hypothesis for UFOs, various national governments have taken the topic quite seriously which is explored in the next chapter.

Chapter 9

The Government's Response to UFOs

Not long after Kenneth Arnold's alleged sighting, the U.S. government began to take a keen interest in UFOs. As the Cold War began to take hold, the initial concern was that the anomalous flying objects could be Soviet in origin. In a secret memo dated September 23, 1947, Lieutenant General Nathan Twining, the commander of the Air Force Material Command, wrote about "Flying Discs" to the Pentagon and recommended that a detailed study into the topic was warranted. He remarked that in his "considered opinion," the rash of UFO sightings had been "something real and not visionary or fictitious."[1] By September 1947, the Air Material Command unofficially began a flying disk project, and on December 30, 1947, the Pentagon gave an order to commence an official study of UFOs.[2] The issue attained greater salience soon thereafter on January 7, 1948, when a large number of witnesses spotted a UFO in proximity to Godman Air Force Base, near Louisville and Fort Knox, Kentucky. In that incident, as explained in the previous chapter, Captain Thomas Mantell, pursued a UFO in the high atmosphere and died when his plane crashed into the ground.[3] On January 22, 1948, Project Sign (also known as Special Project HT-303, or occasionally Project Saucer) was launched.

Project Sign was exclusively an Air Force investigation under the Air Technical Intelligence Center and conducted at Wright Patterson Air Field in Dayton, Ohio. Lieutenant Colonel George Garrett of the Air Force Intelligence's

Collection Division made the first official estimate on the topic in August of 1948, using 16 or more of the most notable UFO cases. Although he could offer no definitive answer as to the provenance of the UFOs, he concluded that they "were not at all imaginary" and that "[s]omething real is flying around."[4] His estimate was distributed to military technology leaders who vowed that no known U.S. devices could have been responsible for the sightings. At this point, the Pentagon decided to forward this information to the Air Material Command to get another expert opinion. To this end, AMC Intelligence chief, Howard McCoy convened an informal think tank of base experts to examine the issue. The committee concluded that the phenomenon was not illusory but authentic.[5]

Still, the Air Force could not come up with a plausible explanation. At first, investigators believed that the UFOs were probably some unspecified Soviet aircraft. A top secret document, Air Intelligence Report 100-203-79 (dated December 10, 1948), which was later declassified, evaluated the possibility that UFOs originated from the Soviet Union and the implications for national security if that were the case.[6] Three distinct camps emerged in Project Sign. One faction advanced a theory that the Soviets were using captured German rockets or other exotic aircraft to wage a form of psychological warfare against the United States. Some of the Air Force scientists considered the possibility that the UFOs might be atomic powered flying machines based on German prototypes that were developed during World War II and subsequently captured by the Soviets. Although this proposition seemed remote, it was about the only one that was thinkable.[7] The investigators later dropped that idea, reasoning that the Soviets would not allow secret advanced devices to fly over the United States where they ran the risk of crashing and being discovered.[8] Over time, it was clear that the Soviets had made no sort of breakthrough and that the UFOs were not some sort of extrapolation of German designs.[9] What is more, rumors began to circulate that Moscow was also interested in the purported flying disks, suggesting that they were as befuddled as the Americans.[10] A second faction argued that the UFOs could be explained by natural phenomena, such as birds, clouds, weather balloons, and hoaxes. Finally, the third faction believed that the sighted crafts were extraterrestrial in origin. This third group passed an "Estimate of the Situation" through the chain of command to the top brass at the Pentagon. Within a year, sentiments shifted in Project Sign toward the theory that the UFOs were not from Earth.

Project Sign examined 243 sightings and submitted its official report—"The Findings of Project Sign"—in February 1949, which concluded that the UFO sightings were most likely extraterrestrial in origin.[11] Nevertheless, the report called conceded that "All information so far presented on the possible existence of space ships from another planet or of aircraft propelled by an advanced type of atomic power plant have been largely conjecture."[12] Project Sign's director, Captain Robert R. Sneider endorsed the report.[13] For his part, the Air Force Chief of Staff, General Hoyt S. Vandenberg disagreed with this conclusion, and as a consequence, the report was returned, declassified, and destroyed.[14] Unpleased with the new slant, General Vandenberg decided to cancel Project Sign.[15]

Working with the Office of Naval Intelligence, Air Force intelligence, produced an opposing report that concluded that the extraterrestrial hypothesis should not be taken seriously. The report recommended that investigative efforts should focus on all things domestic or foreign. The primary author of the report was Major Jere Boggs. The harsh rebuke against Project Sign amounted to a "woodshed whipping."[16] The Project Sign committee was instructed to send copies of the cases it examined to Boggs, the Office of Naval Intelligence, and the Air Force's Scientific Advisory Board. As a consequence, two parallel UFO investigative programs developed; one operated by the Air Force and another by the Navy. Cooperation between the two services on this issue was often contentious; the Navy even questioned the Air Force's competence in this area.[17]

Amidst all the controversy, Project Sign became bogged down in internal dissention over the topic of UFOs. To break the deadlock, the Air Force sought advice from the wider scientific community. To that end, in 1948 the Air Force employed J. Allen Hynek to look into the sudden rash of UFO sightings. At the time, he was a junior faculty member who taught astronomy at The Ohio State University at the nearby Wright-Patterson Air Force Base. The Air Force hired him to review UFO reports and determine if they originated from misperceptions of astronomical objects or events.[18] As an outsider, Hynek was a bit naïve concerning the internal politics of the U.S. government's UFO investigation program. At the outset, he pursued his investigations with the a priori assumption that UFOs could not be extraterrestrial in origin, as he believed no location within the solar system—except for Earth—could sustain advanced life. Furthermore, anywhere outside of our solar system was so impossibly distant as to be unreachable.[19] Initially a skeptic, Hynek predicted that by 1952, the whole UFO craze would be forgotten.[20]

As the security of Project Sign had been compromised by numerous stories in the press, it was officially cancelled. The Air Force publicly announced that Project Sign had been terminated and let the media and the public assume that official investigations into UFOs had ceased. That was not really the case, however, as a new program was launched on December 16, 1948, which proceeded in secrecy under a new title—Project Grudge.[21] At this time, the government could not cavalierly dismiss UFO sightings. On February 16, 1949, a meeting was held in Los Alamos, New Mexico that included Edward Teller, also known as the father of the hydrogen bomb. In a report of the meeting, Navy Commander Richard Mandelkorn wrote that "there is cause for concern of the continued occurrences of unexplainable phenomena of this nature in the vicinity of sensitive installations."[22]

Skeptics dominated Project Grudge.[23] In August 1949, Project Grudge compiled a 600-page secret document—Technical Report #102-AC 49/15–100—which examined 244 UFO sightings covering Europe and the United States for the previous three and a half years.[24] The report concluded that UFOs presented no direct threat to the national security of the United States, but conceded that roughly 23 percent of the cases had not been satisfactorily explained. Nevertheless, it dismissively attributed the UFO sightings to mass hysteria, hoaxes, individual psychosis, and misidentification of various conventional objects.[25] The general thrust of the report was that those who believed in flying saucers suffered from some mental pathology.[26] On December 27, 1949, it was announced that Project Grudge was shut down. The records of the program were placed in storage and most of its personnel were transferred to other jobs, yet the Air Force remained interested in the topic and launched Project Twinkle soon thereafter.[27]

By the early 1950s, opinion was split in the Pentagon over UFOs. One school of thought believed that there was no such thing as flying saucers and that the reports thereof were merely misidentifications, outright frauds, or the result of human error. As such, UFOs presented no demonstrable threat to American national security. The other school of thought maintained that the flying disk technology was likely to be real. Within this latter group there were three theories. One insisted that the disks were Soviet in origin. Another suspected that they were caused by some secret American technology undisclosed to most of the defense and intelligence communities. Finally, another opinion attributed the UFOs to extraterrestrial aliens. An operational gap between these two major schools of thought—the skeptics and the believers—widened over time. The skeptics dismissed the study of UFOs as a waste of time and effort. The other camp urged vigilance for no other reason that the UFOs could present a possible threat to U.S. interests if they were of Soviet or extraterrestrial provenance.[28]

The public's interest in UFOs continued to grow. In April of 1951, the legendary newsman, Edward R. Murrow, produced a television program called "The Case of the Flying Saucer" in which he examined a number of UFO incidents. Surprisingly well-balanced, Murrow nevertheless concluded that he could find no pattern to the sightings and opined that he was still skeptical. This policy of strong public debunking worked well with the general public, but not with those who claimed to have seen UFOs.[29] In the year 1952, hundreds upon hundreds of UFO sightings were reported and media coverage was intense.[30]

In light of the new wave of sightings, the Air Force was compelled to resume investigations into UFOs. On October 27, 1951, Project Blue Book was officially launched with Captain Edward J. Ruppelt as its head.[31] The project's name sought to create an aura of unimpeachable honesty into the UFO controversy (similar to how college students often sign a pledge of honor in their blue book exams).[32] It was Ruppelt who coined the term "UFO" in 1952, feeling that the term "flying saucer" did not account for the full range strange objects spotted in the sky.[33] Ruppelt was convinced that UFOs were worth investigating, but was determined to be absolutely even-handed in his approach. He retained Hynek as his chief scientific consultant. Initially, the Air Force did not disclose the disposition of inquiries into UFOs, but Ruppelt began a system of cooperation with the media by issuing regular releases about sightings and investigations.[34] In essence, Project Blue Book became a repository for UFO cases and a place for people to call and file reports of sightings.[35] Top Air Force officials publicly supported the new program. Lieutenant General Nathan Twining informed the public that he took the UFO phenomenon quite seriously, telling an audience in Amarillo, Texas, on May 15, 1954, that the "best brains in the Air Force" were trying to solve the mystery of flying saucers.[36]

UFOs became a frequent theme in popular culture in the 1950s. Hostile extraterrestrial aliens invading the Earth featured prominently in science fiction films. Films such as *War of the Worlds* and *The Day the Earth Stood Still* captured the public's imagination on the prospect of extraterrestrial visitation. Some observers wondered if an incident that began at 11:40 P.M. on Saturday, July 19, 1952, portended a real-life invasion scenario when seven blips appeared on the air traffic control radar screen at Washington National Airport (now Reagan National Airport). According to some accounts, seven flying disks were seen passing over the White House and the Capitol Building. An intercept aircraft was dispatched from Newcastle Air Force Base in Delaware, but the intruders had disappeared before it arrived at roughly 3 A.M.[37] The following day, Americans read headlines in newspapers proclaiming "Saucers Invade DC" and "Saucers Swarm Over Capital."[38] A few days later, on July 23, 1952, a fleet of flying saucers were alleged to have invaded the air space of Washington, D.C., during which they hovered over military installations.[39] Reportedly, the military commander in Washington was so perturbed by the reports that on July 29, he issued a "shoot down/talk down order.[40] Supposedly, Albert Einstein protested the order, counseling the White House that it should be rescinded, not only to preserve intergalactic peace, but self-preservation on the part of humans as well. The order was later voided.[41]

On July 29, 1952, a press conference was held in which Major General John A. Samford, the director of Air Force Intelligence, told reporters that he was convinced that all of the sightings over Washington, D.C. in the preceding two weeks had been caused by temperature inversions. For the most part, the media accepted this explanation and dismissed the sightings as mass hallucinations.[42]

Nevertheless, President Harry S. Truman was so concerned about these sightings that he instructed the National Security Council to do something about UFOs. CIA director General Walter Bedell Smith, who at the time chaired the National Security Council, formed a scientific advisory panel to evaluate the potential threats related to the UFOs. Nicknamed the "Robertson Panel," Dr. Howard P. Robertson, a physicist and weapon systems specialist from the California Institute of Technology, served as the chairman. Also on the panel were Luis Alvarez, who played a leading role in the Manhattan Project, and Samuel Goudsmit, an associate of Albert Einstein who made numerous contributions to physics.[43] The panel met over four days in January 1953.[44] According to files released under the Freedom of Information Act, the CIA was concerned that public interest in UFOs could undermine national security. From the perspective of the CIA, the UFO phenomenon presented only two possible threats. First, it could create a dangerous panic among the American public. During a spurious UFO wave, communications channels could be clogged, making the United States vulnerable to a Soviet invasion.[45] Second, flying saucer "nuts," scam artists, and cultists could be manipulated into spreading anti-American, pro-Communist, anti-war, and anti-bomb messages to the public. For instance, a celebrated UFO "contactee" of this era, George Adamski, opined that alien civilizations probably had communist forms of government and that the United States would fall just as the Roman Empire did.[46] Fearing that UFO research groups might be used for subversive purposes, the CIA suggested that they should be placed under surveillance. The CIA encouraged all agencies within the intelligence community to work with the media and infiltrate civilian research groups for the purpose of debunking UFOs. By doing so, public interest in the topic would diminish.[47] Under the aegis of the CIA, the Robertson panel proposed the creation of a broad educational program that would integrate the efforts of all concerned agencies. To allay public anxiety over UFOs, around 1953–1954, the CIA instructed Project Blue Book to explain away all UFO cases at all costs.

In accordance with these instructions, the Air Force was reluctant to give its imprimatur to any media project that seemed to suggest the extraterrestrial hypothesis for UFOs, even science fiction entertainment. For instance, when the well-known movie director Howard Hawks petitioned the Air Force for assistance in using military locations, personnel, and equipment for the making of the science fiction film *The Thing*—a story about an ancient shipwrecked alien discovered frozen, but still alive, in Antarctica—his request was rejected by the Pentagon, which maintained that the flying saucer was a "myth" and that its policy was "not to participate in any proposal that [would] perpetuate this hoax."[48] As a consequence, the Air Force refused any cooperation in the production. Furthermore, the Air Force objected to any mention or pictorial display of its personnel or equipment in the film.[49]

By the late 1950s, some private researchers believed that Project Blue Book was not receiving information on the most important UFO cases. A very restrictive information management style was implemented with the Pentagon ordering Wright-Patterson to be silent on the issue. Project Blue Book was closed to inquisitive reporters, as all inquiries were directed to the Public Information desk at the Pentagon. Increasingly, Blue Book became irrelevant.[50] More and more, citizens were seen as unhelpful actors in UFO investigations.[51] The relationship between the federal government and private UFO groups, such as the National Investigations Committee On Aerial Phenomena (NICAP), was strained. A Cold War of sorts broke out between the NICAP and the Air Force, which made it all the way to the U.S. Congress. Beginning in 1956, NICAP's director, Donald Keyhoe, pressured Congress to hold hearings, but these requests were denied.[52] The Air Force consistently applied pressure to squelch formal congressional committee inquiries into the topic.[53] On occasion, though, the NICAP garnered support from some high-level government officials. For instance, in 1958 Senator Barry Goldwater (R-AZ) appeared on the front page of the NICAP's *UFO Investigator* as part of a story in which he spoke of his interest in the topic.[54]

The largest scientific study of UFOs undertaken by the U.S. government was reported in *Project Blue Book Special Report No. 14*, which was completed in 1955.[55] It remains the most exhaustive study of the topic. The study was conducted by engineers and scientists at the Battelle Memorial Institute in Columbus, Ohio, a highly respected research and development organization.[56] Some observers criticized the Air Force for giving too much weight to the literal statements of the witnesses.[57] Nevertheless, the report concluded that it was "highly improbable that any of the reports of unidentified aerial objects [. . .] represent observations of technological developments outside the range of present-day knowledge."[58] For his part, Blue Book's director Captain Ruppelt believed that the Air Force was not taking the issue seriously enough. Rather than playing an educational role, Project Blue Book had been downgraded to a virtual non-entity.[59] In 1953, Ruppelt left the Air Force and was replaced by Major Hector Quintanilla. In frustration, Ruppelt began a series of exchanges with Donald Kehoe.[60]

Two figures—Major Donald Kehoe of NICAP, a leading civilian research group, and Dr. James E. McDonald, a senior atmospheric physicist from the University of Arizona—brought credibility and knowledge to the UFO subject while challenging the approach of Project Blue Book.[61] With the publication of best-selling books and magazine cover stories about UFO stories in 1966, public interest in the topic was at its peak. In 1950, Kehoe published a book titled *The Flying Saucers Are Real*, in which he asserted that for over the past 175 years, Earth had been under systematic close-range observation by extraterrestrial aliens.[62] McDonald accused the government of no "grand cover-up;" rather, he thought it could be more aptly described as a "grand foul-up."[63] Emerging as a leading advocate of the alien spacecraft hypothesis, McDonald worked to bring greater attention to and disclosure on the topic of UFOs after viewing the files of Project Blue Book, whose explanations he found unconvincing. In fact, he even cajoled then-President Lyndon B. Johnson to commission a study into UFOs. Eventually, Johnson relented and authorized a study that included numerous leading researchers committed to UFOs, including Kelly Johnson of the aero-engineering "Skunk Works," and Art Lundahl of the National Photographic Interpretation Center. The study concluded that UFOs most likely involved advanced technology created outside of Earth.[64]

In 1956 Ruppelt published his own book, an insider's view titled *The Report on Unidentified Flying Objects*, but it was unclear as to where Ruppelt actually stood on the issue. In his book, he vacillated between the skeptic and believer camps. The Air Force was relieved that except for a few public lectures and TV appearances, Ruppelt did not become a public figure or an ongoing nuisance.

Nevertheless, the Air Force was not pleased with his book.[65] Published in 1959, the second edition contained three new chapters. In the final chapter, Ruppelt unequivocally stated his position: there was nothing to UFOs. They were not real after all, as he wrote: "I have visited Project Blue Book since 1953 and am now convinced that the reports of UFOs are nothing more than the reports of balloons, aircraft, astronomical phenomena, etc. I don't believe they are anything from space."[66]

Why did Ruppelt change his position? Some people believe that this was his original position all along—he just did not articulate it clearly enough in the first edition. Another school of thought thinks that the Air Force pressured him to renounce the extraterrestrial hypothesis for UFOs. Finally, another group speculates that he was angered by Keyhoe's attacks on his old service, and when informed of how his book was being used toward this end, Ruppelt decided to make amends.[67] The Air Force monitored NICAP director Donald Keyhoe and his media appearances very closely. In 1957, the Secretary of the Air Force, Richard E. Horner, issued statements to the media in an effort to undermine Keyhoe's media appearances.[68] Ruppelt died of a heart attack in 1960.[69]

The Strategic Air Command (SAC) Commander at the time, General Curtis LeMay, dismissed the invasion of space reports as a diversion that should not be taken seriously. The launch of the Soviet Sputnik satellite on October 4, 1957, presented a concrete challenge to the United States. To meet this challenge, the technology and intelligence communities created the Advanced Research Project Agency (ARPA, later DARPA) to secretly develop advanced weapons systems to prevent a technological surprise.[70] The Soviet nuclear arsenal, now equipped with intercontinental missiles, was his overriding concern. These were real world threats; UFOs were far too speculative for LeMay to worry about. Other Pentagon personnel, however, were alarmed by the concentration of sightings over major nuclear installations.[71] During the so-called "Great Fireball era" from 1948–1950, more than 200 such events were logged the great majority of which were in the vicinity of Albuquerque, White Sands, and western Texas military bases.[72]

Despite the public's growing interest in UFOs, the major media seemed to comply with the CIA's debunking program. In 1966, a CBS television two-hour special was organized around the Robertson Panel conclusions. Hosted by Walter Cronkite and titled "UFO: Friend, Foe, Or Fantasy?" the program debunked UFOs from all angles, some of which included false claims, such as there was no radar or photographic evidence to support the reality of UFOs.[73] In a private letter, Dr. Thornton Page, an astronomer and military intelligence expert who worked with the CIA as part of its program to debunk UFOs, disclosed that he had served as a technical consultant to Cronkite's television special and influenced its direction so that it was structured along the lines of the Robertson panel.[74]

The CIA's efforts to diminish interest in UFOs, notwithstanding, the issue would not go away. The precipitating factor for renewed attention was a new wave of UFO sightings which occurred from 1965 to mid-1967.[75] Two high-profile incidents in particular spurred the government to take UFOs more seriously. The first—what came to be called the "Exeter incident"—began on September 3, 1965, in southeastern New Hampshire, where Norman J. Muscarello, an 18-year-old man, was hitchhiking home to the small town of Exeter from Amesbury, Massachusetts. While walking, he observed an enormous red sphere that rose like a red moon from behind some trees. What appeared to be some sort of craft careened through the sky, but generated no engine noise. Badly shaken by the incident, Muscarello stumbled into the Exeter police station and reported the incident. Minutes later, officer Eugene Bertrand entered the station and reported that about an hour earlier, he had spotted a car parked alongside the highway whose occupant—a woman too distraught to drive—claimed that she had been followed for 12 miles by a glowing red object. Listening to Muscarello, Officer Bertrand began to wonder if there was something to the two stories. Now curious, he drove Muscarello to the field where he claimed to have seen the UFO. Not long after their arrival, they spotted a brilliant round object rise from behind the trees. After the incident, more witnesses reported seeing the object. Eventually, John Fuller, a columnist for the magazine *Saturday Review*, published his own carefully researched version of the affair. He later parlayed the article into a book titled *Incident at Exeter*.[76]

Philip J. Klass—an abiding "debunker" of UFO claims—attributed the whole incident to "ball lightning," a rare phenomenon in which lightning assumes an oval shape, takes on an intense red color, and moves with erratic vigor. A major objection to his explanation, though, was that no thunderstorm, which is necessary to produce ball lightning, occurred in the area at the time of the incident. Unwilling to give up on his thesis, Klass countered that high-voltage power lines in the area where the incident took place were capable of producing plasma—a region of ionized gas created by a strong electrical charge—that could take the form of ball lightning. *Newsweek* magazine called his theory "one of the most persuasive explanations" for the incident, but added that the Air Force was "noncommittal." For their part, UFO enthusiasts remained unimpressed with Klass' explanation.[77]

The second major UFO incident commenced on March 14, 1966, when numerous citizens and police officers around Ann Arbor, Michigan, reported seeing what appeared to be objects flashing across the predawn skies. On March 20, near the town of Dexter, a truck driver and his family reported seeing a pyramid-shaped object land in a field. As they approached the site, a white light the object emitted turned red, at which point the object disappeared. The next day, over 50 persons saw an object matching the same description near Ann Arbor. And that same evening, over 80 female students at Hillsdale College, along with a local civil defense director and the college dean, reported seeing an object shaped like a football that swayed, wobbled, and glowed while in flight.[78]

From 1965 to 1969, at least a few scientists began to take the subject of UFOs seriously in part because of pressure from the U.S. Congress.[79] Moreover, the events that occurred in Michigan generated more public concern among residents there that could not be so easily dismissed. Now under the leadership of Major Hector Quintanilla, Project Blue Book adopted a policy of rapid public disposition of UFO cases.[80] To that end, the Air Force sent Dr. J. Allen Hynek, the astronomer who worked for a number of years as a consultant to Project Blue Book, to investigate the aforementioned UFO incidents in Michigan. His tentative explanation was that swamp gas was responsible for the anomalous sightings. As people clamored for a plausible explanation, Hynek's thesis was met with derision and attained infamy in the annals of Ufology.[81] The swamp gas explanation provoked outrage from much of the public, especially in Michigan, where so many witnesses claimed to have seen the UFOs. Representative (and future president) Gerald Ford (R-MI) called for an apology to his constituents and an investigation into the Air Force's UFO procedures.[82] The NICAP took advantage of this public sentiment to blast the Air Force, even going so far as to suggest a cover-up. Within a week after the swamp gas explanation, Hynek, Quintanilla, and the Secretary of the Air Force, Harold Brown, were called to testify on the issue by the House Armed Services Committee. Testifying at the hearings on April 5, 1966, Secretary Brown showed statistics on 10,147 UFO sightings reported to the Air Force from 1947 to 1965. Out of these sightings, the Air Force identified 9,501 as natural phenomena that had been erroneously interpreted. Carefully choosing his words, the remaining 646, Brown noted, were "those in which the information available does not provide an adequate basis for analysis, or for which the information suggests a hypothesis but the object or phenomenon explaining it cannot be proven to have been here or taken place at that time."[83] The implication Brown made clear was that neither he nor anyone affiliated with Project Blue Book believed that UFOs were extraterrestrial in origin.[84]

Out of these hearings, the Air Force agreed to initiate an extensive study on UFOs that would be conducted by a major university. Lieutenant Colonel Robert Hippler of the Air Force's Directorate of Science and Technology was given the task of recruiting a university for the study. He created a panel to recommend an institution and a number of universities were proposed, including the University of Dayton, The Massachusetts Institute of Technology, Harvard, the University of California, Berkeley, and the University of North Carolina. However, at the time, UFOs were regarded as leprosy to academia, so the project was a hard sell.[85]

In 1966, the University of Colorado agreed to host a government-funded study of UFOs. The Air Force's contract for the study amounted to $313,000. The grant was highly unusual in the sense that normally grants are awarded to scientists with experience and interest in the topic with which they contract to study. In this case, however, the grant was given to a scientist with little interest and no experience in the topic, who essentially had to be pushed into the study.[86] Edward U. Condon, a prominent physicist and head of the National Bureau of Standards, agreed to lead the effort. After receiving his Ph.D. in physics from the University of California, Berkeley in the 1920s, Condon spent two years in Germany working with some of the world's most prominent physicists. During World War II, he served as a district head for the Manhattan Project at Los Alamos. He was once the target of Richard Nixon for an alleged, but bogus, accusation of being a pro-Communist security risk. Now at the end of his career, Condon had held numerous organizational presidencies and was much honored by his peers. Nevertheless, he had a reputation for ruffling feathers and doing things his way. An inveterate joker, the opportunity to study UFOs seemed like an amusing diversion.[87] The Condon committee, as it came to be known, included a preponderance of psychologists which caused somewhat of an uproar among the hard scientists.[88] The committee met five times between October 14 and 31, 1966, before the contract officially began.[89]

In January of 1967, a representative of the Air Force suggested an anti-extraterrestrial hypothesis so that the Air Force could cease its UFO program. Condon was happy to oblige.[90] Three months after his appointment to lead the committee, he announced at a public meeting: "It is my inclination right now to recommend that the government get out of this business. My attitude right now is that there's nothing to it, but I'm not supposed to reach a conclusion for another year."[91] Even more damning, two concerned project members of the study unearthed a memorandum by the project coordinator Robert J. Low, which was addressed to two university deans. In the memo, Low discussed problems associated with undertaking the project. Most notably, he opined that by conducting a study on UFOs, it could put the researchers "beyond the pale" and they could stand to lose "prestige in the scientific community." One way to counter this problem, he suggested, was to have the study conducted almost exclusively by "nonbelievers who, although they couldn't possibly *prove* [italics in original] a negative result, could and probably would add an impressive body of evidence that there is no reality to the observations." Further, he counseled that "The trick would be, I think, to describe the project so that, to the public, it would appear a totally objective study but, to the scientific community, would present the image of a group of nonbelievers trying their best to be objective, but having an almost zero expectation of finding a saucer."[92]

According to some UFO researchers, the fact that he used the word "trick" suggests that he gave his game away. In July of 1967, Roy Craig, a physical chemist who served on the committee, passed the memo along to the rest of the members, who were shocked and concerned over its verbiage and tone. Even Condon conceded that the memo was inappropriate. As a consequence of the memo, the project personnel lost confidence in Low.[93] Moreover, the publication of Low's memo damaged the credibility of the committee. The pro-UFO community saw the memo as proof positive that Low was irrevocably biased against giving the case for UFOs a fair consideration. Consequently, the NICAP formally withdrew its support for the effort, and Donald Keyhoe called for a government inquiry.[94] Conversely, the anti-UFO camp saw the memo as an innocuous missive; the word "trick" was merely used in a colloquial sense, without any prejudice implied.[95] For his part, Condon was infuriated when the Low memo was made public. He fired the two staffers who leaked the memo the day after he heard about it.[96]

Before the conclusion of the Condon Committee investigation, the U.S. Congress made yet another inquiry into UFOs. On July 29, 1968, the Committee on Science and Astronautics of the U.S. House of Representatives held a Symposium on Unidentified Flying Objects in Washington, D.C. Dr. James E. McDonald was the chief organizer for the event. J. Edward Roush (D-IN) chaired the symposium. Numerous scientists testified at the symposium. The proceedings were compiled in a volume called *Symposium on UFOs*.[97]

Finally released in early 1969, *The Scientific Study of Unidentified Flying Objects*—popularly known as the Condon Report—concluded that "Nothing has come from the study of UFOs in the past twenty years that has added to scientific knowledge."[98] Furthermore, the Condon Report opined that the evidence indicated that UFOs posed no threat to national security. The Condon committee was divided. One faction, which included James E. McDonald, denounced the project. John Fuller, a journalist who identified with those who found UFO sightings credible, blasted the study in an article titled "Flying Saucer Fiasco," which appeared in a May 14 issue of *Look* magazine. The major media outlets, however—including *The New York Times*—and also the National Academy of Sciences—essentially rubber-stamped the Condon Report's findings.[99]

The Air Force wanted out of the UFO business and used the Condon Report as its excuse to pull the plug on Project Blue Book.[100] All totaled, Project Blue Book investigated 12,618 sightings.[101] The vast majority of the cases—roughly 90 percent—were dismissed as misidentified aircraft or explainable by natural causes; however, 701 remained as "unidentified."[102] These findings are typical of other UFO investigations in that the residual of unexplained cases typically account for about 5 to 10 percent of the total.[103] Finally, on December 17, 1969, the Secretary of the Air Force Robert C. Seamans Jr. announced that the Air Force would terminate Project Blue Book, thus ending all official investigations into the topic of UFOs.[104] As a consequence, the UFO question shifted to the margins.[105] Increasingly, scientists refused to take the issue seriously.

The Condon Report was a major setback to Ufology. A sense of betrayal set in the UFO community, which contributed to the "us-versus-them" mentality that would characterize its relationship with the federal government.[106] Critics claimed that Condon had come to his conclusions before the study began. With his bias, he was disinterested in the facts—even those contained in his own report.[107] What might be described as a minority report—*UFOs? Yes! Where The Condon Committee Went Wrong*—was later published by Dr. David Saunders. Saunders, whom Condon fired, pointed out what he saw as important aspects that were not addressed in the study. In an earlier article titled "Flying Saucer Fiasco" published in *Look Magazine* in 1968, he mentioned the memo from Robert J. Low, which described what some inferred as a strategy of subterfuge that would make the study look scientific.[108]

Over time, some skeptics reversed their position on UFOs. Most notable, was J. Allen Hynek, who became more amenable to the extraterrestrial hypothesis.[109] In 1972 he released a book titled *The UFO Experience: A Scientific Inquiry*, in which he acknowledged that the Air Force followed the CIA's lead in debunking UFO claims. Moreover, he complained that the best UFO cases did not go to Project Blue Book.[110] Increasingly frustrated by the lack of official interest, in 1973 he founded the Center for UFO Studies (CUFOS), which is based in Illinois. After his death in 1976, it was later renamed the J. Allen Hynek Center for UFO Studies.[111]

Why does there appear to be such a strong taboo against taking the UFO subject seriously when there is considerable evidence for it? Rather than an intentional conspiracy, perhaps the U.S. government might be as baffled as everyone else on the UFO question. In an interesting essay titled "Militant Agnosticism and the UFO Taboo," two political scientists, Alexander Wendt and Raymond Duvall, advance a theory as to why the U.S. government has supposedly been less than forthcoming on the UFO question.[112] As they point out, skeptics cite a number of seemingly intractable obstacles to interstellar travel to argue against the extraterrestrial hypothesis. Nevertheless, Wendt and Duvall argue that the origins of the UFO taboo are political, not scientific. As they see it, the prospect of UFOs presents three major challenges to the sovereignty and credibility of the state. First, if UFOs are accepted as truly unidentified, then that proposition would acknowledge a potential threat that could undercut the legitimacy of the state insofar as protection against potential threats is the most elemental function of the government. Second, a confirmation of the presence of UFOs would create tremendous pressure for a world government, which today's nation states would be reluctant to form. Third, and most important, the extraterrestrial possibility would call into question the anthropocentric model of modern sovereignty, which as they explain, forms the basis of authority of states to command the loyalty of their subjects. The arrival of extraterrestrial aliens, they point out, would be something analogous to the Christian "Second Coming." In such a scenario they ask, to whom would people give their loyalty? Could states survive if such a question became real, not just theoretical?

In sum, Wendt and Duvall argue that the presence of UFOs creates a deep, unconscious insecurity in which certain possibilities are unthinkable because of their political implications. As a consequence, the taboo emerges not so much from a vast conscious conspiracy seeking to suppress "the truth" about UFOs. Rather countless undirected practices that help us "know" that UFOs are not extraterrestrial in origin and can therefore be disregarded are carried out by the government, but not in the style of a covert conspiracy.

Still, some officials have taken UFOs seriously and have attempted to get to the bottom of the issue. Presumably, the executive branch would be the most important arm of government for investigating UFOs because of the numerous agencies at its disposal. Former President Jimmy Carter once stated that in 1969, he and a group of businessmen spotted a UFO while attending a Lions Club meeting in Leary, Georgia.[113] Shortly after assuming the presidency, he instructed his science advisor, Frank Press, to write a memo to the National Aeronautics and Space Administration (NASA), recommending that the agency set up a panel of inquiry on UFOs. However, after a number of letters, memos, and inquiries made their through the various levels of the hierarchy of the NASA bureaucracy, the agency turned down the president's request in December 1977.[114] For his part, Senator Barry Goldwater (R-AZ), a former pilot and retired major general in the Air Force Reserve, was convinced that a secret UFO program did exist. He once called the commander of the Strategic Air Command, Curtis LeMay, and asked permission to access the room at Wright-Patterson where information on UFOs resided—at which point LeMay got angry and cursed him out, exclaiming, "Don't ever ask me that question again!"[115]

Former President Ronald Reagan wrote that he had witnessed at least two UFO sightings. According to one account, in June of 1982, after a screening of the soon-to-be released motion picture *E.T.* at the White House, President Reagan confided to Steven Spielberg, "You know, there aren't six people in this room who know just how true that [the *E.T.* movie] really is." But before the president could elaborate, the conversation was interrupted by well-wishers at the screening.[116] While president, he occasionally speculated about the prospect of extraterrestrial aliens and their impact if they ever visited Earth. Speaking before the United Nations General Assembly in September 1987, Reagan remarked: "Perhaps we need some outside, universal threat to make us recognize the common bond that unites all humanity. How quickly our differences world-wide would vanish if we were facing an alien threat from outside this world. And yet I ask you, is not an alien threat already among us?"[117]

Although the U.S. government ceased formal investigations into UFOs with the dissolution of Project Blue Book, official interest in the topic did not completely go away. The UFO researcher Kevin D. Randle claims that the Air Force continued to investigate UFO sightings even after Project Blue Book was terminated. According to some claims, the Aerospace Defense Command, which is now inactive, took over responsibility for reports on UFOs.[118] Randle notes that the publication of Air Force Regulation identified the 4602nd Air Intelligence Service Squadron as the unit that would conduct the in-field UFO investigations. The regulation required the unit to report its findings to the Air Technical Intelligence Center at Wright-Patterson Air Force Base, but there was no requirement to pass findings on to Project Blue Book. Randle sees this as proof that the Air Force maintained a classified study of UFOs that was not part of Project Blue Book. Over the years, he claimed that the designation of the 4602nd Air Intelligence Service Squadron had changed several times and eventually evolved into the Detachment 4696th Air Force Intelligence Group headquartered at Fort Belvoir, Virginia. No records of these units are publicly available. According to Randle, the new codename for UFO investigations was Project Moon Dust, but that name was later changed too.[119]

Ostensibly, the U.S. government appears to be aloof toward UFOs despite the documentary evidence. But in recent years, some members of Congress have taken up an interest in UFOs. For instance, in July 2005 Congress held hearings that were initially supposed to include discussion on unidentified aerial phenomena (UAP). The hearings were sponsored by Representative Tom Davis (R-VA). Although not publicly disclosed, Davis was supposed to have had his own UFO sighting, which prompted his interest in the hearings. However, the hearings ended up taking another direction—counterterrorism. Representatives from the Department of Homeland Security, Department of Defense, and the Federal Aviation Administration (FAA) testified that the skies were safe, but made no mention of UFOs, which had initially prompted the hearings.[120]

More recently, John Podesta, who served as President Bill Clinton's White House chief of staff, wrote a foreword to Leslie Kean's 2010 book, *UFOs: Generals, Pilots, and Government Officials Go on the Record*, and publicly supported her efforts to bring more information on the topic of UFOs. For her part, Kean characterizes the U.S. government as a "pariah" on the international scene concerning official UFO investigations for its lack of transparency. In contrast, several European and South American governments have taken the issue seriously and established agencies to investigate UFOs.[121] In particular, Kean commends the French government for its open-mindedness on UFOs as evidenced by its creation of GEIPAN, a unit in the French space agency (CNES) that investigates unidentified aerial phenomena.[122]

In February 2012, the FAAs Air Traffic Organization Policy stated that persons who wish to report UFO sightings should contact Bigelow Aerospace Advanced Space Studies, a division of Bigelow Aerospace.[123] The firm was founded by Robert Bigelow, who grew up in Las Vegas, Nevada. When he was about 10 years old, his grandparents who lived next door to him, told him about their own experiences in which they witnessed UFOs. Bigelow's father became an avid UFO researcher. Piqued by these stories, by the time he was an early teenager, Bigelow vowed to himself that he would someday be involved in some type of space activity. For many years, he kept his goal a secret, not even telling his wife.[124] In a sense, his pursuit was not unlike that of the nineteenth-century businessman, Heinrich Schliemann, who after reading a copy of Ludwig Jerrer's *Illustrated History of the World* at the age of eight, was determined to one day find the lost city of Troy, which at the time was thought to be only a myth. Like Schliemann, Bigelow worked many years to save money to pursue his dream. After amassing his fortune in real estate and the hotel chain Budget Suites of America, he launched Bigelow Aerospace, a Nevada-based space technology startup company in 1998. To date, the firm's most ambitious project has been the launching of inflatable space station modules. Bigelow has given millions of dollars of grant money to UFO research groups such as the Mutual UFO Network (MUFON). According to some sources, Bigelow hopes to reengineer alien technology.[125] Bigelow occasionally talks about his interest in UFOs, but for the most part, he is tight-lipped on the topic. His involvement in collecting UFO reports was the subject of an episode of Jesse Ventura's *Conspiracy Theory* television program.[126]

On occasion, foreign governments and even international governmental organizations have expressed interest in UFOs. For instance, in 1978 the United Nations General Assembly invited interested member states to coordinate research on extraterrestrial life, including UFOs. The General Assembly encouraged researchers to inform the Secretary General on their findings, but little seems to have come of this invitation.[127]

After a wave of UFO sightings were recorded in Belgium in 1989–1990, the Belgian government took an interest in the topic as well. Approximately,

2,000 cases were reported, of which 650 were investigated and 500 remain unexplained. Some observers described the aircraft as triangular, while others described them as massive inverted aircraft carriers. A high proportion of these sightings involved lights in a triangular formation on the underside of a huge craft that hovered over villages and towns at low altitudes.[128] Several witnesses observed the craft making tilting maneuvers that revealed a dome at the top. Belgian Major General Wilfried De Brouwer maintained that the maneuvers of the UFOs were beyond the capacity of even experimental aircraft, and no "black program" could have been responsible for the sightings. Moreover, U.S. government authorities assured the Belgian Air Force that there were no U.S. military aerial test flights that could have occasioned the reports.[129]

The French government began to investigate UFOs in 1974. Authority for investigating UFOs was initially placed with the National Gendarmerie. On May 1, 1977, Yves Sillard, the president of CNES (Centre National d'Étude Spatials)—the French equivalent of NASA—created a research arm for of the study of UFOs.[130] On July 16, 1999, a private French group—COMETA (the committee for In-Depth Studies)—released a 90-page document that concluded that the best explanation for the small number of seemingly inexplicable UFO cases was the extraterrestrial hypothesis. The COMETA report was written under the auspices of the Institute of Higher National Defense Studies, a private think tank. Nevertheless, the report was based on a study that was carried out by numerous retired generals, scientists, and space experts who spent three years analyzing military and pilot encounters with UFOs, which prompted supporters to defend its credibility. The report concluded that the UFOs "could be the work of a craft of extraterrestrial origin."[131] Going further, one of the report's authors, Jean-Jacques Velasco, who from 1983 to 2004 led GEIPAN, concluded that the existence of UFOs was without question. Critics of the COMETA report point out that its bibliography contains a long list of popular books in the UFO literature. Furthermore, they argue that the release of the report was manipulated to make it appear as if were an official French government study. Finally, the original COMETA report was published in a tabloid, which led some critics to impugn the document's authority.[132] This proved to be an embarrassment to Institute of Higher National Defense Studies.[133]

Until recently, the British government maintained an official project to investigate UFOs. The origins of the British program can be traced to a 1950 initiative by the chief scientific adviser to the Ministry of Defense, the great radar scientist Sir Henry Tizard. He set up a small group called the Flying Saucer Working Party. In its final report issued in 1951, the group concluded that all UFO sightings could be explained as misidentifications, ordinary objects, optical illusions, psychological delusions, or hoaxes.[134] But after a series of high-profile sightings that included incidents in which UFOs were seen by Royal Air Force (RAF) pilots, or tracked by military radar, the Ministry of Defense reconsidered its original skepticism and decided to further investigate the phenomenon. Over the course of its lifetime, the Ministry of Defense's UFO project received more than 12,000 sighting reports.[135]

The most celebrated of the official British UFO investigators is Nick Pope, who in 1991 began his position at the UFO desk in the British Ministry of Defence. Describing himself as "the real Fox Mulder"—the character from the TV Series *The X-Files*—he claimed that he was viewed as somewhat of an iconoclast by his peers in the Ministry. He sensed that his bosses did not want him to delve too deeply into the subject.[136] To his surprise, Pope found that the RAF personnel were quite open-minded to the topic of UFOs. As he explained, the RAF is responsible for protecting the skies over England, and was alert to the potential threat of a superior technology. Furthermore, he noted that many RAF and commercial pilots have observed UFOs.[137] During his experiences investigating UFOs, Pope found that the witnesses often told convincing accounts of their sightings. They did not seem to crave publicity; on the contrary, they often seemed embarrassed and sorry for taking up his time.[138]

After examining a myriad of UFO cases and immersing himself in the literature on the topic, Pope concluded that he was left with no doubt in his mind that some of the objects are extraterrestrial in origin.[139] He speculates that aliens began taking an interest in humans after World War II, when atomic weapons and new technology could someday threaten their planets.[140] Generally speaking, Pope is not sanguine about the alien presence. Citing incidents such as cattle mutilations, alleged abductions, and occasional downed human aircraft, he fears that the aliens are "carrying out what can only be described as crimes against humanity."[141] To better prepare ourselves for a possible hostile confrontation with extraterrestrial aliens, he calls on all national governments to "take the issue seriously, investigate sightings, and announce their conclusions, however bizarre or unsettling they might be."[142] Still, Pope is adamant that there was no official cover-up of the UFO phenomenon by the government in Britain.[143]

Despite numerous recorded incidents, many people dismiss the subject of UFOs as something that is unserious and unworthy of study. Be that as it may, there is still substantial interest in the topic; at one time, it was estimated that roughly half of all Freedom of Information requests in the United States concerned UFOs.[144] When the French government released its UFO files in one fell swoop in 2007, the demand was so high that its dedicated website crashed.[145] Still, official inquiry into the topic appears limited, as public officials rarely discuss the topic in a serious manner. Even more incredible than the UFO sightings are the reports of direct alleged contact with extraterrestrial beings. Despite the sensational nature of some claims, Nick Pope finds it implausible that all of the alleged abductees are lying. The next chapter explores this controversy.

Chapter 10

Close Encounters

THERE IS A DEARTH OF convincing trace evidence to make a solid case for the extraterrestrial hypothesis, but with the multitude of UFO eyewitness accounts that have been reported over the years, many people remain open to the idea. More controversial is a small subset of UFO sightings in which some people claimed to have had intimate contact with beings that are presumably from other worlds. The veracity of their claims is a contentious issue. Some testimonies appear to be highly dubious, while others have a quality of verisimilitude.

A leading ufologist, Jacques Vallée, created a taxonomy of human contact with extraterrestrial aliens. According to his classification, Close Encounters of the First Kind involve sightings without any physical contact. Close Encounters of the Second Kind consist of those cases in which aliens left behind physical traces, such as indentations in the ground or heightened radioactive readings. Close Encounters of the Third Kind are those cases in which alien beings have been seen, usually in the vicinity of the UFO. Finally, Close Encounters of the Fourth Kind are alien abductions.[1] This chapter focuses on the abduction phenomenon. Is it real or fabrication? To be sure, some claims strain credulity and seem patently false. For instance, in 1952 George Adamski regaled people with stories that he had encountered and communicated with an alien from the planet Venus.[2] Subsequent probes of Venus indicate that the planet is inhospitable to life, thus belying his fantastic claims. Other accounts, however, cannot be dismissed so easily.

Reported cases of direct contact between humans and aliens began not long after the contemporary UFO era commenced in the late 1940s. In some instances, humans alleged to have had intimate relations with their alien abductors, implying the existence of some sort of hybrid breeding program involving humans and extraterrestrials. For example, in October 1957, in one of the first recorded alleged abductions, a Brazilian cowhand named Antônio Villas-Boas reported being kidnapped from his tractor.[3] According to his account, he had sexual intercourse with a four-and-a-half foot, blond-haired, blue-eyed alien that appeared to be female.[4] After she made love to him, she pointed to her belly, then to the sky and then left him.[5] Over the next few days, Villas-Boas discovered unusual wounds on his body. A doctor who examined him a few months later recorded that his symptoms resembled those typical of radiation poisoning.[6] After the incident, Villas-Boas claimed that Brazilian authorities interrogated him quite harshly, yet he never wavered from the details of his story.[7]

On August 1 of the next year, Albert Bender told about being taken to an ice cave in Antarctica, where he observed numerous alien aircraft. While there, he saw aliens in their original form, which he described as outlandishly horrible. Their assumed leader, whom Bender referred to as the "exalted one," answered his questions about cosmology, religion, science, and medicine.[8] According to Bender, the aliens were primarily interested in extracting an element from Earth's sea water and shipping it back to their home planet.[9] In 1962, Bender published a book—*Flying Saucers and the Three Men*—in which he expounded on his alleged encounter, claiming that he arrived to the secret Antarctic underground base by way of astral projection and he met male, female, and hermaphrodite creatures. Not long after the publication of the book, he retired from Ufology and managed a hotel in California.[10]

Some abduction accounts have clear political overtones, as was the case with Louis Farrakhan, the controversial leader of the Nation of Islam, the most prominent Black Nationalist organization in the United States. He once told a bizarre story of an incident in September 1985, which took place on a mountaintop in Mexico. According to his account, he had a vision of being taken to a massive UFO, the "Mother Plane," on which he encountered the late Elijah Muhammad (who once led the Nation of Islam).[11] During the ordeal, Muhammad warned minister Farrakhan of a plot by then-President Ronald Reagan and the Joint Chiefs of Staff to wage war in the Middle East. Farrakhan concluded that the target of the supposed aggression was Muammar Gadhafi. So moved by the experience, he traveled to Libya in March of 1986 to warn Gadhafi of Reagan's alleged plan. The next month, American warplanes bombed Tripoli. Farrakhan saw the first Gulf War as further confirmation of Elijah Muhammad's message.[12]

Understandably, skeptics impugn the veracity of the aforementioned abduction cases. One of the more convincing abduction cases involved a married couple, Barney and Betty Hill. Their story began outside of Portsmouth, New Hampshire, in September 1961, when they were driving home late at night from a vacation. As they travelled along U.S. Route 3 near Indian Head in North Lincoln, they spotted a strange light in the sky. When they stopped the car, the light seemed to get closer. They soon discerned that the light was an object that appeared to be a huge flattened circular disc with a row of intensely illuminated blue-white windows along its front edge. The object hovered above them about 200 feet in the air.[13] Even stranger, they observed a group of 11 "little men" standing in the road. Soon, the creatures surrounded them, at which time their memories stopped.[14] According to their accounts, which were given under hypnosis, once they made sight of the strange beings, they were placed in some kind of trance that made them lose control of their movements. From there, they were led into a spacecraft.[15]

Initially, the Hills had no immediate recollection of the encounter after the incident. But what most puzzled them was a two-hour gap for which they could not account in their trip home from Canada.[16] For years, their ordeal was suppressed, but under the treatment of a Boston psychiatrist, Dr. Benjamin Simon, the Hills recounted amazing stories of their experiences while under hypnosis. They recalled being taken on board a spacecraft and placed in separate rooms, whereupon they were subjected to physical examinations. To them, the ordeal was akin to the experience of laboratory animals. In Betty's case, samples were taken of her hair and skin. A long needle was inserted into her navel, presumably, as some sort of pregnancy test. During her experience, Betty Hill discussed the universe and the human concept of time with the leader of the crew.[17] Interestingly, in 1964 she recounted a memory of a star map that was shown to her while she was onboard the spacecraft. Her extraterrestrial examiner pointed to an area on the map of the planetary system from which the aliens originated. As Betty Hill explained, the aliens told her that the lines on the star map represented heavy trade routes, lighter lines were normal trade routes, and the dashed lines were occasional expeditions.[18] At the end of their ordeal, the alien beings administered hypnotic suggestions to the Hills so that they would not recall anything about the incident.[19]

New astronomical data available for the first time in 1969 seemed to support part of Betty Hill's story. Subsequent measurements and designs made of the planetary and star system she described later matched a twin-star system which was not known to astronomers at the time of the alleged abduction.[20] The researcher Marjorie Fish found a configuration in the sky that closely matched the pattern as described in Betty Hill's star map. Based on Fish's analysis, the aliens appeared to be from Zeta Reticuli in the constellation Reticulum, which is about 37 light years from Earth.[21] Moreover, from the perspective of the Hills, who resided in New Hampshire, the star system would not be visible in the night sky. Zeta 1 and Zeta 2 cannot be seen by observers north of Mexico City's latitude.[22]

The Hills had an unassuming background, which lent believability to their story. Betty was a highly regarded social worker for the state of New Hampshire. Her African-American husband was a postal employee and a civil rights worker. At the time, interracial couples were somewhat of a novelty and lacked acceptance in mainstream American society, which some observers cite as all the more reason for the couple to avoid publicity. Although their therapist Dr. Benjamin Simon later wrote an article stating that he did not believe in extraterrestrials, he was nevertheless convinced that the Hills were sincere in their own accounts and were not fabricating them.[23] The story was unknown to the general public until 1966, when *Look* magazine published a two-part excerpt of a book by John G. Fuller titled "The Interrupted Journey," which went on to become a national best-seller.[24] In 1975, the book was made into an NBC-TV movie called *The UFO Incident*, starring James Earl Ray as Barney Hill and Estelle Parsons as Betty Hill.[25] In 2007 Betty's niece, Kathleen Marden, along with the noted UFO researcher Stanton Friedman, coauthored a book—*Captured*—which provided a full account of the Hills' experience, along with supporting scientific information. Friedman suspects that the aliens he believes visit the Earth originate from the star system Betty Hill identified during her alleged abduction ordeal.[26]

Another notable abduction case occurred on November 5, 1975, when Travis Walton, a 22-year-old man, and six of his colleagues were returning to their homes in Snowflake, Arizona, in a large truck after a hard day of clearing brush in the Sitgreaves National Forest. At around 6:00 P.M., as they drove through a on logging road, they claimed that a strange craft began hovering above them about 100 feet away. Intrigued by the object, they stopped their vehicle to take a closer look. Soon after they exited the truck, a bolt of bluish white light flashed from the UFO, knocking Walton to the ground. His companions fled from the scene in a state of panic. After the object departed, they returned to the scene, but Walton was gone. They duly reported the incident to the Navajo County sheriff. Walton was missing until November 11, when he called home from a telephone booth near Heber, Arizona. After his ordeal, he was dehydrated and had lost a significant amount of weight.[27] According to his version of the story, after regaining consciousness from the bolt of light, he found himself aboard an alien craft, where he was subjected to a number of experiments. In 1978, Walton wrote his account of the episode in a book titled *The Walton Experience*. [28] In 1993, the book was turned into a film—*Fire in the Sky*—starring D.B. Sweeney as Walton. Considerable controversy surrounds the case. On the one hand, those who support Walton's story point out that he passed two polygraph examinations—one in 1976 and another in 1993.[29] On the other hand, some skeptics, for instance Philip Klass, suspect that Walton and his colleagues concocted the story as a way to get their employer out of a Forest Service brush-clearing contract that was running behind schedule and was facing heavy penalties.[30]

Over the years, subsequent abductee accounts have echoed those experienced by Walton and the Hills. One of the most celebrated abduction stories comes from the novelist Whitley Strieber, whose 1987 book, *Communion*, became a *New York Times* bestseller and was later made into a film in 1990 by the same name starring Christopher Walken. Strieber claimed to have had elaborate personal encounters with nonhuman beings over the course of his life. He first noticed something strange on December 26, 1985, while in his cabin located in a secluded corner of upstate New York. In the middle of the night, he heard a peculiar swooshing sound coming from his living room downstairs. He checked the living room, but found no evidence of anything untoward. After he returned to bed, a small being, perhaps four feet tall, entered his bedroom. Paralysis overcame him, and he reported being subjected to a type of clinical examination by small beings who wore dark blue coveralls. A female among them seemed to be guiding the procedure. During the examination, he claimed that a needle was inserted into his brain and a device into his rectum. Soon thereafter, he found himself in a room elsewhere, presumably onboard some alien ship. After that event, he recalled previous "missing time" episodes in his life, some occurring when he was a child. This experience is consistent with the reports of other abductees who recount a feeling of disorientation followed by "missing time" for which they cannot account. To bolster his credibility, Strieber subjected himself to a battery of psychological tests and a neurological examination. He was determined to fall within the normal range in all respects. To further prove the veracity of his abduction experiences, he underwent a polygraph examination that was administered by an operator with 30 years' experience, which he passed.[31]

Throughout his ordeal, Strieber observed four different types of alien figures. One type was small and robot-like in demeanor. Members of another group, attired in dark-blue coveralls, were short and stocky. They had wide faces and were either dark gray or blue, with glittering deep-set eyes, pug noses, and broad, somewhat human mouths. Representatives of a third type were about five feet tall, very slender, with extremely prominent and mesmerizing black slanted eyes. Finally, those in the last group were somewhat smaller, with similarly large heads, but round black eyes like large buttons.[32] He described their movements as choreographed as if their every action seemed decided elsewhere and then transferred to the individual being.[33] Strieber described the beings he encountered as insect-like in appearance and deportment. He conjectured that they belonged to a hive-like community and thought with a group mind. Although they might have superior intellect, he wondered if they thought slowly when compared to humans.[34] Although seemingly outlandish, his descriptions are markedly similar to those people who have undergone the dimethyltryptamine (DMT) experience, which will be discussed shortly.

Strieber advanced a number of explanations as to where the beings originated. In addition to the extraterrestrial hypothesis, he conjectured that they could hail from another dimension. He held out the possibility they could be space travelers from Earth's future. Perhaps, he wondered if the beings were some kind of manifestation from his own psyche but were nevertheless real and corporeal.[35] As an author of a number of horror novels, including *Wolfen*, which became a motion picture, Strieber obviously had an interest in the unexplained. However, before the recollection of his abduction experiences, he claimed that he did not believe in the extraterrestrial hypothesis for UFOs.

As Nick Pope once noted, the highly emotive and controversial nature of the UFO topic makes it inevitable that a certain lunatic fringe would be drawn to it.[36] Consequently, many scientists do not take the topic seriously. For instance, the noted astronomer Neil de Grasse Tyson dismisses the extraterrestrial hypothesis for UFOs, finding the abduction stories to be particularly ludicrous. To date, he does not believe that the UFO cases examined satisfy the standards of evidence that any scientist would require for any other claim.[37] Furthermore, he notes that the abductees have not related much interesting information from their alleged conversations with aliens, as they tend to consist of vague allusions, generalities, and platitudes on the dire need to achieve world peace.[38]

It has been estimated that the percentage of abduction cases that have been brought forth under hypnosis ranges from 80–90 percent, and in some casebooks reach 100 percent.[39] But hypnosis can be unreliable, and some subjects are highly suggestible under hypnosis. It is possible for sincere people to develop fake memories concerning topics such as alien abductions, past-life regression, and ritualistic abuse. Such memories can result when a researcher or therapist coaxes an account that fits expectations.[40] What is more, abduction cases are often surreal and dreamlike in character and often end up as hallucinogenic journeys. In many cases, humanoid aliens shape shift into birds, giant insects, and other phantasmagoric creatures.[41] One explanation is that hypnagogic (toward sleep) and hypnopompic (awakening from sleep) anomalies induce fantasies in some individuals, who then attribute the experience to an alien abduction.[42]

The late Dr. John E. Mack, who was professor of psychiatry at the Cambridge Hospital, Harvard Medical School, conducted the most extensive research on alien abduction cases. He founded the Center for Psychology and Social Change (now called the John E. Mack Institute) in 1993, which, *inter alia*, examined subjects who claimed to have been abducted by aliens. With his impressive academic credentials, his research was not to be taken lightly. Initially very skeptical, over time, Mack concluded that the vast majority of those subjects that he interviewed were psychologically healthy persons that were trying to contend with an extraordinary experience.[43] Reviewing the research of Budd Hopkins, Mack was intrigued by the consistency of the abduction accounts.[44] Mack argued that hypnosis was a credible tool in the study of the abduction phenomenon.[45]

In his books, *Abduction* and *Passport to the Cosmos*, Mack referred to the abductees as "experiencers." As the abduction experiences began, he found that the subjects were often disturbed by humming sounds, strange bodily vibrations, or paralysis. The experiencers often reported that they were placed on some type of examining or treatment table. Some abductees claimed to have received implants that were designed to monitor them, not unlike those that terrestrial scientists use to track bears, wolves, and deer.[46] Mack and other researchers found that some of the abductees bore small scars that corroborated their claims of implantations.[47] They commonly reported seeing energy-filled tunnels and cylinders of light during their experiences. The typical alien was usually described as short, with a large head, skinny body, big eyes, gray skin, and a small mouth. Some, however, were described as reptoid and instectoid. Not unlike the ordeal of Betty Hill, many abductees reported that the aliens appeared to be interested in the reproductive apparatus of humans, possibly to breed some kind of human-alien hybrids. The abductees recounted that the aliens attempted to communication with them, often by way of telepathy, as if they heard alien voices speaking inside their heads. It is not uncommon for abductees to feel that there is one alien in particular with whom they have a special relationship.[48]

According to Mack, interactions with extraterrestrials have often had a transformative effect on the abductees. After their experiences, many of them became committed to a number of progressive political causes, including environmentalism, and peace and disarmament movements.[49] Mack speculated that the aliens performed a role not unlike shamans insofar as they come to Earth to open the minds of humans and awaken them, opining that the encounters seemed "almost like an outreach program from the cosmos to the spiritually impaired."[50]

Some of Mack's colleagues at Harvard looked askance at his work, and in 1994 a "Special Faculty Committee" was formed to assess the validity of his research. Abduction researchers, along with Mack's attorney, Eric MacLeish, mobilized widespread support for Mack, but by August 1995, the committee closed shop without any official censure of Mack or his research.[51] That same year, Mack co-hosted a conference on the topic of alien abductions that was held at the Massachusetts Institute of Technology.[52] On September 24, 2004, while attending a conference in London, Mack was killed when he was run over by a truck while crossing a street.[53]

What accounts for the alien abduction phenomenon? Are extraterrestrials really responsible? One of the more bizarre theories put forward is that aliens reside in unseen dimensions and occasionally interact with humans. J. Allen Hynek coined the term "metaterrestrial" to denote something that originates outside of the ordinary three-dimensional space-time framework in which we normally function.[54] Likewise, Jacques Vallée came to believe that aliens originated from other dimensions beyond space-time from a multiverse that is all around us.[55] He once suggested that the UFO experience could involve non-human consciousness or intelligence that seems to shape-shift according to the culture with which it comes into contact. That is, whoever is behind the UFO enigma socially manipulates the minds of the contactees according to their individual beliefs and the cultural context in which they live.[56] The British Ufologist Jenny Randles once theorized that in some rare cases a witness's subjective impression of a UFO is so strong that it can manipulate objective reality. Referring to this as the "Oz factor," Randles conjectured that alien beings could be contacting humans through consciousness alone and are able to induce a subjectively real incidence of an encounter. A particularly sensitive person might serve as a radio receiver of cosmic messages.[57]

Some researchers have conjectured that aliens could originate from alternative dimensions rather than distant planets. So-called "Transcendentals" are disincorporated energy beings that manifest themselves in any way, shape, or form. As they exist in dimensions outside of our time, they are capable of "popping" into time at any point they so choose, perhaps even carrying out operations in different times contemporaneously. Some remote viewers claimed to have encountered Transcendentals who seem to have access to all levels of human consciousness, and at times are willing to reveal some details of their operations.[58] Remote viewers were part of a CIA and Defense Intelligence Agency project called Stargate. These "psi spies" were trained to locate various military intelligence and targets using telepathy. Their minds become disembodied, enabling them to visit distant locations in both time and space. A few remote viewers have gone so far to claim that during some missions, they have actually visited other planets.[59] But even its proponents concede that remote viewing is far from 100 percent reliable.[60]

As bizarre and implausible as these claims might seem, contemporary theoretical physicists hypothesize that it is quite likely that several more spatial dimensions exist, but we are unable to see them. Since the end of the twentieth century, research in string theory has dominated theoretical physics. As string theory posits that there are other dimensions that have gone undetected, it is germane to the abduction issue, and a brief exposition is in order. String theory is an effort to obtain the holy grail of physics—a so-called unified theory that would unite the two pillars of physics of the twentieth century—that is, Einstein's theory of relativity and quantum mechanics. These two theories embody virtually all knowledge concerning the fundamental forces of nature. Both have been repeatedly validated in laboratory experiments; unifying the two, however, has proven problematic. The theory of relativity describes the nature of gravity and the physics of large entities. By contrast, quantum theory deals with the behavior of small particles in the subatomic realm. As they are currently formulated, though, general relativity and quantum mechanics cannot both be right. This is why string theory holds out great promise insofar as it has the potential to explain all the physical laws of the universe in one master equation.[61] In fact, in string theory, general relativity and quantum mechanics require each other in order to make sense. Upon Einstein's death in 1955, work on unification ground to a halt, but decades later, string theory would emerge as the leading candidate for a unified field theory.

Antecedents to string theory, however, can be traced back much earlier. In 1919, Theodor Franz Kaluza from the University of Konigsberg in Germany sent a paper to Einstein that proposed a five-dimensional theory of gravity. What intrigued Einstein was that with this added fifth spatial dimension, James Maxwell's electromagnetic force, which was a theory of light, could be combined with the theory of gravity. In 1926 a Swedish physicist, Oskar Klein, conjectured that this fifth dimension was extremely small and "curled up" like a circle.[62] The Kaluza-Klein theory, as it came to be known, posited that light was a vibration of a fifth unseen dimension.[63] Although compelling, the Kaluza-Klein theory languished as physicists remained unconvinced that a fifth dimension really existed.[64]

Over the following decades, physicists discovered a plethora of subatomic particles, which suggested that some elemental property to matter and energy had yet to be discovered. In 1968, while thumbing through old math books, Gabriele Veneziano and Mahiko Suzuki independently stumbled upon the Beta function written down in the nineteenth century by Leonhard Euler. They found that the Beta function satisfied almost all the stringent requirements of the scattering matrix, or S-matrix, describing particle interactions.[65] Veneziano believed that a string picture could describe the interaction of interacting particles. In 1970, Yoichiro Nambu of the University of Chicago elaborated on this concept and proposed the idea of strings to make sense out of the chaos resulting from the hundreds of hadrons (composite particles made of quarks that are held together by the strong nuclear force) that were discovered in laboratories.[66] Around that same time, Keiji Kikkawa, Bunji Sakita, and Miguel A. Virasoro applied mathematical formulas to explain how these strings interacted.[67]

It was not until 1984, however, when Michael Green and John Schwarz announced that sting theory was free of anomalies, that interest in the subject really took off.[68] According to string theory, the building blocks of nature consist of extremely tiny vibrating strings. The subatomic particles that make up the universe are similar to the notes played by a symphony orchestra. The vibration of the string determines the characteristics of the subatomic particle, such as a photon or a neutrino.[69]

Over time, five major variants of string theory were formulated. The main differences among them were the number of unobservable dimensions in which the strings developed. These curled up dimensions are purportedly infinitesimally small, perhaps the size of the Planck's length.[70] As these strings and dimensions are so minute, today's instruments cannot detect them. In 1994, the Nobel laureate Edward Witten and Paul Townsend announced that the ten-dimensional string theory was actually an approximation to a more mysterious eleven-dimensional model, which they christened M-theory.[71] M-Theory unified the five variants of string theory and added one more dimension, for a total of eleven.

Life as we know it would be profoundly different in an alternative dimensional world. Back in 1884, a Shakespearean scholar, Edwin Abbott, published a novel—*Flatland*—which told a story about life in a two-spatial dimensional world. Although written as a satire of the Victorian society in which Abbott lived, the tale gives insight into the lives of denizens in realms far different than ours. As narrated by Mr. A. Square, Flatland is a world populated by people who are geometric objects. In this stratified world, the more sides on a person, the higher his social rank. For instance, women are Straight Lines, workers and soldiers are Triangles, professional men and gentlemen (like Mr. A. Square) are Squares, and the nobility are Pentagons, Hexagons, and Polygons. One day, a strange being from Spaceland (a three-spatial dimensional world like ours), Lord Sphere, visits the world of Flatland and introduces Mr. Square to the marvels of another dimension. With his different perspective, when Lord Sphere looks at beings in Flatland, he can see the inside their bodies and view their internal organs. By contrast, when Lord Sphere enters the lower-dimensional realm, Mr. Square, the resident of Flatland, only sees circles of ever-increasing size penetrating his world. Lord Sphere invites Mr. Square to Spaceland, which involves a harrowing journey in which he is peeled off of his two-dimensional world and deposited in a three-dimensional realm. Excited by his experience, upon his return to Flatland, Mr. Square decides to tell his fellow Flatlanders of his remarkable journey. However, his strange tale is perceived as seditious by the authorities. As a consequence, he is tried before the Council and sentenced to prison, where he lives out the rest of his life as a martyr.[72]

In addition to different dimensions, there are the possibilities of other realms existing in the properties of dark matter and dark energy. Although these properties cannot be directly seen, their effects can be measured. Contemporary astrophysicists now believe that the attractive force of dark matter is responsible for holding the stars together in galaxies. Otherwise, there would not be enough gravity to keep the stars from wandering away. Although we cannot see it, some physicists believe that it is ubiquitous in our galaxy. Conversely, it is believed that dark energy is responsible for driving the galaxies away from one another. Collectively, dark energy and dark matter account for about 96 percent of the "stuff" in the universe. By contrast, conventional matter and energy consist of the remaining 4 percent.[73]

Could there be a terrestrial explanation for alien abductions? More specifically, could the human mind be a portal through which to encounter beings residing in other dimensions? Some people wonder that perhaps the secret gateways to alien worlds may be hidden in our own minds. As incredulous as this might sound, some researchers have speculated that the hallucinogenic drug DMT (dimethyltryptamine) could be used as a possible gateway to make contact with alien entities. Shamans have long used variants of the drug in rituals to enhance their mystical experiences. A physician, Dr. Rick Strassman, conducted a government-sanctioned study of DMT. To his amazement, he found that almost 50 percent of his subjects reported encounters with bizarre figures, which they perceived as residing in different dimensions.[74] Initially, Strassman thought that these encounters were entirely subjective experiences that his subjects had conjured from their psyches. But as his research continued, he began to suspect that these entities were "autonomous noncorporeal beings."[75] Because of the lucidity of his subjects' accounts, he decided to act as if the realms his volunteers described and the inhabitants with whom they interacted were genuine.[76]

Adhering to a painstaking protocol, Strassman conducted his study in the early 1990s at the University of New Mexico's School of Medicine in Albuquerque. He recruited 60 volunteers with prior experience using hallucinogenic drugs to participate in his experiment. DMT is an extremely powerful, yet short-acting psychedelic drug. Some scientists believe that DMT is endogenously produced in the pineal gland in the brain.[77] Supposedly, production is at its peak during critical periods in one's life including birth and death.[78] The drug can also be synthetically produced and administered in large doses, which maximizes the effect. Among other things, Strassman believes that DMT is responsible for the "near-death experience." Under the drug's influence, many of his subjects reported experiences not unlike those detailed by alien abductees. To his subjects, their experiences were lucid and did not seem dream-like.[79]

Strassman found striking commonalities between the reports of John Mack's subjects and those of his DMT volunteers.[80] For instance, sound and vibration would build before the encounter just before the scene explosively shifted to an alien domain. The ambience suggested that the realm they entered was inhabited by a technologically advanced civilization. The "autonomous noncorporeal beings" were usually described as instectoid or reptoid in appearance. In some instances, the beings were described as bearing a resemblance to praying mantises, albeit in humanoid form. Not unlike Mack's subjects, Strassman's volunteers often found themselves on a bed or landing bay, or in some high-technology room. Often, they found themselves placed on an examination table. Interestingly, the beings seemed unsurprised when meeting the human subjects, as if they had the capability to observe humans at will. Nevertheless, they seemed to have a sense of urgency, as if they anticipated the limited time with which they had to work on the examinations. Usually, one particular alien seemed to be in charge. The aliens sought to communicate with the subjects using gestures, telepathy, or visual imagery. Some abductees reported a feeling of neuropsychological reprogramming or a transfer of information involving unusual symbols, rather than words or sounds.[81] The purpose of the contact was uncertain, but some subjects believed that the aliens were attempting to improve humans individually or as a species.[82]

Is there a reason for the similarities in the reported experiences of abductees and DMT subjects? Strassman speculates that magnetic fields may affect the functioning of the pineal gland and thus alter consciousness. He point outs that several reported abduction experiences occurred near high-intensity power lines, which produced powerful magnetic fields. Possibly, this precipitated an increased level of endogenous DMT in some people, thus occasioning the abduction experience.[83]

While some of the subjects found the experience to be pleasant and transformative, others found it frightening and disconcerting. Many people report the DMT experience as an event which brings about a total awareness, a cosmic consciousness if you will; however, this profound sense of knowledge and awareness is often fleeting and dissipates after a day or two. Similar experiences have been reported from people under the influence of so-called magic mushrooms, which have been consumed for millennia by various indigenous tribes in Latin America.[84]

Strassman conjectured that there are other "channels" of existence that are always present; however, our hard-wiring prevents us from experiencing them. Under the influence of DMT, some people are able to access these different planes of existence. Perhaps, Strassman wonders, these realms might reside in dark matter. He speculates that humans may actually have parallel "dark matter bodies." Based on this assumption, subjects could indeed undergo surgical procedures, such as implants, during alien encounters, but there would be no physical evidence of the procedure in our world.[85]

The DMT explanation appears to have some merit for the alien abduction phenomenon. According to many accounts, the abductee is awoken in his or her bedroom in the presence of some humanoid entity and is then transported through solid surfaces and onto an awaiting craft.[86] Strassman speculates that perhaps DMT provides access to "other" channels, or dimensions that we cannot usually perceive with our five senses.[87] In addition to Strassman's study, numerous testimonials of persons who have undergone the DMT experience can be found on the Internet on sites such as YouTube. Many of their experiences are consistent with those of Strassman's subjects. Following up on Strassman's research, in 2006, Marko A. Rodrigues published a scientific paper in which he discussed ways to possibly determine if the entities experienced by the persons under the influence of DMT do have an autonomous existence, rather than being the projections of hallucinating brains. For future research, he suggested a test that could be administered that would involve asking the entities to perform a complex mathematical task involving prime numbers to verify their independent existence.[88]

Collectively, the abductees' claims are far from convincing. Although some abductees appear to be sincere, others are obvious frauds, which ultimately undermine the credibility of the abduction subset of Ufology. The cases involving DMT, however, appear compelling, but legal restrictions on the drug's usage make research in that area problematic. In the United States, DMT is designated as a Schedule I drug under the Controlled Substances Act of 1970. If UFOs are indeed extraterrestrial in origin, it stands to reason that alien visitors would occasionally interface with Earthlings. According to some researchers, there is already much evidence to support this proposition, but it is suppressed primarily by the U.S. government. This and other issues are explored in the next chapter.

Chapter 11

UFO Conspiracy Theories

As NASA discovers more and more exoplanets, hope continues to grow that someday we might make contact with some form of extraterrestrial intelligence. Ufologists, however, maintain that extraterrestrial aliens have already visited Earth on numerous occasions. Many people believe that the U.S. government (and other governments around the world for that matter) is holding back on its knowledge of UFOs. According to a 2002 poll conducted by the Roper organization, 72 percent of the respondents reported that they were convinced that the U.S. government had knowledge on extraterrestrials but withholds the information from the public.[1] The critical nature of extraterrestrial visitation, so the argument goes, is too sensitive for the government to release to the populace. Knowledge that the government cannot protect its skies from alien intruders would call into question its power and competence. For this reason, the government would have a strong motive to keep silent on the issue.

Although the U.S. government officially investigated UFOs for over 20 years, some researchers believe that it has been less than forthcoming in the information it has shared with the public. Chapter 10 dealt with the documentary record of the U.S. government's investigation of UFOs. This chapter addresses the controversial proposition that a certain segment of the U.S. government knows far more about UFOs than what has been made available to the public.

Practically from the start of the contemporary UFO era, government insiders have advanced accusations of official cover-up. For example, a retired Marine Corps major and pilot, Donald E. Keyhoe (1897–1988), did much to popularize the fledging UFO hysteria in the early 1950s with his first book, *Flying Saucers Are Real* (1950), in which he argued that there was a massive cover-up of an alien invasion by the government.[2] Although he offered no direct evidence for the extraterrestrial hypothesis, he inferred from the general disarray of the Air Force's investigations that UFOs were authentic. He suggested that certain unnamed sources had confirmed to him the existence of extraterrestrial crafts.[3] Some conspiracy theorists go further and claim that the U.S. government actually conspires with the alien visitors. Some versions of this genre insist that governments began colluding with extraterrestrial aliens on the cusp of World War II.

According to some accounts that are most likely apocryphal, the Germans retrieved a flying saucer that had crashed in the Black Forest in 1936.[4] German scientists attempted to reverse-engineer the downed craft, with limited success.[5] Myths surround the Third Reich's involvement in the occult, including efforts to make contact with extraterrestrial aliens who would provide knowledge on advanced technology that could be harnessed to further Germany's geopolitical ambitions. Adding to the flying saucer mystery was a group of Russian émigré women led by an enigmatic medium and Nazi sympathizer named Maria Orsic, who allegedly channeled messages from aliens on a planet in the Aldebaran star system on how to construct anti-gravity discoid aircraft.[6]

During the 1920s and 1930s, some Germans did indeed experiment with disk-shaped craft. These crafts, however, were highly unstable and difficult to navigate.[7] But over time, a legend took hold that German scientists during World War II worked on a flying saucer program that would employ antigravity methods for propulsion. Before the war, a brilliant German forester, Viktor Schauberger, developed a bizarre collection of theories about the complex interaction of forces of nature that constantly created or reinvigorated matter. He theorized that a vortex could be used as a conduit for the transmission of energy. In his spare time, he dabbled in designing vacuum devices and flying saucers. In April 1944, as Germany's predicament in the war became increasingly desperate, Schauberger was drafted into the Waffen-SS at the age of 59.

According to some versions of the story, Heinrich Himmler's SS assumed stewardship of the program. An enigmatic figure, SS General Dr. Hans Kammler, supervised the program.[8] One of the devices created as part of the project was *Die Glocke* (the Bell), a bell-shaped device powered by rotating cylinders. The machine was supposed to have been based on descriptions of a flying saucer from the Hindu Vedic records, specifically, the descriptions of the Vimana. In the Vedic worldview of eighth- to twelfth-century India, there was hierarchy of planets encompassing 400,000 humanlike races and 8,000,000 other lifeforms.[9] According to some interpretations of the text, ancient aliens used exotic craft in battles that they fought in the air against each other. The conspiracy narrative contends that German scientists sought to reverse-engineer this craft from the descriptions in the ancient texts. Kammler's whereabouts after the war is uncertain. Some accounts claim that he committed suicide at war's end, while others assert that he was executed by the Soviets. Still others claim that he traded his knowledge of the flying saucer program for his freedom and made his way to the United States under the aegis of Operation Paperclip. He then lived out his life and died in obscurity in either Texas or Virginia.[10]

The Germans were not the only ones to have allegedly experienced UFOs during the war. According to a letter from a scientist whose grandfather served as one of Winston Churchill's bodyguards, the British prime minister conspired with Dwight D. Eisenhower to suppress the truth about a spectacular UFO sighting that was witnessed by the crew of a Royal Air Force aircraft returning from a reconnaissance mission. Supposedly, Churchill feared that exposure would cause the public to panic and shatter their faith in religion.[11] Just two years after the war, the United States would become the focal point for UFO sightings, the most spectacular of which is the alleged Roswell affair.

Conspiracy theorists contend that the U.S. government became involved with extraterrestrial aliens shortly after World War II. The impetus for the working relationship was an apparent crash of an alien craft that occurred near Corona, New Mexico. What became known as the "Roswell incident" occurred around the same time that stories of flying saucers were sweeping the country. Military pilots were placed on alert and radio operators were on standby as they looked skyward for signs of unusual aircrafts. The state of New Mexico in particular was a sensitive and guarded area, as the first atomic bombs were built at the research center at Los Alamos. Furthermore, captured German V-2 rockets were launched just south of White Sands near Alamogordo, and not far from there was the Trinity site, where the first atomic bomb was detonated in 1945.[12]

The Roswell story begins on the evening of July 3, 1947, when a severe thunder and lightning storm occurred in central New Mexico. Local ranchers later described hearing a loud explosion that did not sound like other thunderclaps.[13] On July 4, 1947, W.W. "Mac" Brazel, discovered some unique debris on the ranch where he worked. According to newspaper reports of the period, the debris included strips of rubber, pieces of paper, shiny foil, and bits of wood. Three days later, he made the 75-mile trek to Roswell, which is about 60 miles to the southeast of the ranch, to report his fantastic discovery to the authorities. When he arrived at the sheriff's office, Brazel showed the materials to Deputy Sheriff B.A. "Bernie" Clark. Puzzled by the find, Clark phoned the Chaves County Sheriff, George M. Wilcox, who after some discussion on the incident with Brazel, decided to call the Army Air Base and inform the military of the situation.[14] Soon thereafter, the sheriff contacted Major Jesse Marcel at the nearby Roswell Army Air Field, which was the home of the 509th Bomber Squadron. On Monday, July 7, Major Marcel and Captain Sheridan Cavitt went to Corona with Brazel to examine crash site.[15] Allegedly included in the recovered materials, were the bodies of strange grayish creatures, one of whom was alive and taken back to the base.[16]

Shortly thereafter, the commanding officer of the 509th Bomber Squadron, Colonel William "Butch" Blanchard, announced through his office that his unit had captured a crashed flying saucer on a ranch outside of Roswell, New Mexico. Accompanying the press release was a photo showing some of the pieces of the saucer.[17] The story was quickly picked up by news services around the country.[18] The public relations officer, First Lieutenant Walter Haut, issued the following statement:

> The many rumors regarding the flying disc became a reality yesterday when the intelligence officer of the 509th Bomb Group of the Eighth Air Force, Roswell Army Air Field, was fortunate enough to gain possession of a disc through the cooperation of one of the local ranchers and the sheriff's office of Chaves County.
>
> The flying object landed on a ranch near Roswell sometime last week. Not having phone facilities, the rancher stored the disc until such time as he was able to contact the sheriff's office, who in turn notified Major Jesse A. Marcel of the 509th Bomber Group Intelligence Office.
>
> Action was immediately taken and the disc was picked up at the rancher's home. It was inspected at the Roswell Army Air Field and subsequently loaned by Major Marcel to higher headquarters.[19]

The story was quickly picked up by news services around the country.[20] On July 8, the *Roswell Daily Record* announced in a dramatic headline: "RAAF Captures Flying Saucer on Ranch in Roswell Region." But soon the story mutated. The next day, General Roger Ramey of the 8th Air Force announced on local radio that the original report had been a mistake. He retracted his initial claim and now averred that the wreckage, which consisted of a box kite and a balloon, was from a "high-altitude weather device" and not an extraterrestrial spacecraft.[21] A press conference was hastily put together in Fort Worth, Texas, at which photographers were permitted to take a few shots of some twisted wreckage that was supposed to be nothing more than a misidentified weather balloon.[22] General Ramey ordered one of the intelligence officers who went to the crash site, Jesse Marcel, to admit that he had mistaken a weather balloon for a flying saucer.[23] After the press conference, Marcel was abruptly removed from Wright Field and ordered to return to Roswell.[24] After some of the photographers complained that they were not allowed close enough to the recovered debris, a second press conference was held; however, some of the cameramen claimed that the wreckage was not the same and that fragments had been switched.[25] Years later, the UFO researchers, Kevin D. Randle and Donald Schmitt, claimed that the original wreckage of the purported craft was replaced with balloon debris just minutes before newsmen were ushered inside the press conference.[26]

Allegedly the same day that he arrived at the base in Roswell, General Ramey told assembled officers in a secret meeting that two additional crash sites were discovered and that the army would admit to one, but keep the others classified. A second crash site was supposed to be located just north of Roswell. A third crash site was alleged to have been approximately 175 miles northwest of Roswell, in the area of New Mexico known as the San Agustin Plains.[27] The military quickly sought to confiscate all traces of physical evidence from the crash, and the debris was flown back to the 8th Air Force headquarters at Fort Worth, Texas.[28] From there, the retrieved materials were sent to the Foreign Technology Division located at Wright Field in Dayton, Ohio, a key intelligence organization, which was responsible for the breakdown and analysis of captured enemy weapons and equipment.[29] According to some accounts, three alien bodies, described as humanoids about four feet tall, were transported to a facility at Wright Field.[30]

Despite official denials, there appeared to me more to the story than the government was letting on. Frank Joyce, an announcer from the local radio station KGFL, interviewed Mac Brazel soon after he came forward to report the incident. Brazel told Joyce that he found what appeared to be the wreckage of a flying saucer, with "little people" not far from the crash site. But soon thereafter, Brazel was detained by the military and remained in custody for the better part of a week.[31] After he was released, he recanted his original story and agreed that what he found was consistent with the weather balloon explanation. He told the Associated Press that he was "sorry he told about [the incident]."[32] In retrospect, it seems odd that the Air Force would detain him for simply misidentifying a weather balloon. After all, there is no evidence that the military personnel who did the same faced any disciplinary action.[33] Some people believe that he was pressured to change his story. A few of his neighbors claimed that shortly after he was released from custody, he purchased a brand-new pickup truck and then left his employment at the Foster ranch to start his own business in Alamogordo, New Mexico, thus suggesting that he received a bribe.[34] Likewise, members of his family claimed that he made the retraction under duress. Brazel's son Bill and daughter Bessie, recounted that their father was bitter and humiliated after the incident. His good natured-spirit was gone, seldom to be seen again. Bessie recalled an incident in which her father took her aside and warned her, "Don't believe everything you read about me in the papers. The government is going to use me to keep something secret!"[35] According to some of his family members and associates, Brazel remained convinced until his dying day that what had crashed on the ranch was indeed a flying saucer.[36] Mack Brazel died of a heart attack in 1963. Adding more suspicion is that some of Brazel's descendants died or disappeared under strange circumstances.[37]

According to some accounts, Chaves County Sheriff George Wilcox was used by the military as an enforcer to dissuade civilians in Roswell from talking about the incident.[38] Army Intelligence personnel fanned out through Roswell and neighboring communities to suppress information on the crash.[39] Allegedly, military men visited Wilcox and his wife and warned them that not only they, but also their children and grandchildren would be eliminated if the couple told anyone the truth about the incident.[40] After the initial flurry of news reporting, which lasted about a week, the Roswell story quickly faded into obscurity and would not be resurrected until three decades later.

The Roswell story was all but forgotten, but in early 1978, Jesse Marcel, the former intelligence officer of the 509th Bomb Group at the time of the Roswell incident, broke his silence. He told the UFO researchers Stanton Friedman and Len Stringfield that he was the man who picked up the pieces of the flying saucer over 30 years ago.[41] Supposedly, he was resentful for the embarrassment the U.S. government had caused him because he was essentially made a scapegoat for the Roswell affair. After more than 40 years, he wanted to get it all off of his chest.[42] A seemingly credible figure, Marcel had a distinguished military career and was awarded five air combat medals in World War II.[43] He held a number of very important positions during his Air Force career. For example, he was in charge of Army Air Force's security and intelligence briefings at Kwajelein Base in the Pacific, at which two atomic bombs were test-fired. At another time, he was assigned to the top secret Armed Forces Special Weapons Project—a military agency that was responsible for nuclear weapons after the Manhattan Project was succeeded by the Atomic Energy Commission. Until the day he died in May 1986, Marcel swore that what he recovered in the desert north of Roswell was not the same debris that was displayed for the Fort Worth *Star-Telegram* news photographer James B. Johnson in General Ramey's office on July 8, 1947. He claimed that he was instructed to pose with a substituted weather device and ordered not to say a word to any of the press waiting outside of the room.[44] In his estimation, the debris recovered at the site was no weather balloon, but rather, something that was "not of this Earth."[45] Adding credence to his story is that fact that Marcel, with his extensive military experience, would seem unlikely to have misidentified the debris as a flying saucer if it were only a weather balloon. As Stanton Friedman pointed out, Marcel was very familiar with all kinds of weather balloons and would not have mistaken such ordinary debris for a downed aircraft.[46] Furthermore, Marcel was never publicly reprimanded for the fiasco.[47]

Soon after Marcel's revelations, the story began to crack open. As a consequence of Marcel's revelations, Roswell has become the focal point of UFO research in America, spawning dozens of books, television documentaries, and videos.[48] Over the next decade, Stanton Friedman and Bill Moore interviewed 62 persons from the area who were involved in the incident, thus bringing attention to an affair which had been all but forgotten.[49] Numerous witnesses came forward and corroborated the crashed saucer version of the Roswell incident.

Frank Joyce, the radio announcer who had interviewed Mac Brazel, claimed that the military applied pressure to silence him as well. As a consequence, his interview with Brazel was never aired. Allegedly, Senator Dennis Chavez (D-NM), then-chairman of the Senate Appropriations Committee, phoned Walter Whitmore, Sr., the majority owner of the KGFL radio station, and strongly advised him to do as he was instructed by the Federal Communications Commission and not broadcast a recorded interview with Mack Brazel.[50] If Joyce's account is to be believed, Senator Chavez threatened KGFL's majority owner, Walt Whitmore, Sr., that his station would lose its broadcasting license if it went ahead with its plans to air the Brazel interview.[51] Soon after he interviewed Brazel for his program, Joyce claimed that he was hauled off, against his will, to a military hospital in Texas. When Joyce was released, he was told not to return to Roswell. He complied and went off to Albuquerque to start a new life. There, he became a well-known newsman and features reporter for KOB radio and television, where he retired in 1997. In May of 1998, two UFO researchers, Thomas J. Carey and Donald R. Schmitt, visited Joyce in his Albuquerque home, where he broke his silence and revealed his story about Mack Brazel finding the strange wreckage and the small bodies.[52]

Adding more drama to the case were the allegations of Glenn Dennis who worked at the Ballard Funeral Home. A 22-year-old embalmer at the time of the Roswell incident, he received a call from the base inquiring about the availability of "children's caskets." Later that afternoon, he received another call to transport an airman who had been injured in a motorcycle accident in town to the base hospital. (Besides providing mortuary services, the Ballard Funeral Home also had the base contract for ambulatory services when requested.) Upon escorting the injured airman to emergency entrance of the hospital, Dennis observed an ambulance whose door was opened. Inside there appeared to be the front end of a canoe-like structure with strange writing and symbols along its side. After entering the hospital, he claimed to have seen a nurse he knew who excitedly told him to leave, or else he would "be in a lot of trouble." On his way out of the building, Dennis claimed that he was accosted by a tall red-haired captain and a black sergeant who told him to leave right away and keep his mouth shut. Allegedly, the black sergeant who had accompanied the officer remarked that Dennis would "make real good dog food."[53] (For its part, the Air Force found this allegation spurious insofar as the pairing of a white officer and a black NCO was unlikely prior to the desegregation of the armed forces in May 1949.[54]) A day or two later at the Officer's Club, Dennis claimed to have met with the nurse to ask what all the fuss was about. Nervously, she told him that she had seen three dead "foreign bodies" lying on gurneys. Two of them were in bad shape and appeared mutilated. She described the creatures as being 3 to 4 feet tall, with long arms and frail bodies. Their heads were oversized. Each had only four fingers, but no thumbs. Their eyes were sunken and oddly shaped. She complained of an awful smell emanating from the bodies. Like so many of the other persons involved in the incident, Glenn Dennis claimed that military personnel threatened him with death if he ever revealed the story.[55]

Many years later, the Air Force reviewed its record of the Roswell affair and concluded that the "missing nurse" described by Dennis was probably First Lieutenant Eileen M. Fanton. When investigators questioned her about the incident, she failed to corroborate Dennis' version of events. The Air Force surmised that the "red-headed captain" was probably Captain Joseph W. Kittinger (now a retired Colonel), who was stationed at the base at the time of a balloon mishap in May 1959. Kittinger often accompanied balloon launch and recovery crews. He had a very laudable service record, receiving the Distinguished Flying Cross. The Air Force categorically rejected the accusation that Kittinger, or any other personnel, threatened civilians at Roswell.[56] As an aside, Kittinger was associated with Dr. J. Allen Hynek from 1958 to 1963, during which they frequently discussed UFOs at length. Kittinger said that he was "flabbergasted" when he learned that Hynek later reversed his position on UFOs and endorsed the extraterrestrial hypothesis.[57]

Skeptics often dismiss the Roswell incident as a myth because of a lack of trace evidence. But as the UFO researchers Thomas J. Carey and Donald R. Schmitt point out, eyewitness testimony is considered credible evidence in a court of law and can even be used to sufficiently convict someone of murder, leading to the death penalty.[58] Through the course of their investigation into Roswell, they claimed to have amassed a roster of more than 600 people, directly or indirectly associated with the incident, who support the claim of a flying saucer recovery.[59] Perhaps the most important testimony the authors received came from First Lieutenant Warren Haut, the public relations officer who initially announced that the Air Force had captured a flying saucer. In a sealed statement released after his death, he confirmed many of the elements of the Roswell story, including the recovery of a wrecked flying saucer, "memory metal," and alien bodies.[60] Carey and Schmitt believe that he waited so long to disclose the "truth" because he wanted to keep his word of silence to his close friend, the 509th Bomber Squadron, Colonel William "Butch" Blanchard, who ordered base personnel not to disclose details of the incident. Haut would go on to become one of the greatest advocates of the Roswell alien crash story. He even founded the International UFO Museum & Research Center in Roswell.

In the course of their investigations, Carey and Schmitt found that with a few notable exceptions, there was a direct correlation between military rank and reluctance to talk about the incident, that is, the higher the rank, the more disinclined that person was to say anything. This suggested to the researchers that the career military veterans feared losing their pensions.[61] However, there were exceptions. For instance, Arthur E. Exon, a retired brigadier general became the highest ranking military office to come forward and support the crashed saucer version of the Roswell story. He claimed that a quantity of the material retrieved from Roswell arrived at Wright Field for testing by a special team of lab workers. According to his account, the material appeared "unusual," like foil, but could not be dented by hammers.[62] This description was consistent with numerous eyewitness accounts that claimed that very thin sheets of metal recovered at the site when crumpled would amazingly resume its flat shape with no wrinkles or creases, leading some UFO researchers to christen it "memory metal."[63] Exon was convinced that the materials were not something manufactured on Earth.[64]

In the 1990s, the UFO community sought to generate congressional interest in the Roswell incident. Representative Steven Schiff (R-NM) (1947–1998) worked hard to get answers for his constituents and others about what exactly happened at Roswell. At his behest, both the General Accounting Office (GAO) (now the Government Accountability Office) and the Air Force commenced investigations into the incident. In the fall of 1993, Schiff broached this topic with Charles A. Bowsher, the Controller of the GAO. On January 12, 1994, Schiff announced that he was requesting a GAO investigation. Soon thereafter, on February 9, 1994, Richard Davis, the Director of the National Security Analysis group at the GAO, wrote a letter to William J. Perry, the Secretary of the Department of Defense, that the GAO was initiating a review of the Department's involvement in the Roswell incident. As a result of the inquiry, on July 28, 1995, the GAO released a report titled "Results of a Search for Records Concerning the 1947 Crash Near Roswell, New Mexico." Only 20 pages long, the document concluded that a thorough search of the records concerning Roswell found no evidence of a crashed alien craft. The records search did, however, find an FBI teletype that mentioned a weather balloon launch in the area.[65]

Some critics found discrepancies in the government's explanation. For instance, there was a suspicious FBI memorandum dated July 8, 1947, and discovered under the Freedom of Information Act. Sent from an FBI agent in the Dallas, Texas office to an FBI office in Cincinnati, the memo suggested that a lie should be perpetrated upon the public at the staged press conference held in Fort Worth. Some researchers infer that the FBI teletype implies that a clandestine flight carrying the recovered Roswell material went to Wright Field, where the Foreign Technology Division made preparations for its arrival while the Army was in the midst of spinning its weather balloon story.[66] A second document, sometimes referred to as the "smoking gun" of the Roswell case, was a telex held in the hands of General Ramey—the architect of the weather balloon story—during a press conference. Supposedly, the telex told a different story than the one he gave to members of the press.[67] The telex appeared in one of the photos taken of the debris in 1947. When the photo was enlarged, some researchers claim to have seen the phrases "the victims of the wreck" and "in/on the disc." Some UFO researchers have interpreted the blurred image of this text as a description of victims and some sort of recovered wreckage.[68]

The Air Force conducted its own separate investigation. When queried by the GAO, the Secretary of the Air Force ordered a complete inquiry to examine all information and records on the subject. To encourage openness, he went so far as to issue a proclamation absolving anyone with knowledge about the Roswell incident who believed that they were still subject to security or secrecy oaths.[69] The result of the official investigation was a publication titled *The Roswell Report: Fact versus Fiction in the New Mexico Desert*. Released to the public on September 8, 1994, the report weighed in at approximately 1,000 pages and was dubbed the "phone book" by Ufologists. According to the Roswell Report, nothing important occurred in New Mexico in early July of 1947. The Air Force did concede, however, that a highly classified operation—Project Mogul—could have led some witnesses to misidentify a remnants of a weather balloon for a crashed saucer. Project Mogul was a mission to detect hydrogen bomb tests from afar using instrumented balloons. The project was the brainchild of an American physicist, Maurice Ewing, who contended that "sound channels" in both the ocean and atmosphere could be used to acoustically detect Soviet test explosions. His scheme was to launch a long train of balloons that were outfitted with sensitive microphones so that they could detect the noise of a mushroom cloud as it rose through the layer of cold air.[70] A similar project was undertaken by the Navy. On September 25, 1947, the Office of Naval Research began launching the large "Skyhook" balloon, which was designed to test the atmosphere high above the limits of conventional aircraft.[71] From the late 1940s to the 1950s, polyethylene balloons were often misidentified as flying saucers. Often laminated with aluminum, the polyethylene balloons had a shiny appearance that could resemble a flying saucer.[72] Despite these investigations, Schiff remained unconvinced and accused the government of a "massive ongoing cover-up."[73]

Critics of The Roswell Report charged that the Mogul theory lacked strong documentary evidence. No document was uncovered that confirmed the connection between the wreckage found at Roswell to any specific balloon flight; rather, the Air Force used speculation and inference to reach this conclusion. Nevertheless, as the UFO researcher Jan Aldrich conceded, government agencies usually do not devote high priority to records preservation. Consequently, huge amounts of records are routinely destroyed and do not necessarily imply a cover-up.[74]

Two years later, the Air Force released a revised version of the report titled *The Roswell Report: Case Closed.* The report was authored by Captain James McAndrew, an Intelligence Applications Officer assigned to the Secretary of the Air Force Declassification and Review Team at the Pentagon. Some important questions were left unanswered in the original report, most notably, the issue of the alien bodies. The revised report noted that anthropomorphic dummies were used in high-altitude parachute tests that were conducted during the 1950s for scientific research.[75] The report concluded that witnesses had likely mistaken these anthropomorphic dummies for alien beings.[76] Besides Project Mogul, the report explained that more balloons were launched as part of the Constant Level Balloon Project.[77] Occasionally, civilians would pass through a drop zone and head over to see what had landed. Inasmuch as the project operated in secrecy, recovery teams behaved in a furtive, scurrying manner. Furthermore, the dummies used for the drops had fused fingers and no ears or noses, which might have led to rumors that UFOs carrying aliens had crashed in the scrubland outside of Roswell. In one incident in 1959, two parachutists,

Dan Fulgham and Joe Kittinger crashed as their balloon came down in a field on the outskirts of Roswell. Suffering head injuries, Fulgham was taken to the hospital at Walker Air Force Base, which was staffed in part by civilians. To some observers, the injured Fulgham with his face so severely swollen that he could not open his eyes, may have been mistaken for an extraterrestrial alien.[78]

Unconvinced, many UFO researchers impugn the Air Force's explanation because typically crash dummies are the size of adult humans and not four-and-a-half foot, child-sized persons. The Air Force conceded that the high-altitude tests that used anthropomorphic dummies did not commence until June 1954 and concluded in February 1959.[79] Hence, the starting date for this test program was seven years before the Roswell incident.[80] To explain the temporal discrepancy, the Air Force claimed that the witnesses were the unwitting victims of a mental process affliction known as "time compression," whereby recollections of past events tend to contract the time frames in which they took place as a person ages.[81]

The Roswell incident continues to intrigue the public and cries out for explanation. In a bizarre theory advanced in 2011, the investigative journalist Annie Jacobsen maintained that the alien beings were actually deformed patients of the infamous Nazi physician, Dr. Joseph Mengele. According to her account, he was captured by the Soviets at war's end. The Soviet leader Joseph Stalin ordered that a craft—a World War II vintage flying wing—crash in the United States to frighten Americans.[82] Besides a dearth of documentation, Jacobsen's story is implausible because there was no aircraft in the world at that time that could have flown the distance described in her book.[83] Similarly, the British UFO researcher, Nick Redfern advanced the theory that the U.S. military in Roswell was actually testing Japanese planes developed during World War II for dropping bombs on distant targets using manned balloons. He maintains that the scheme was a suicide mission from which the pilots were not supposed to return. Supposedly, the Air Force conscripted midgets, which accounts for the "alien bodies" described in testimonies of the witnesses to the crash site.[84]

The Roswell incident remains one of the most controversial UFO cases in history. Capitalizing on the incident's notoriety, in 1992 the Roswell UFO museum opened, followed four years later by the annual Roswell UFO Festival.[85] With all that has been written about the topic, readers are caught in a conundrum. So far, there has been a dearth of documentary and physical evidence put forward to make a convincing case that a flying saucer crashed in New Mexico. However, there is a wealth of testimony from key figures involved in the incident. Moreover, there is a remarkable consistency in these accounts, which lends credibility. There seems to be a convergence of testimonial evidence from those who were involved in the incident. As a result, it is difficult for the skeptics of glibly dismiss the incident as a myth.

Adding grist to the Roswell conspiracy mill is the alleged MJ-12 (Majestic-12) group. Purportedly, MJ-12 was founded on September 24, 1947, by a special classified executive order issued by President Harry S. Truman.[86] The story first "leaked" in 1984, when an American television producer, Jaime Shandera, received a roll of film that was sent anonymously through the mail and postmarked in Albuquerque, New Mexico.[87] The film contained pictures of various government documents that detailed the machinations of MJ-12—a secret government body established to handle relations with extraterrestrial aliens who had arrived on Earth. In June 1987, the UFO researcher Bill Moore publicly announced the Majestic-12 conspiracy to the public. According to the purported top secret records, indeed a crashed alien aircraft was recovered at Roswell, New Mexico.[88] Documents explained that "four small human-like beings had apparently ejected from the craft at some point before it exploded." All four were reported dead and badly decomposed because of predators and exposure to the elements during the approximately one-week time period that elapsed before their discovery. An analysis of the bodies conducted by Detlev Bronk, a prominent American scientist, tentatively concluded that the biological and evolutionary processes responsible for their development was quite different than for *Homo sapiens*, thus suggesting that they were "Extraterrestrial Biological Entities" or "EBEs." A memo detailed how news reporters were given a cover story to convince them that the whole incident involved only a misguided weather research balloon. A covert analysis organized by General Twining and Vannevar Bush concluded that the disc was most likely a short-range reconnaissance craft. The memo speculated that it was possible that the craft originated from Mars, but that Dr. Donald H. Menzel (publicly, a prominent UFO debunker) considered it more likely that it belonged to beings from another solar system. There was also brief discussion of another alien craft that had crashed on December 6, 1950, at the El Indio-Guerrero area of the Texas-Mexican border. However, the object was almost totally incinerated by the time the security team arrived at the scene. To avoid public panic, the memo recommended that the MJ-12 Group operate under the "strictest security precautions."

Dr. Vannevar Bush, the head of the Office of Scientific Research, which had recently developed the atom bomb, was identified as the leader of MJ-12. Besides him, the original members of MJ-12 were said to be the following: Rear Admiral Roscoe Hillenkoetter, the third director of Central Intelligence; Lloyd Berkner, a former secretary of the Joint Research and Development Board; Dr. Detlev Bronk, an eminent scientist who chaired the National Research Council; James Forrestal, a former secretary of defense; Gordon Gray, Truman's special assistant; Dr. Jerome Hunsaker, the head of the department of Mechanical and Aeronautical Engineering at the Massachusetts Institute of Technology; Dr. Donald Menzel, a noted astronomer and UFO skeptic; General Robert Montague, commander of the Sandia military base in Albuquerque; Rear Admiral Sidney W. Souers, secretary to the National Security Council; General Nathan Twining, the commander of the Wright-Patterson Air Force Base; and General Hoyt Vandenberg, the director of Central Intelligence.[89] Other scientists that were later allegedly involved in MJ-12 included John von Neumann and Robert J. Oppenheimer.[90] Eisenhower's appointment of Nelson Rockefeller as his "Special Assistant for Cold War Planning" was said to be a cover for his true role in directing U.S. foreign policy with the aliens.[91]

The strange death of James Forrestal (an alleged member of MJ-12) has led some conspiracy theorists to believe that he intended to break the MJ-12 story to the public. In 1948, Forrestal suffered a nervous breakdown. On May 22 of that year, he jumped to his death from his sixteenth-floor room at the Bethesda Naval Hospital. Rumors persist that his condition deteriorated as a result of intimidation by "men in black."[92] Supposedly, he had come to favor a policy of public disclosure on the alien presence, and as a consequence, paid the ultimate price.[93]

According to some versions, MJ-12 was later reorganized into two layers. The outermost layer included up to 40 persons who formed a Study Group that was designated PI-40. Their main function was to conduct specialized studies and provide policy recommendations concerning extraterrestrial issues for the smaller decision-making group, MJ-12, which would implement the policies after gaining executive approval from President Eisenhower.[94]

Some conspiracy theories assert that President Eisenhower actually met with extraterrestrial emissaries. According to this narrative, on February 24, 1954, while he was vacationing near Palm Springs, California, Eisenhower was apprised of a recent crash of an alien craft. As a ruse, a story was floated that he complained of a toothache and was whisked away to visit his dentist, but in reality, was taken to Edwards Air Force Base, where he viewed the wreckage. More fanciful accounts claim that live alien beings with "white hair, blue eyes, and colorless lips" were kept there as well.[95] Another version claims that Eisenhower met an alien delegation on February 5, 1955 at Holloman Air Force Base in southern New Mexico. The actual meeting was said to have taken place on board an alien spacecraft.[96] At the meeting, the aliens identified themselves as originating from a planet near a red star in the Constellation of Orion. Supposedly, their planet was dying. A formal treaty between the alien nation and the United States was signed by Eisenhower.[97] Eisenhower agreed to stop sending scramble jets to fire on the spacecraft in exchange for the aliens not making public displays of their presence on Earth.[98] In the most nefarious version of the story, Eisenhower granted permission to the aliens to abduct humans for medical experiments, with the proviso that they be returned to their abodes unharmed after the aliens were through with them.[99]

Interestingly, in February 2010, Laura Magdalene Eisenhower, a great-granddaughter of the president, posted a document on the Internet that described her alleged contacts with aliens.[100] Some former government employees have also claimed that there was collusion between aliens and the U.S. government. For instance, Victor Marchetti, the former executive assistant to the Deputy Director and special assistant to the Executive Director of the CIA, once claimed that the U.S. government had established contact with aliens, but was determined to keep the information secret.[101]

Not surprisingly, the fantastic allegations surrounding MJ-12 have occasioned considerable interest among the public. The MJ-12 story was featured in *The New York Times* and the *Washington Post*. In addition, it was the subject of an ABC-TV news program, *Nightline*.[102] The controversy surrounding the story attained such notoriety that in 1987, the FBI commenced an investigation to determine if whether there had indeed been a breach of security with the release of top secret documents, or were they merely fabricated. After over a year of effort, the FBI could not find any agency or person willing to admit that the documents had been purloined from the government or make a formal complaint that they had been forged.[103]

Not all of the MJ-12 documents have held up well under close scrutiny. Evidence later transpired that at least some of the MJ-12 documents were forgeries. For example, the UFO skeptic, Philip Klass, uncovered the signature of President Harry Truman on one of the MJ-12 documents, that was identical to another signature of his on a totally unrelated document. Inasmuch as Graphologists maintain that no two signatures can be exactly the same, skeptics dismiss the MJ-12 documents as a fraud.[104] Some UFO enthusiasts go so far as to claim that disinformation specialists created the MJ-12 controversy to divert attention away from the true oversight committee.[105] A committed researcher, Kevin D. Randle, argued that the documents have distracted the UFO community for too long. Most damning, he finds a dearth of provenance for the documents—that is, no compelling chronology and actual agencies that could have produced them.[106] Randle suspects that the MJ-12 documents were created by some persons in the UFO community to advance their personal agendas. Nevertheless, he finds it highly probable that a secret committee was indeed formed to address the issue of extraterrestrial aliens. And, like the much-vaunted MJ-12, he reasons that such a committee would include political and military leaders along with top scientists.[107] Randle speculated that the alleged crashed ship at Roswell might even have been a test to determine the scientific and technological sophistication of Earthlings. According to this line of reasoning, when humans succeed in reverse-engineering the craft, they would have demonstrated their maturity as a species.[108]

For years, the MJ-12 story lacked credibility because of a lack of corroborating evidence. As all of the alleged members were dead, they could not comment on the affairs. But finally in 1997, a government insider—retired Army Lieutenant Colonel Philip J. Corso—came forward to substantiate the story. He claimed that Majestic-12 indeed existed as a body to manage the extraterrestrial issue.[109] In the early 1960s, he was head of the Foreign Technology desk in Army Research and Development at the Pentagon. Ostensibly, his main duty was to evaluate foreign technology acquired by the Army; however, under the cover of this program, he claimed that he coordinated the development of new technologies based on recovered alien technologies at Roswell. In his book, *The Day After Roswell*, he maintained that that the prerequisites of modern technologies, including lasers, fiber optics, Kevlar bulletproof vests, and even transistors, were all reverse-engineered from the debris recovered at the 1947 crash in Roswell.[110] Corso had a veneer of credibility; previously, in 1992, he had testified before a congressional panel investigating the fate of American service members who were declared missing in action and prisoners of war.[111]

Corso claimed that there was an ongoing war between Earthlings and the alien invaders. Several weapons systems developed after World War II were not directed so much at the Soviets, but rather extraterrestrial interlopers. According to Corso, the Cold War was used as a cover for the United States to develop numerous high-tech weapons systems.[112] Sophisticated tracking and defense systems were developed to look into space as well as spy on the Soviet Union.[113] Corso provided details of a plan formulated by General Arthur Trudeau, the Army Chief of Staff for Research and Development—Project Horizon—to fortify the moon and turn it into a military installation. Not only would the base serve as a deterrent against the Soviet Union, it would also deny aliens from using the moon as a station from which to attack Earth. President Ronald Reagan's strategic defense initiative, when operational (popularly known as Star Wars), would be aimed not only at Soviet missiles, but also at alien spacecraft.[114] Based on this line of reasoning, Reagan believed that not unlike his arms control summits with the Soviets, it was better to negotiate with the aliens from a positions of strength.[115] Despite their technological superiority, Corso believed that the aliens did not have the raw power to launch a global war against the Earth.[116] Furthermore, on occasion, alien aircrafts were vulnerable to terrestrial counterforces.[117] Corso went so far as to claim that he actually viewed a deceased alien recovered from Roswell. Based on its unusual anatomy, he speculated that the aliens may have been genetically engineered to travel long interstellar distances.[118]

Written with great specificity, Corso's book has a quality of verisimilitude that some people find convincing. For example, Senator Strom Thurmond (R-SC), for whom Corso once served as aide, even wrote a foreword, but later retracted it after controversy surrounding the book began to arise.[119] Other readers, however, are far more skeptical. Critics have examined Corso's book and found it to be riddled with errors.[120] John Alexander, a retired U.S. Army colonel who has extensively researched UFOs, found many discrepancies in his story. Nevertheless, he was impressed by Corso's narrative when he discussed the topic with him. During their conversations, he observed that Corso was remarkably consistent with his story, noting that "Usually when someone has made up a story from whole cloth, they get tripped up when retelling the story over time. With Phil, it was almost as if the incidents were recorded on tape and you could push the start button at any time and hear the identical report."[121]

Some researchers go far beyond Corso and claim that extraterrestrials are quite active in our political affairs. So-called "grays"—large-headed aliens with pallid skin—are believed to harbor malevolent intentions toward humans and are most often reported to be involved in human abductions and experimentation.[122] Allegedly, representatives of the U.S. government entered into an agreement with the grays, allowing them to build an underground base in New Mexico in return for advanced space weapons and technology. At times this relationship has deteriorated.[123] By contrast, a taller phenotype of aliens—"Nordics"—are thought to be magnanimous toward humans and have been locked in a struggle with the grays for many years.[124] Some UFO researchers believe that the Nordics seek to direct the misguided *Homo sapiens* species along a more spiritual path.[125]

In his book, *Exopolitics: Political Implications of the Extraterrestrial Presence*, Michael E. Salla, a former professor at the School of International Service at American University in Washington, D.C., advanced the case that aliens have "multiple agendas, activities and histories that interact with global humanity's own subjective experiences, aspirations and belief systems."[126] According to his narrative, the history of collusion began in 1934, when then-President Franklin Delano Roosevelt signed a secret deal with extraterrestrial aliens.[127] Salla identifies three alien subgroups that are active on Earth. The first two subgroups sought to establish a world government that would be led by their human proxies. The third subgroup consists of two alien races: the Nephilim (a giant humanoid species) and a reptilian species. The latter subgroups, have a pessimistic view of humanity, fearing its ability to maintain proper stewardship over the Earth.[128] Salla asserts that an alien craft was shot down either during the first Gulf War in 1991 or around 1998 when the U.S. military was enforcing a strict no-fly zone over Iraq. The Iraqis recovered the craft and gave sanctuary to the downed alien beings. Saddam's weapon experts then reverse engineered the alien technology, thus creating advanced weapons of mass destruction.[129] Salla argues that President George W. Bush's decision to launch Operation Iraqi Freedom in 2003 was not to dismantle Saddam Hussein's WMD program, but rather, acquire secret alien technology held by the Iraqi government.[130] Even by the standards of Ufology, Salla's research is on the fringe and appeals only to the most credulous of readers.

Assuming that aliens have visited our planet, surprisingly very little has been written about their home planets. In *Secret Journey to Planet Serpo: A True Story of Interplanetary Travel*, Len Kasten detailed with great specificity the secret diplomacy between a race of extraterrestrial aliens and representatives of the U.S. government. His "research" demonstrates the proclivity of Ufology to synthesize various conspiracies into one grand narrative.

His story begins in November 2005, when Victor Martinez, a host of an Internet forum titled "The UFO Thread List," received a series of emails from a sender known only as "Anonymous" who claimed to have inside knowledge of a top secret government program. According to his account, "Project Crystal Knight" was an interstellar exchange program between the U.S. government and a race of extraterrestrial aliens. As Kasten explains, the U.S. government entered into an alliance with a race of extraterrestrial aliens for defensive purposes to offset a rival bloc involving another alien race and Nazi Germany. The Nazi-alien axis can be traced back to 1908, when an officer in the German Army—Karl Ernst Haushofer—was sent to Tokyo to serve as an artillery instructor to the Japanese Army. The enigmatic Haushofer is best known as the original theoretician of geopolitics. While in Tokyo, he was one of only a few westerners to be inducted into the Black Dragon Society. A smaller inner circle of the group—the Green Dragon Society—concentrated on harnessing psychic and occult power to further Japan's imperial ambitions. Legend has it that members of the Green Dragon Society established contact on an astral level with the Society of Green Men, who lived in a remote monastery and underground community in Tibet. Kasten avers that the Green Men were connected to a huge underground empire created by Reptilian extraterrestrials from Alpha Draconis. Impressed with German-Prussian militarism, the Green Men convinced the Green Dragons to invite Haushofer into their secret society and initiate him into some of their mysteries. It was prophesized that a Japanese-German alliance would conquer Russia and rule the massive Eurasian landmass. With hegemony over Eastern Europe and much of Asia, the Japanese-German alliance would one day be in a position to confront the Western European and Anglo-American alliances.

Haushofer was also involved in a Munich-based secret society known as the Thule Gesellschaft, or the Thule Society. Founded by Rudolf von Sebottendorf (1875–1945), the group advanced a theory on the origins of the Aryan race, claiming that it emerged from a long-lost Atlantis-type civilization known as Thule, located somewhere far in the northern hemisphere. In an effort to win the support of the broad masses, the Thule Society launched the German Workers' Party (DAP) in 1919. That same year, an obscure Army corporal investigating nationalist groups attended a meeting of the fledging political party. Adolf Hitler would eventually assume leadership and change the name of the party to the German National Socialist German Workers Party (NSDAP). Through his student and fellow Thulist Rudolf Hess, Haushofer met Hitler in the Landsberg Prison, where the future German führer was serving a sentence for a failed putsch attempt against the Bavarian government. Haushofer tutored Hitler on geopolitics and the need for *Lebensraum* so that Germany could achieve its manifest destiny.

Although Germany did not win the war, Kasten claims that some surviving

Nazis established a base in Antarctica. To be sure, there was indeed a German expedition to the frozen continent. In 1938, Captain Alfred Ritscher of the Kriegsmarine claimed a small area of land that was christened "Newschwabenland" (New Swabia). The more fantastic accounts maintain that the Germans built a submarine base at that location and later, a base for flying saucers. A myth was created that in 1946, Admiral Richard E. Byrd discovered a hidden civilization while leading a U.S. Naval excursion to Antarctica. According to an apocryphal account, Byrd's fleet was repulsed by the "last German battalion" and flying saucers. With the help of a Reptilian race that once resided in Atlantis, the Germans made great advances in aircraft technology. According to Kasten, a joint German-Japanese crew piloted a flying saucer—the Haunebu IV—which crashed landed on Mars in January 1946, after a perilous eight-month journey.

Soon thereafter, so the narrative goes, the U.S. government established contact with extraterrestrial aliens. If the Roswell legend is to be believed, on July 4, 1947, an alien aircraft crashed in central New Mexico. Supposedly, the aliens took a keen interest in the U.S. atomic bomb program then underway in New Mexico. Five of the aliens were found dead near Roswell, but a surviving crewmember was rescued by U.S. military personnel from a nearby Army Air Base. Later at a second crash site nearby, four additional dead alien bodies were recovered. The U.S. government personnel working on case referred to the aliens as Extra-terrestrial Biological Entities, or "Ebens" for short. The surviving Eben remained alive until 1952. He was extremely intelligent and learned English quickly by listening to military personnel who were responsible for his care.

As Kasten explains, the U.S. government decided to enter into an alliance with the Ebens to guard against the threat of a Fourth Reich in Antarctica. Eventually, after much effort, the humans and Ebens were able to roughly translate each other's languages and begin an interstellar dialogue. Around the time that John F. Kennedy assumed the presidency, the Ebens and the U.S. government agreed on an exchange program. Because of Kennedy's deep mistrust of the CIA, he decided that the Defense Intelligence Agency would manage the project. The Air Force was the lead agency that took responsibility for finding the 12 volunteers for the mission. All of the recruits were unmarried at the time of the mission. None had children. Preference was given to those candidates who were orphans. The concern was that family members might somehow learn of the top secret program. The 12 selected consisted of ten men and two women. Each crewmember was designated with a three-digit number.

On the afternoon of April 24, 1964, two alien aircrafts landed in New Mexico just as scheduled. One of the visitors, a female known only as "Ebe2," presented the astronaut team with "the Yellow Book"—a type of encyclopedia that contained numerous facts about the Earth and the galaxy. Unlike a standard book, when the reader viewed the transparent surface of the Yellow Book, words and pictures would magically appear. The two parties agreed on a return visit for July 16, 1965. Right on schedule, the team boarded the Eben shuttle craft on that date. Despite the vast distance between the two destinations—38.42 light years, or 240 trillion miles—it took only nine months for the Eben spaceship to make the long journey from Earth to their home planet, Serpo. By traveling though "space tunnels," they were able to take shortcuts to their destination. On board the ship was a sophisticated device that enabled the Ebens to communicate with their home planet at a rate faster than the speed of light.

Upon the team's arrival on Serpo, the same Ebe2, whom the crew depicted as warm-hearted and friendly, served as the team's escort on Serpo. During their stay, the team was granted a considerable degree of license to travel on the planet as they saw fit; however, they were forbidden to enter private living houses, which were described as standardized dwellings in the style of the adobe domiciles that predominate in the American southwest.

Located in a binary star system, the planet Serpo would never get completely dark. The planet was estimated to be about 3 billion Earth years in age, while its two suns were roughly 5 billion years old. The crew learned that the Ebens had originated on another planet but left when it was threatened with extreme volcanic activity. To avoid the disaster, the Eben population relocated to Serpo and colonized the planet roughly 5,000 years ago.

Numbering only about 650,000 inhabitants, Eben society was described as peaceful, but also regimented. The population was homogenized, not only in conventions, but in physical appearance as well. Each male had a mate. Couples were allowed no more than two children. In their daily lives, the Ebens were spartan and disciplined. There were no stores, shopping malls, or money. Rather, Ebens obtained the items they need at central distribution centers. Although the Eben populace was highly intelligent, they displayed little individualism and creativity. The concept of individual liberty was not operable on Serpo. Be that as it may, the team noted that one of the more intelligent hosts that followed them around—Ebe5, or the "Ebe Einstein"—gradually began to appreciate the concept of liberty and wanted to know more about the principles of freedom experienced on Earth.

Earlier in their history, the Ebens fought the "Great War" against a hostile race of alien invaders. Although many hundreds of thousands of Ebens died in the conflict, which lasted about 100 years, they eventually prevailed with the help of particle-beam weapons. The Ebens warned the team that several hostile alien races exist within the Milky Way. The war so traumatized the Eben population that they were willing to acquiesce to a strong central authority.

Eben society is described as a "military-industrial oligarchy." The Council of Governors exercised complete authority and control over every aspect of Eben life. The population was divided into the elite controllers who jealously guarded the secrets of science and technology, and the masses that were manipulated and generally kept in the dark. The actions of the Ebens were controlled by a device worn on their belts. The military, which also served as a police force, was ubiquitous, although the team never saw the police carry guns or weapons of any type.

To be expected, the Ebens had developed several advanced technologies and techniques, including particle-beam weapons, anti-gravity vehicles, cloning, the ability to view and record past events, interstellar communication, and sophisticated translation devices. They also learned how to extract free energy from the quantum vacuum. The fact that they had long ago mastered genetic engineering could perhaps explain the homogeneity of the population. Anti-gravity technology enabled them to effortlessly move massive objects, which explained how ancient megalithic monuments on Earth— including the pyramids of Egypt and Stonehenge—were built on Earth with the assistance of Ebens, thousands of years ago.

For their religion, the Ebens practiced a form of monotheism that worshiped a supreme being who was conceptualized as a deity that was connected to the universe. The religion also venerated numerous "subentities" analogous to saints in Christianity. They believed in an afterlife in which the soul of the departed was taken to a midpoint realm not unlike the Christian purgatory. Once the soul was ready, it then proceeded to join the "Supreme Plateau," where it remained for eternity. But in some instances, souls were reincarnated. These souls returned back to our plane of existence to perform some specific act, not unlike the notion of karma in Eastern religions.

After a 13-year tour of duty, on August 18, 1978, seven of members of the team returned to Earth. Three had died—one on the spaceship en route to Serpo and two on the planet—while two others decided to remain on Serpo. According to Kasten, the last survivor of the mission died in Florida in 2002. Several members of the team died of cancer believed to have been engendered by exposure to the intense radiation they absorbed while on Serpo.

Kasten asks: Why was such sensitive classified information concerning the top secret project released? He suspects that an aging, highly placed intelligence operative, inspired by the transparency policy instilled by then-President Kennedy, decided near the end of his life that the public had a right to know about this fantastic program and the U.S. government's special relationship with extraterrestrials. Kasten speculates that the famous film director Steven Spielberg had inside information about Project Crystal Knight and used it for the basis of his film *Close Encounters of the Third Kind*.[131]

As Kasten's "exposé" illustrates, much of the literature in Ufology makes fantastic claims backed up by highly dubious sources with murky provenance. What is more, it seems puzzling that authors who expose such closely guarded secrets suffer no apparent repression from the government or any other agency. However, some researchers claim that this is not always the case. Inasmuch as the government allegedly maintains secrecy on the UFO topic, it takes harsh measures to prevent people from disclosing its true nature.

Some researchers insist that there is an active campaign to suppress the truth about UFOs. Toward this end, so-called men in black (MIB) are alleged to visit UFO witnesses. Described as robotic in demeanor, they are neither warm nor benign; yet, they are nevertheless menacing to their subjects whom they seek to dissuade from talking about their UFO experiences.[132] Episodes of alleged men-in-black encounters commenced with the contemporary UFO era, but earlier accounts have been recorded.[133] The men-in-black legend was first popularized by Albert Bender, a resident of Bridgeport, Connecticut, with an abiding fascination with the occult and unexplained phenomena. He claimed to have been visited by three men in black in 1953 who warned him not to reveal the truth that they conveyed to him about UFOs.[134] Coincidentally, as explained in Chapter 9, that same year, the CIA convened a select group called the Robertson Panel, which examined the national security implications of the UFO issue and recommended a policy of discouraging UFO discourse in America. The UFO researcher, Jerome Park, suspects that Bender may have been interviewed by FBI agents, possibly as a result of the recommendations of the CIA's Robertson Panel. The encounter may have so upset Bender that he may have transformed a seemingly normal but traumatic incident, into a bizarre tale involving alien men in black.[135] Bender recounted his experience to Gray Barker, whose 1956 book, *They Knew Too Much About Flying Saucers*, gave Bender's saga widespread publicity in the UFO community.[136] In 1952 Bender founded the International Flying Saucer Bureau, but soon after his alleged men-in-black encounter in 1953, he desisted in his public interest in UFOs.[137]

Very often, men in black are reported to appear in groups of three. In the United States, they are said to travel in 1950s-style black Cadillacs. In the British Isles, they prefer 1960s-era black Jaguars. Both models are almost always described as looking curiously brand new.[138] Men in black are alleged to visit witnesses even before they have had a chance to report their UFO sightings. They are often described as attired in clothes that have been out of fashion for some time. At times, they appear to have trouble breathing. Their speech is laced with archaic expressions. They are especially interested in photographs and any physical evidence that witnesses possess concerning their UFO encounters. To keep the witnesses silent, men in black are said to make threats against them and their family members.[139]

Some researchers speculate that the men in black occasionally follow through with their threats, citing the suspicious deaths of UFO researchers and witnesses. For instance, Morris K. Jessup, an early researcher who did much to popularize UFOs, was found dead in his car, ostensibly from self-inflicted asphyxiation. Some observers, however, impugn the official version of his death. Interestingly, Jessup was also connected to the shadowy Carlos Miguel Allende, who was mainly responsible for the creation of the "Philadelphia Experiment" legend.[140] Another UFO researcher, D. Scott Rogo, was stabbed in his home by an unknown assailant.[141] Dr. James E. MacDonald, who was discussed in Chapter 8, was the apparent victim of a bizarre suicide.[142]

Another suspicious death was linked to a UFO incident in Pennsylvania. On December 5, 1965, residents in Kecksburg reported seeing something fall into a heavily wooded area around 6:30 P.M. According to witnesses, military personnel flooded the area and sealed off the woods. Soon thereafter, a flatbed truck was brought in, only to be seen later carrying something covered by a tarpaulin. Adding to the mystery, a local radio reporter, John Murphy, planned to broadcast an investigative report on the incident, but before it was aired, he was visited by mysterious men in black. According to his colleagues at the radio station, he was visibly shaken by the visitation and his report was heavily censored, as it did not even mention any object in the woods. The incident came to be known as "Pennsylvania's Roswell." Increasingly despondent, Murphy refused to talk about the incident anymore. Years later, he was killed in a hit-and-run accident as he walked across a street while visiting Ventura, California. The driver involved in the accident was never caught.[143]

On some rare occasions, multiple persons have men-in-black encounters in a single community. Allegedly, some residents of Point Pleasant, West Virginia, received visits from men-in-black between 1966 and December 1967, just before the Silver Bridge, which spanned the Ohio River and connected Point Pleasant to Gallipolis, Ohio, suddenly collapsed. The accident claimed 46 lives. The events surrounding the event were the basis for the 2002 film *The Mothman Prophecies*, starring Richard Gere.[144]

As the saying goes, extraordinary claims require extraordinary evidence. To date, Ufology has not lived up to this standard. Nevertheless, there are serious and sincere researchers in the field who deserve to be heard. Since the cancellation of Project Blue Book in 1970, the scientific community has generally been uninterested in UFOs. Although skeptics concede that some UFO sightings have no immediate explanation, they counter that this is really no different than any other field of science in which a residue of anomalies is left unexplained by the dominant theory.[145]

Still, some researchers are adamant that UFOs are real and originate from outside of our planet. Foremost among them is Stanton Friedman. His trajectory into Ufology began in 1958, when he needed one more volume to round out an order from a discount mail order bookseller so that he would not have to pay postage. Edward J. Ruppelt's *The Report on Unidentified Flying Objects* caught his attention. After reading the book, he was intrigued, but not entirely convinced. Still, he was curious enough to keep UFOs on his reading list.[146] At the time, he worked as a nuclear physicist for General Electric near Cincinnati.[147] Later in his career, he worked on nuclear rockets at the Westinghouse Astronuclear Laboratory in Pittsburgh, Pennsylvania. While there, he and a few of his colleagues founded a UFO group and established a 24-hour hotline so people could call to report sightings. When he was laid off by Westinghouse in 1969, he decided to enter into Ufology full-time. Over the years, he has made more than 700 lectures that are well received by audiences around the world.[148]

Friedman argues that there is overwhelming evidence to support the extraterrestrial hypothesis: He estimates that there are billions of extraterrestrial civilizations in our galaxy alone.[149] But how could extraterrestrial aliens make the long journey to Earth? As Friedman correctly points out: "Technological progress comes from doing things differently in an unpredictable way. The future is not an extrapolation of the past."[150] When I asked him the most convincing evidence that he would adduce to support the extraterrestrial hypothesis, he replied:

> It would be the consistent testimony and all of the large-scale scientific studies indicating that reliable, respectable people, all over the world and associated instrumentation have been observing the same strange things—manufactured objects behaving in ways that we cannot duplicate. What's most impressive to me is that if you look at the physical trace cases, you look at the abduction cases, you look at the radar visual sightings, if you look at just the 41 cases described

> by James E. McDonald in his congressional testimony of 1968, there's an incredible variety [of sources] and consistency. Multiple witness radar visual sightings over big cities. Sightings out in the country. Day time. Night time. Sightings by pilots. Sightings by astronomers. It's a plethora of evidence.[151]

Why would extraterrestrials visit Earth? Friedman proffers a few reasons why. First, Earth is the only planet in the solar system mostly covered by water—more than likely, the *sine qua non* for life. Second, Earth is the only planet in the solar system that has a high level of oxygen in its atmosphere. Third, Earth is the densest planet, as it is made up of heavy metals that are quite rare in the solar system. Fourth, a variety of life flourishes on Earth, which could make the planet attractive to interstellar researchers.[152] Finally, aliens might take an interest in humans for no other reason than security. In the twentieth century, humans developed lethal weapons of mass destruction and the capabilities to leave Earth, albeit with limitations. Interestingly for Friedman, at the end of World War II, there were three signs that humans would be moving out into the galaxy—V-2 rockets, atomic bombs, and radar—and all three of these technologies were located in the area of New Mexico where Roswell is located.[153] This would explain why the alleged aliens were reconnoitering the area.

Why would the government conceal the true nature of UFOs? Friedman answers that the advanced flight technology of such crafts would have significant national security implications that the government would prefer to keep secret. If this flight technology could be duplicated, the government could easily defeat the military forces of any country. There is also the issue of societal panic in the style that supposedly occurred in 1938 during Orson Welles' broadcast of H.G. Wells' *War of the Worlds*.[154] Furthermore, if extraterrestrial aliens ever decide to attack us, the government would be unlikely to protect us from such an advanced civilization. As a consequence, this could undercut confidence in the government.[155]

As an example of how the government is able to keep secrets, Friedman cites the case of the Manhattan Project, noting that then-President Harry S. Truman was completely unaware of the effort to build an atomic bomb until 13 days after the death of President Franklin D. Roosevelt.[156] While that is true, he fails to mention that Soviet spies had infiltrated the Manhattan Project almost from its inception.[157] Citing another example, Friedman points out that the "Ultra" project to decipher the German secret communications code was kept secret during World War II, despite the fact that the effort involved 12,000 people who worked at Bletchley Park in England. Moreover, the program was kept secret for 25 years after the war had ended.[158]

In Friedmann's opinion, MJ-12 is a real body that is still extant today. He refers to the U.S. government's silence on this alleged program as the "Cosmic Watergate."[159] A "new Woodward and Bernstein team," Friedmann says, is needed to expose this story to the general public.[160] He suspects that Dr. Donald Menzel, a Harvard astronomer who wrote several books and articles debunking UFOs was a member of MJ-12. After his death, Friedman even speculated that Carl Sagan took his place on the secret panel.[161] Nevertheless, the two were classmates while at the University of Chicago and occasionally corresponded with each other. Friedman even visited Sagan before he died.[162]

Not surprisingly, relations between Friedman and the SETI community are cold. Derogating SETI as the "Silly Effort To Investigate" Friedman characterizes the enterprise as a cult led by charismatic leaders who operate on a strong dogma. As he explains, cults do their best to ignore or repress testimony that is contrary to their beliefs, in this case, the extraterrestrial hypothesis for UFOs.[163] He derides the so-called debunkers and skeptics for their seeming narrow-mindedness, as he explained to this author:

> Well, they [the skeptics] are not looking in the right place [for evidence]. Physical evidence covers a lot of ground. There are roughly four thousand physical trace cases from more than eighty countries, which include changes in the dirt, and so forth. What they are really saying is "gee nobody has turned over a body or a crashed saucer to us and without that, we won't accept the reality."
>
> I differentiate between debunkers and skeptics. The skeptics say, "well let's see." "I don't know, but let's find out." The debunker does

> know—ABA—"anything but aliens" is his answer. They [the debunkers] do their research by proclamation. Attack the data. Attack the people. And if it were true, I would already know it.[164]

Skeptics point out that the aliens are often described as having too human-like characteristics, which one would not expect in a truly extraterrestrial life form. Moreover, UFOs do not maneuver in ways one would expect if they were actual physical objects. On radar screens, they seem to make 90-degree turns at high speeds that would rip a physical object apart.[165] The SETI scientist Seth Shostak finds it unlikely that virtually all governments of the world could conspire to hide the truth about alien UFOs. Furthermore, he cites a lack of credible evidence that has been adduced by the UFO enthusiasts. Why, he asks, have the many satellites orbiting the Earth not spotted UFOs pirouetting above our planet? Finally, he notes that despite more than a half century of presumed alien visitation, the sum total of UFO visits appear to have had no significant consequences.[166] Instead of landing in a high-profile public place, aliens have a penchant for appearing to farmers and stranded motorists. For his part, Friedman counters that negotiations usually occur between groups with roughly the same level of power. As we would not be their technological equals, they would feel no compelling reason to "land on the White House lawn" to meet with the president of the United States.[167] Supporting Friedmann's analysis, the Nobel Laureate George Wald once opined that contact with an extraterrestrial civilization would produce "the most highly classified and exploited military information in the history of the earth."[168] He feared that if communication between Earth and an extraterrestrial civilization was established, humans could become dependent on their alien benefactors.[169] Consequently, details of any collaboration would remain highly classified.

Other UFO researchers are highly skeptical of any notion of a government conspiracy involving extraterrestrial aliens. While still in the military, U.S. Army Colonel John B. Alexander formed the Advanced Theoretical Physics Project (ATP) to determine what role the Department of Defense played in UFO investigations. Membership in the group included persons from the Army, Navy, and Air Force as well as persons from the defense aerospace industry and some members of the Intelligence Community.[170] Alexander had his own UFO sighting on May 31, 1972, while driving northbound on I-135 in Kansas. In the predawn hours, he noticed a brilliant, hovering, cylindrical object in the sky.[171]

Alexander believes that MJ-12 might have existed, but not for the purpose of extraterrestrial aliens. According to a confidential source, he claimed that MJ-12 was created to ensure the continuity of government in the event of a national emergency.[172] With his numerous contacts in the black world of the defense department, Alexander concluded that there was no conspiracy of silence on UFOs in the government. He pointed out the so-called "capabilities paradox" with respect to UFOs:

> The capabilities paradox refers to the perceived inability of the U.S. Government to manage things efficiently, except for on distinct area—UFOs. It is this unique area in which the government is perceived as near omniscient, well organized, and highly focused. Many people believe the government can't secure our borders, especially those with Mexico. However, let a UFO fly by, let alone crash, and the responsible agencies are instantly involved in chasing the intruders or recovering material.[173]

Although Alexander rejects the notion of a government conspiracy to suppress the truth about aliens, he finds numerous UFO cases convincing, most notably the incident at the Rendlesham Forrest in England discussed in Chapter 8, which support the extraterrestrial hypothesis.

In recent years, several books on UFOs have added new vigor in the debate over the extraterrestrial hypothesis. For instance, in *Witness to Roswell: Unmasking the Government's Biggest Cover-Up* (2009), Thomas J. Carey and Donald R. Schmitt conducted extensive interviews with a number of the key figures in the case that lend credence to the theory that there is more to the story than what the government has let on. Likewise, Leslie Kean's *UFOs: Generals, Pilots, and Government Officials Go on the Record* (2010) relied on numerous seemingly credible people, including high-ranking military officers, commercial pilots, and notable public officials, to make the case that there is something real behind UFO sightings. Seeking to separate herself from the lunatic fringe in ufology, Kean calls for a "militant agnosticism" in pursuit of UFO investigations—that is, not automatically attributing UFOs to extraterrestrial sources.[174] If, however, the eyewitness accounts presented in her book are to be believed, then one would be hard pressed to conclude otherwise. Other explanations proffered in the past would be at least, or even more mind-boggling, such as time travelers, Nazi flying saucers from underground bases in Antarctica, or visitors from other dimensions.

Be that as it may, there appears to be little to no corroborative evidence outside of UFO studies that would support the extraterrestrial hypothesis. As Leslie Kean conceded, there have been no deathbed confessions or willed documents from any government scientists or officials that have revealed the truth about special access programs on UFOs. Moreover, she points out that we have not seen any fantastic reverse-engineered military technology that might have been retrieved from captured UFOs, despite rumors to the contrary. Numerous memoirs of past presidents and other prominent leaders often reveal embarrassing details of their lives, but so far, none provide any insight about their secret relationships with extraterrestrial representatives. And although some researchers dismiss the notion of a "vast conscious conspiracy" on the part of the U.S. government that prevents information on UFOs from reaching the public, it is hard to see how an issue of such earth-shaking significance could be withheld if it were true. Instead, the world's leaders seem to go about their day-to-day business and decision-making without considering the influence of extraterrestrial visitors, who, if they were so magnanimous, might at least give us some advice on how to find alternatives to fossil fuels, considering their ability to master interstellar travel. As others have noted, contact with extraterrestrials would be a transformative event, yet if it has already occurred, the world's leaders seem startling aloof and amazingly tight-lipped—a remarkable adherence to security protocol in an age of exposé journalism. At least from surface appearances, public officials do not seem to take into account the presence of aliens, with few exceptions. For instance, far from shunning the prospect of extraterrestrial life, in August 1996, then-President Bill Clinton enthusiastically announced that NASA scientists had found evidence for life on Mars in the form of microscopic features inside a meteorite recovered from Antarctica in 1984, although subsequent analysis chipped away at that conclusion.[175]

As far as UFOs are concerned, the U.S. government's position seems to be one of disinterest. As the machinations of the Condon Committee, which was discussed in Chapter 9, made clear, the U.S. government wanted out of the business of investigating UFOs. The Air Force officially shut down Project Blue Book at the end of 1969. Although rumors surface of a successor program, there does not appear to be any documentary evidence of any serious effort to look into UFOs. In fact, the Federal Aviation Agency's official guidance recommends that UFO sightings be reported to a private entity, Bigelow Aerospace Advanced Space Studies in Nevada. All of this would seem to militate against conspiracy theories claiming that the U.S. government is deliberately concealing information on UFOs and extraterrestrial aliens.

If extraterrestrial aliens have visited Earth, to date they do not seem to be overtly hostile. Knock on wood, there has been no alien armada that has attacked our planet to plunder our resources. Thankfully, that scenario remains confined to Hollywood. Even if extraterrestrial life exists elsewhere in the galaxy, many people believe that they have no way of ever making direct physical contact with Earth. The laws of physics as we currently conceive them would seem to rule out the prospect of interstellar travel. But perhaps for a civilization that is millions of years more technologically advanced than ours, this could well be within the realm of possibility. As such, it behooves us to consider the consequences of direct contact as detailed in the next chapter.

Chapter 12

Preparing for Direct Contact

Popular culture continues to reflect our fascination with extraterrestrial aliens. In May 2012, a science fiction film—*Battleship*—was released whose plot centered on a confrontation between the U.S. Navy and a fleet of extraterrestrial invaders in the Pacific Ocean. The story begins in 2005, when NASA scientists discover an exoplanet with similar characteristics to Earth. Assuming that there could be intelligent life on the planet, NASA transmits a powerful signal from a communications array in Hawaii that is boosted by a satellite in orbit. Remarkably, the alien recipients of the message are able to mount an interstellar invasion a mere seven years later.

The above scenario, although implausible because it ignores the cosmic speed limit (i.e., the speed of light, which according to Einstein's special theory of relativity, nothing can exceed), nevertheless illustrates the potential perils of attempting to communicate with extraterrestrial aliens. Hostile alien invasions are a staple in Hollywood films. For instance, in March 2011, a film—*Battle: Los Angeles*—was released that depicted yet another hostile alien invasion of Earth. The purpose of the offensive is to eradicate the human population so that the aliens can harvest our planet's resources. Just a few months before, in the film *Skyline*, aliens used massive force fields to vacuum humans into their craft, whereupon the space invaders removed their victims' brains. Adding a sense of verisimilitude to the film's trailer was a clip of a television news program featuring Dan Rather, who reported on a recent warning by the eminent British astrophysicist Stephen Hawking that communicating with extraterrestrial civilization could be potentially dangerous. In April 2010, the Discovery Channel broadcasted a documentary in which Hawking speculated on the existence of extraterrestrial life. In his mind, the multitude of billions of stars and galaxies suggests that life in other solar systems is almost a certainty. Hawking pondered an important question: Could visitors from extraterrestrial civilizations pose a threat to Earth? In the documentary, an armada of massive space ships roams the galaxy as interstellar nomads, who having exhausted all the resources on their home planet, search for other planets to conquer and colonize. He concluded that making contact with aliens is "a little too risky." Moreover, drawing upon the experience on Earth, he mused that "if aliens ever visit us, I think the outcome would be much as when Christopher Columbus first landed in America, which didn't turn out very well for the Native Americans."[1] Years before, he issued a similar warning in his classic text, *A Brief History of Time*.[2]

Hawking is not the first authority to warn of the potential hazards of alien contact. For instance, in his book, *The Third Chimpanzee*, the noted historian Jared Diamond offered an essay on the perils of attempting to contact alien civilizations, observing that whenever a more advanced civilization encountered a less advanced one, the results have often been catastrophic for the weaker party, including slavery, colonialism, or extinction.[3] He once described the 1974 Arecibo message, which was aimed at the globular star cluster M13 some 25,000 light years away, as suicidal folly, comparing it to an Incan emperor describing the wealth of his kingdom to the gold-crazed Spaniards. Sending out radio signals, he advised, was "naïve, even dangerous."[4] Likewise, the Czech astronomer Zdnek Kopal warned in 1972 that "Should we ever hear the space-phone ringing, for God's sake let us not answer, but rather make ourselves as inconspicuous as possible to avoid attracting attentions."[5] Even Carl Sagan, who was generally sanguine about the prospects of interstellar comity, once counseled that a relatively young civilization, such as Earth's, should listen quietly, "before shouting into an unknown jungle that we don't understand."[6]

How should we prepare for the prospect of direct contact with an extraterrestrial civilization? Presumably, any civilization that could send a spacecraft across interstellar distances would be far more technologically advanced than we are. Such overwhelming technological superiority would presuppose military superiority as well. First and foremost, pessimists warn that technologically superior aliens could pose an existential threat to humans. Hence, it would be prudent to follow the precautionary principle, which counsels for the avoidance of any action that will create a risk of harm.[7] While this argument seems reasonable at first blush, upon further reflection these fears would seem to be greatly overstated for a straightforward reason related to the physics of interstellar travel. Nevertheless, because of the potential dangers we should seriously consider various scenarios.

As will be discussed in much greater detail in the next chapter, when considering interstellar travel, two important factors to consider are the vast distances between solar systems and the enormous energy requirements that would be necessary to fuel spacecraft that could traverse such distances. The closest solar system, Proxima Centauri, is about 4.2 light years away from Earth. To put that in perspective, if a vessel from Earth could somehow travel the speed of light, it would take over four years to reach that solar system. According to Einstein's Special Theory of Relativity advanced in 1905, interstellar travel is seemingly impossible because no vessel could travel faster than speed of light; it would take centuries or millennia to travel the distances between solar systems. For example, at the speed of the Voyager spacecraft launched in 1977, it would take over 70,000 years to get to Proxima Centauri. But in 1915, Einstein argued in his General Theory of Relativity that under some circumstances, faster-than- speed-of-light-travel is theoretically possible under some extreme conditions. For instance, by using large amounts of energy, an advanced civilization might be able to warp space, thus enabling faster-than-light travel. Alternatively, a wormhole could be used to create a shortcut across space and time. A functioning wormhole might be able to serve as a bridge between two distant points in the universe. Assumedly, practical interstellar space travel would only be viable for very advanced civilizations that have learned how to harness enormous amounts of energy.[8] For an extraterrestrial civilization to conduct interstellar travel, most probably it would have to obtain at least Type I status on the Kardashev scale as discussed in Chapter 4.

Opinion has long been divided on the outcome of contact between aliens and earthlings. Optimists conjecture that an alien civilization that takes the trouble and expense of actively trying to contact us would have altruistic motives.[9] Pessimists, on the other hand, warn that interstellar travel might be possible and that extraterrestrials could be aggressive, so it behooves us to be wary of making contact.[10] On that note, a 1977 U.S. Congressional Research Service report noted that although it was tempting to hypothesize that any extraterrestrial civilization that mastered interstellar travel would have overcome differences that spawn wars, it was far from certain that it would be peaceful toward us. As a consequence, the aliens could arrive on our planet "well armed and ready for combat."[11] Alien visitors might initially be neutral, but all the while, "size up" Earth for a subsequent takeover depending upon the planet's defense and desirability.[12] Could our civilization mount an effective defense? Ruefully, Seth Shostak opined, "Any earthly defense against such a society would be like the armies of the American Civil War making plans to battle the U.S. military of today."[13]

So why would aliens visit the Earth? As Stephen Hawking mused, an alien civilization might exhaust its home planet and search for other planets from which to extract resources. Would an alien civilization make the long trip to Earth to plunder our resources? Conceivably, one could envisage a resource war in the sense of trying to grab up all the planets in a galaxy, not unlike nations and corporations on Earth seek to secure supplies of raw materials.[14] This is a recurrent theme in science fiction. For instance, the blockbuster 1996 film, *Independence Day*, depicted a large-scale, coordinated attack on Earth by aliens who travel from planet to planet to harvest their resources.

Upon closer examination, though, this scenario seems unlikely. It stretches the bounds of credulity to believe that an advanced alien civilization would come all the way to Earth for energy products. It's a safe bet that any civilization capable of traveling such long distances either by way of spacecraft and/ or wormholes would not be using oil and other pre-Type I civilization energy sources. What is more, the transportation costs to bring the energy products back to the mother planet would not be economical.[15] Possibly, fusion reactors, or even anti-matter reactors, could be used to fuel such space vessels, in which case hydrogen, or some isotope thereof, would be required.[16] Hydrogen, however, is one of the most abundant chemical elements in the universe and would not require interstellar travel to obtain. Conceivably, an alien civilization might want to extract minerals from other planets. Yet, it would not be practical to come all the way to Earth for minerals. After all, they could more than likely be found on planets in much greater quantities in their own solar system or in nearby solar systems. And it would be far more practical to conduct strip mining on planets on which they would not have to deal with restive denizens, such as Earthlings. In short, the rarity of advanced life and the tremendous distances between civilizations suggest that there would be plenty of planets and stars for all those that were capable of exploiting such methods of resource extraction.

Still, others point out that the way in which humans treat other animals is not reassuring. Higher-level mammals, such as dolphins and even chimpanzees, are often mistreated for commercial or experimental reasons.[17] Moreover, evolutionary biologists point out that altruism occurs with decreasing intensity as individuals grow more distantly related.[18] Nevertheless, examples of interspecies altruism are practiced on Earth. For instance, in numerous cases, dolphins have rescued shipwrecked humans. Likewise, humans enjoy certain animals as pets and treat them well. Furthermore, there are numerous organizations that advocate on behalf of animal welfare.

Perhaps, aliens would crave us as food. There is a good possibility that alien visitors would be descended from carnivores.[19] Based on this assumption, the biologist Michael Archer feared that aliens would more likely be predatory than benevolent.[20] Invoking evolutionary theory, it stands to reason that extraterrestrial alien visitors are most likely to be descended from predators insofar as they have traditionally been more intelligent than prey. And any species smart enough to master interstellar travel must be highly intelligent and would probably come from a predator lineage.

The food scenario has some intuitive appeal. After all, many humans consume animal flesh. In a classic episode of the science fiction program *The Twilight Zone*—"To Serve Man"—seemingly magnanimous aliens called the Kanamits come to Earth and share their advanced technology, which solves the planet's greatest woes, including eradicating hunger, disease, and the need for warfare. Humans are encouraged to take trips to the Kanamits' home planet, which is supposedly a paradise. However, a female code breaker eventually deciphers a Kanamit book, ironically titled *To Serve Man*, and determines that it is actually a cookbook on how to serve humans as meals.

When examined more closely, however, this scenario lacks credibility as well. Presumably a Type I, or Type II, or Type III Civilization capable of interstellar travel would also have mastered agricultural engineering and would have solved problems in farming and livestock a long time ago. Even on Earth, which has not yet attained Type I civilizational status, there have been marked improvements in nutrition on a global scale. A portion of the human population still suffers from hunger, but there has been a steady reduction in that segment over the years. In fact, more people are obese today than are undernourished.[21] But even if aliens did find us tasty, they could determine the sequence of our amino acids that makes us appetizing and then reconstruct the relevant proteins on their home planet. Furthermore, as Carl Sagan pointed out, the high freightage costs would make shipping humans en masse across interstellar distances impractical.[22] Although it is certainly reasonable to assume that aliens would consume animal flesh, after all, most humans do that too—creating a wormhole just to go to Kentucky Fried Chicken seems rather far-fetched.

Some commentators fear that aliens might try to subjugate us. If they should ever arrive on Earth, the argument goes, they would most likely see us as competitors. Moreover, just because a civilization has a high level of technological development, it does not necessarily follow that it would be altruistic to others. As Michael Michaud pointed out, in the 1930s, Germany led the world in science and technology, yet engulfed the planet in a world war. Drawing on these historical analogs, one observer mused that aliens are "more likely to come bearing the outer space equivalent of swastikas, and alien/human contact plans that involve boxcars and camps."[23] But once again, this scenario seems unlikely. The prospect of advanced alien civilizations coming to Earth to enslave humans seems implausible. A civilization capable of interstellar travel would not have a labor shortage; it could more easily build robots and bio-machines to do the necessary grunt work.[24]

Could aliens use humans to help biologically propagate their species? Many of the abductee accounts imply that aliens are acutely interested in humans' reproductive organs. As discussed in Chapter 3, some scientists, including the co-discoverer of the structure of DNA Francis Crick, have speculated that aliens may have actually "seeded" planets in the universe by sending microbe-laden probes out into space in a process called "directed panspermia" as a way in which to spread the building blocks of life.[25] In the film *Mars Needs Women*, Martians suffer a genetic deficiency that produces only male offspring. However, a civilization that could travel great distances would almost certainly have mastered bioengineering as well and could ameliorate such a deficiency. Moreover, inasmuch as alien life-forms would probably be based on entirely different DNA and protein molecules, they would probably have no interest in either eating us or mating with us.[26]

Perhaps an advanced martial extraterrestrial civilization might use Earth as a type of training ground for their warriors. Such was the plot of the *Predator* film series, in which alien creatures visit various hot spots and war zones on Earth to hone their martial and hunting skills. This could be exciting for members of some alien civilizations. After all, at one time, big game hunting was the province of distinguished gentlemen in the West. The scenario has some plausibility until one considers the frivolousness of such a trip. Would aliens be willing to spend 1,000 years in suspended animation for such an excursion? Would aliens create a wormhole just to go big game hunting?

More seriously, territorial motives could inform interstellar colonization. Based on assumptions of terrestrial life, we would assume that life has a natural tendency to expand. Should extraterrestrial life be any different?[27] Some SETI researchers argue that insofar as life evolves through competition, we should be wary of making contact with extraterrestrials who might view us as competitors. Because organisms self-replicate, they will be subjected to the dynamics of natural selection, with the implication that those that replicate themselves more effectively will come to predominate over those that replicate themselves less effectively. To survive for so long, they would have been successful in their competition with other life forms. For that reason, to quote David Swift, it might be "prudent to remain silent in the jungle."[28] This sentiment is an enduring theme in science fiction. For instance, in H.G. Wells' classic novel, *The War of the Worlds*, Martians invade Earth to take over the planet. The novel was written at the height of the British Empire, when power was often measured in land.[29] Eventually, an advanced extraterrestrial civilization would be forced to embark on interstellar travel if it wanted to survive, as its sun would have a limited life.[30] In such a scenario, an extraterrestrial civilization might want to colonize so-called Goldilocks planets, that is, those that that fall within a star's habitable zone and would be roughly in size of the planet Earth, so that it could have an atmosphere. Such a planet would avoid overly hot or cold temperatures so that it could retain liquid water on its surface, assumedly a *sine qua non* for the emergence of life. As explained in Chapter 2, the so-called "Rare Earth hypothesis" implies that habitable planets are extremely uncommon in the universe because a number of unlikely events and conditions would be necessary to give rise to these worlds. Such prime real estate, so the argument goes, would be highly coveted by aliens, not unlike the American continents were for European settlers centuries ago.

The Columbus analogy, though, would probably be inapplicable for alien encounters. Judged by contemporary standards, the Spanish conquistadors were not much more technologically superior to their Native American hosts. There was no great capabilities gap between the two parties. The Spaniards coveted the gold of the latter, and eventually their territory, both of which were limited commodities that could only be obtained on Earth for both cultures. Moreover, it was foreign diseases to which Native Americans had no immunity that decimated that population, rather than a systematic plan of conquest and genocide. Rather than travel to Earth and subdue humans, it would probably be more feasible for a Type II or Type III civilization to artificially create the conditions supportable of life on a much closer planet. This was the scenario of the 1990 film *Total Recall*, starring Arnold Schwarzenegger, in which an alien artifact—a terraforming machine—has the ability to create an oxygen-bearing atmosphere on Mars. Conceivably, a Type II or Type III Civilization might even be able to reconfigure a planetary system so that more planets orbit inside the habitable zone.[31] Perhaps out of a sense of magnanimity, an alien civilization would leave our sun and solar system alone and choose to colonize other solar systems. In that sense, settling other planets would not really be imperialism in the classic sense of the term because, as Frank Tipler and John Barrow opined, the planets would just be "dead rocks and gas."[32] A technologically sophisticated extraterrestrial civilization may also be accompanied with an advanced ethical development that values other life forms and decide to leave them unmolested.[33] Finally, if a Type III civilization sought more stars from which to extract energy, it could find a great abundance of more desirable and larger stars instead of our sun which is average in stature.[34]

Some have argued that superior civilizations would seek out and destroy less-advanced civilizations for their own self-defense. The analog of the Cold War is instructive, as neither superpower was principally concerned about territory, but rather feared the prospect of a preemptive nuclear strike. Preemptive galactic warfare, the cosmologist Edward Harrison argued, would be prudent insofar as a species that has overcome its own self-destructive qualities may view other species bent on galactic expansion as a sort of virus.[35] Ironically, the 1959 science fiction film *Plan 9 from Outer Space*—often cited as the worst movie ever made—contained such a plot that logically explained hostile alien intentions. In the film, extraterrestrial beings seek to stop humans from developing a doomsday weapon that could destroy the universe. One is reminded of the first test of the atomic bomb at the Trinity site at Los Alamos in New Mexico on July 16, 1945: The Manhattan Project scientist physicist Enrico Fermi offered to take wagers that the test would merely destroy the entire State of New Mexico or wipe out all life on Earth.[36] More recently, some observers of projects such as the Relativistic Heavy Ion Collider (RHIC) on Long Island feared that experiments to create a miniature black hole could devour our planet.[37] Conceivably, an alien civilization might seek to cripple another civilization before it developed the capability to retaliate or wreak galactic havoc. The Manhattan Project scientist John von Neumann and the electrical engineer Ronald Bracewell conjectured that extraterrestrial civilizations might send self-replicating robotic probes to explore other solar systems. Some scientists feared that such artificially intelligent automatons might be programmed to destroy other civilizations.[38] Alternatively, some scientists feared that the machines could mutate and change their motivations, thus becoming competitors to their creators.[39]

For an advanced Type III civilization, interstellar colonization would be feasible. Furthermore, aliens facing an imminent disaster on their home planet may be forced to flee and seek a new home elsewhere. Robin Hanson adapted an economic model to speculate on the dynamics of galactic colonization. The basis of his model is that competition inevitably shapes the pattern of growth. Regardless of whatever motives a community might have for spreading, there will be a myriad of diverse cultures vying for planetary *Lebensraum*. The leading edge of this migration wave will be determined purely by competitive selection effects. If a colony has the resources and technological capabilities to migrate out to the cosmos, it could not be stopped unless it ran into another extraterrestrial community doing the same because, presumably, there would be no legal writ that applies galaxy-wide.[40]

In the event of an invasion, Travis S. Taylor and Bob Boan advanced a strategy for a military confrontation with hostile extraterrestrial aliens. They reason that it would be folly to confront extraterrestrial aliens in a conventional military battle plan. Presumably, the aliens would be technologically far more advanced than we if they cold master interstellar travel. Instead, they advocate a form of asymmetrical warfare not unlike the mujahedeen employed against the Soviet Red Army in Afghanistan in the 1980s. By avoiding capture, surviving, and making a general nuisance of themselves, the mujahedeen were able to defeat a far superior conventional force. Likewise, the longer Earthlings could drag out the fight, the better the odds are that they would prevail in a conflict with alien invaders. To ensure manpower strength, Taylor and Boan counsel that women will need to return to their traditional gender roles and concentrate primarily on childbearing. They advise that humans burrow underground and build bases where weapons can be developed and reserve troops can prepare for battle. Taking a page from the contemporary jihadist's playbook, they go so far as to suggest suicide bombing as a possible tactic of resistance. Small tactical nuclear weapons—such as the Davy Crockett launcher—could be used to stealthily strike alien targets.[41]

In planning for this potential conflict, Taylor and Boan call for the creation of a "Sixth Column," a government agency that would have broad autonomy to prepare for such an eventuality including procuring and managing the production of technology for galactic travel, as well as advanced weapons to thwart an invasion. The agency should be highly classified insofar as the dissemination of advanced weapons of mass destruction developed to guard against an alien invasion, could unsettle potential rival governments on Earth and engender a new Cold War and arms race. As a way to develop possible strategies, Taylor and Boan suggest that Sixth Column analysts could publish scenarios in science fiction novels. Feedback from the readers would provide a free sanity check.[42]

Even if extraterrestrial travelers have no ill will toward humans, there could still be a threat of alien microbes harming life forms on Earth, not unlike smallpox, which decimated the Native American population when European explorers landed in America. This prospect might lead governments on Earth to quarantine the interstellar visitors. The Columbus analogy was used by the biochemist Norman Horowitz in 1960 when he warned about the prospect of "back contamination," or the possibility of bringing back extraterrestrial diseases to Earth from samples gathered on the Moon or other planets.[43] Just a few years, before in 1957, Joshua Lederberg—who went on to receive a Nobel Prize for his work in genetics—raised the issue of contamination before the scientific community. At his urging, in 1958 the National Academy of Sciences of the United States recommended to the International Council of Scientific Unions that the problem of contamination be evaluated. The Council complied and created a Committee on Contamination by Extraterrestrial Exploration.[44]

The prospect of contamination proved to be a contentious issue. A dispute emerged between astrobiologists on the one hand, and geologists and engineers on the other, among the scientists and officials preparing the Apollo flights to the Moon from 1960 to 1972. The astrobiologists warned that terrestrial microbes carried to the Moon by the Apollo spacecraft and astronauts could endanger lunar life forms. Furthermore, lunar microbes inadvertently brought back from the Moon, it was argued, could possibly infect the inhabitants of Earth. For their part, the geologists and engineers thought that these fears were overwrought and did not warrant any special precautions.[45]

Some leading scientists, including Carl Sagan, urged that the Apollo spacecraft should be sterilized before traveling to the Moon. NASA ruled out this plan, determining that it would be difficult, if not impossible, to place a sterile lander on the Moon.[46] However, to be on the safe side, the Apollo astronauts were quarantined for several days after their missions to ensure that they could not contaminate Earth with alien microbes.[47] Even today, NASA employs a Planetary Protection system whose goal is to protect planets from any possible contamination of Earthly microbes.[48]

Could extraterrestrial aliens wage biowarfare against the human race? As Paul Davies speculated, rather than offering an incentive to invade, our biosphere might actually be an inconvenience for aliens. The successful colonization of Earth might first require eliminating the indigenous biosphere and replacing it with an alien one. By doing so, the Earth would be terraformed to accommodate the alien interlopers.[49] Instead of a direct military confrontation, alien invaders might introduce deadly pathogens that eliminate the human population. Chandra Wickramasinghe speculated that hostile alien invaders could package their genetic heritage within microbes, including viruses, and launch them into space. They would not need to employ elaborate weapons in the style of science fiction films. With this method, they could take over the Earth without risking a war with human residents.[50] Be that as it may, the microbiologist Abigail A. Salyers maintained that alien microbes would be unlikely to infect humans. As she points out, any bacteria or viruses that evolved on, for example, Mars or Europa would be suited for very cold temperatures. As a consequence, they would probably grow poorly or die in the warm human body. Likewise, a group of scientists with the National Research Council concurred that the probability that alien microbes could contaminate Earth is extremely low.[51]

Much of the speculation on extraterrestrial motives assumes that the aliens will be biological. But if they are able to make the long journey to our planet, there is a strong probability that they could be android or some hybrid thereof. Some SETI luminaries, including Paul Davies, Steven Dick, and Seth Shostak, predict that if we do someday make contact with extraterrestrial intelligence, it is likely to be android, rather than biological. Essentially, they accept the Kurzweil thesis that organic intelligent life will eventually evolve into post-biological entities directed by artificial intelligence. Assuming that we do make contact with an advanced extraterrestrial civilization, it is statistically probable that it will be in the post-biological stage of its evolutionary development.[52] What would be the motivations of machine-based life? Would it have a profoundly different ethical system? Would intelligent machines behave altruistically toward humans both on Earth and elsewhere?

Machine-based life could present potential hostile scenarios between extraterrestrial civilizations. As Seth Shostak mused, because technological improvement would occur on a very short timescale, a "winner take all" contest might apply to thinking machines. The first machines with the complexity to think fast might come to dominate, at least within reach in its portion of the galaxy.[53] As John D. Barrow and Frank J. Tipler pointed out, the biological "exclusion principle," which posits that two species cannot occupy the same ecological niche in the same territory, could be applied to the case of human-extraterrestrial contact. From the perspective of the alien travelers, it might be foolish to wait for possible rivals to get to the "star wars" level of technological sophistication.[54] As regards artificial intelligence, Albert A. Harrison wondered if electric brains could qualify as empathic moral agents. And even if they could, would they be ready to extend friendship to humans? Currently, some computer scientists in the field of artificial intelligence are working to imbue robots with ethical principles and prepare them to resolve ethical quandaries.[55]

Perhaps the most serious danger we face would be if aliens saw Earth as a threat to their galaxy. In his book, *On Aggression*, Konrad Lorenz noted that humans possess some of the same aggressive instincts as animals. But unlike lower species, in which aggression is rarely fatal, humans lack a similar inhibition against fatal violence and are capable of killing their own.[56] Humanity's aggressive tendencies and history of war and conflict could be incompatible with an interstellar association of peaceful aliens. In that sense, Paul Davies argued that "Ironically, the greatest danger from an alien encounter may be ourselves."[57] Aliens might decide to launch a preemptive strike for the greater good of the galactic community. To avert this situation he counsels that we on Earth should signal our best intentions to extraterrestrials, in spite of our penchant for violence at home.[58] In a similar vein, Seth D. Baum, et al. went so far as to counsel terrestrial scientists to avoid sending messages that show evidence of our negative environmental impact. Any messages that indicated the widespread loss of biodiversity could be dangerous if received by an alien civilization with a sense of universalistic altruism toward life in the cosmos. In this scenario, aliens may decide to wipe out humanity to preserve the Earth's ecosystem as a whole.[59] Likewise, the UFO researcher Stanton Friedman mused that aliens could also be concerned about the potential danger that Earth could pose to other civilizations. Aliens might worry that nuclear weapons conjoined with ever more powerful rockets might someday be launched against their own planets. According to Friedman:

> I think every civilization in the local neighborhood, and any local subset of the galactic federation, has put out the word that Earth is a serious threat to the neighborhood. Of course they would be concerned about us. Also, I should note that the transition from low-tech to high-tech can be amazingly short on a cosmic time scale. That means that we may well be the only planet in the local neighborhood at this intermediary stage of development between being able to make a mess of our planet and being able to wreak our brand of havoc elsewhere as well. The bottom line to me is that a major reason for aliens to be checking us out is to quarantine us if we show no progress in acting as responsible galactic citizens.[60]

It is quite possible that an extraterrestrial population could be heterogeneous rather than homogeneous. As a result, an encounter between our civilization and theirs might not follow one general trajectory, but instead could have multiple possibilities. As Seth D. Baum and his colleagues pointed out, an encounter could rapidly change form if a shift in power occurred within the extraterrestrial civilization's leadership. Or perhaps, we might receive mixed messages from an alien civilization that lacks a single unified leadership structure. Maybe several alien factions or nations could originate from the same planet but will make contact with us, each in pursuit of different objectives.[61] In this vein, Michael G. Michaud discussed the fluid nature of conflict on Earth and how it could apply to contact between humans and extraterrestrials:

> We should beware of oversimplifying alien motivations and behavior, for example by proposing that extraterrestrial intelligence is either friendly or hostile. In our own history, nations and other groupings are not always friendly or always hostile to others. Think of US relations with Germany. We fought two wars against that country, yet now Germany is one of our closest allies.
>
> Our own history suggests that intelligent beings may take action for a wider variety of reasons. One is scientific curiosity, which might lead to a visit by an alien probe that may or may not seek to communicate with us. Another possible alien motivation is to check out our civilization to see if we are a potential long-term threat. If the extraterrestrial intelligence decides that we are a threat, we might face danger. If they decide we are not a threat, they might not do anything that we can detect.
>
> I have serious doubts about some of Hollywood's assumptions about alien motivations, such as finding a new home planet or extracting resources from our solar system. A much more advanced species would have easier ways of meeting such needs.[62]

Even if aliens do no seek to eradicate us, they could still limit our cosmic aspirations. In 1973, John Ball proposed the "zoo hypothesis," which argued that extraterrestrial aliens might have quarantined the Earth and followed a directive which proscribed intervention on our planet.[63] Yet another possibility is that aliens might covet humans for entertainment purposes, perhaps even keeping us as pets.[64]

Others, however, are more sanguine about alien intentions. Despite the dire warnings, as Gerard O'Neil pointed out, that fact that Earth has yet to be attacked suggests that nobody out there is hostile to us.[65] Furthermore, optimists contend that the vast distances between solar systems insulate our planet. Frank Drake speculated that advanced civilizations would achieve medical breakthroughs that would drastically lengthen individual life spans, thus rendering them immortal. As he notes, there is nothing in the chemistry of life that requires deterioration and death; rather, death has traditionally been nature's way of making room for the next generation. But through advanced technology and bio-engineering, death could be outsmarted. In a highly advanced society, death would come about only through the physical destruction of the organism either through accident or intentional killing. Such beings, Drake reasoned, would be extremely risk-averse and peaceful. They would feel a desire to spread their goodwill throughout the cosmos and, as a consequence, would probably be very active in detecting and communicating with other intelligent civilizations.[66] It seems reasonable that members of an advanced civilization would be interested in encountering other intelligent sentient beings. Jill Tarter, an astronomer and director of the Center for SETI, speculated that "rather than exploiting us, they might value and support the natural biodiversity of the galaxy."[67]

At our current technological development, we would not be a threat to a Type II or Type III civilization. Moreover, even a military invasion could be problematic for an advanced extraterrestrial civilization. If interstellar travel is exceedingly difficult, then interstellar invasion would even be more challenging.[68] Keep in mind that Operation Overlord, the Allied invasion at Normandy in World War II—a mere 100 miles across the English Channel—required considerable planning and was not launched until the Allies had already attained near complete superiority in the air and in the sea. The Fermi paradox suggests that there would be a lot of empty real estate between civilizations, thus alleviating any kind of galactic impulse for *Lebensraum*.[69] Sagan once argued that it would be more practical for alien civilizations to colonize their own planetary system, rather than endure the costs and hazards of going to other stars.[70] If extraterrestrial civilizations do indeed exist, their aspirations could differ markedly from our own. Nevertheless, as Robert Forward pointed out, even for technologically advanced civilizations, interstellar voyages would probably be justified only for major purposes.[71] From the perspective of representatives of an alien civilization, plundering the Earth for its resources would be neither practical nor desirable.

Historical analogies, Carl Sagan once argued, were probably inapplicable to alien and human contact as he found it unlikely that we would face "colonial barbarity" from a technologically more advanced alien civilization. Quarrelsome extraterrestrials, he reasoned, would probably not last long in interstellar space because they would be eliminated by a more powerful species. According to this line of reasoning, those civilizations that lived long enough to perform significant colonization of the Galaxy would be those least likely to engage in aggressive galactic imperialism.[72] Moreover, extraterrestrials visiting Earth would assuredly be far more technologically advanced than we, so that the former would have nothing to fear from us. And Earth would not likely have anything that they would need.[73]

It is probable that an advanced civilization with a high degree of technical capacity would be based on a high degree of cooperation.[74] Those societies whose members have learned to cooperate have a decided advantage over those that do not. For example, efforts that require collective action, such as defense, require cooperation. But even if an advanced civilization has left war behind, it may not be magnanimous to civilizations elsewhere. As Paul Seabright warned, cooperation within a group can make that group more aggressive in its dealings with outsiders.[75]

If trends on Earth, however, are applicable on other planets, they could augur well for the prospects of interstellar comity. Albert A. Harrison suggests that the ascendance of liberal democracy on Earth could have taken hold among other civilizations in the cosmos.[76] As Francis Fukuyama observed in his 1989 article, "The End of History?," governments around the world were converging on a model consisting of democracy and free markets. According to his analysis, all other ideologies had been effectively exhausted and discredited, thus no credible alternatives remained. Supposedly, all that was left was mere fine-tuning.[77] The end of grand ideological struggles would remove a major cause for war. Research has demonstrated that democracies are unlikely to wage war against one another. Hence, democratization has arguably made our planet a more peaceful place.[78] Likewise, democratization on other planets could by extension make the cosmos more peaceful as well. If extraterrestrial civilizations are democratic, then it could expand the "zone of galactic peace."[79] As civilizations advance, they might develop cooperative tendencies as part of the evolutionary process.[80] Nikolai Kardashev even speculated that alien civilizations might find it advantageous to unite with others to reduce the time delay for communications and other types of activities.[81] On that note, Ronald N.

Bracewell conjectured that the chief commodity of interstellar commerce and exchange would be information.[82]

Could our civilization one day be invited to join a galactic club, not unlike the United Nations on Earth? Ronald N. Bracewell envisaged such a scenario in which members would find developing communities such as Earth and introduce them into the galactic community. Each member would have its own sphere of influence and maintain an ongoing program of launching messenger probes in an endeavor to comb the unexplored frontiers for new technological societies.[83] Some commentators have even speculated on the form that "galactic law" might take to regulate relations between extraterrestrial civilizations.

Adam Korbitz introduced the term *celegistics* to refer to theories of legal systems involving extraterrestrial intelligence.[84] In this framework, Andrew G. Haley coined the term *Metalaw*, referring to fundamental legal precepts that would theoretically apply to all intelligent beings, including human and extraterrestrial. As Haley explained, this Metalaw should be based on an International Golden Rule—that is, Do unto others as they would have you do unto them.[85] Building on this framework, the Austrian space law pioneer Ernst Fasan proposed the following rank order of 11 metalegal principles, starting with the strongest first:

1. No partner of Metalaw may demand an impossibility of another.
2. No rule of Metalaw must be complied with when compliance would result in the practical suicide of the obligated race.
3. All intelligent races of the universe have in principle equal rights and values.
4. Every partner of Metalaw has the right of self-determination.
5. Any act which causes harm to another race must be avoided.
6. Every race is entitled to its own living space.
7. Every race has the right to defend itself against any harmful act performed by another race.
8. The principle of preserving one race has priority over the development of another race.
9. In case of damage, the damager must restore the integrity of the damaged party.

10. Metalegal agreements and treaties must be kept.
11. To help the other race by one's own activities is not a legal but a basic ethical principle.[86]

Fasan later distilled a simpler three-part formula from the above principles:

1. A prohibition on damaging another race.
2. The right of a race to self-defense.
3. The right to adequate living space.[87]

Even though alien visitors may "come in peace," they could still have a disruptive effect on Earth. For instance, their presence would undercut the uniqueness of man, thus posing a major theological challenge to our terrestrial religions. With regard to politics, how might the people on Earth view the viability and legitimacy of their governments when far more intelligent and technologically sophisticated aliens are in their midst? The introduction of new technology could upend industries on which many communities depend. Moreover, there could be a struggle among governments on Earth to harness these new technologies for their own national interests.[88] Roberto Pinotti conjectured that one possible response to alien contact would be a magnified ethnocentrism on Earth. Some people may seek to preserve their identity by espousing it in an exaggerated form to survive a confrontation with alien life. By doing so, however, it could lead to the fragmentation of multicultural societies, causing significant authority problems for federal and supranational governments.[89]

Conversely, the contemporary ideology of multiculturalism might be expanded to encompass the extraterrestrial visitors. By doing so, Earth's civilization may one day amalgamate with an extraterrestrial culture. On that note, Carl Sagan presaged that the less advanced civilization would eventually cease its independent historical development and merge with the more advanced civilization. As a consequence, the number of communicating extraterrestrial civilizations would increase over time.[90]

Why would aliens visit us? In a word, curiosity. As Frank Drake opined, "Many human societies developed science independently through a combination of curiosity and trying to create a better life, and I think those same motivations would exist in other [alien] creatures."[91] Presumably, members of an extraterrestrial civilization capable of reaching Earth would be highly curious—a quality that would be necessary to practice the science that led to their advanced technological development.[92] On Earth, human scientists are intensely interested in lower life forms, such as insects; it would logically seem to follow that aliens would be interested in other life forms as well.[93] Certainly, an alien civilization traveling to Earth would have knowledge of physics that far exceeded our own. Therefore, the aliens might be more interested in areas such as ethics, religion, and art.[94] One reason they might visit Earth is to practice a form of cosmological anthropology.[95]

The record of UFO sightings correspond to the curiosity motivation. According to the alleged eyewitness accounts, UFOs are usually stealthy, avoiding ostentatious displays in favor of short engagements followed by quick disappearances. Likewise, the abduction narratives follow a similar pattern in which subjects are taken in isolation from bystanders and released after a relatively short encounter. Admittedly, if alien encounters were otherwise, it would provide the long elusive smoking gun evidence of their existence. After all, an extraterrestrial spacecraft landing on the White House lawn would settle the issue one and for all. Still, it is noteworthy that UFO sightings would seem to be consistent with the curiosity motivation.

To prepare for the eventuality of alien contact, we must consider alien motives. To that end, the former chairman of the Transmissions from Earth Working Group, Michael Michaud, recommended that the Committee on SETI be broadened beyond astronomers to include philosophers, historians, theologians, etc.[96] Based on his experience as a diplomat, Michaud sets three priorities for mankind with respect to contact with extraterrestrials. First, we must guarantee our safety and security in the event of any alien intrusion onto Earth. Second we need to develop, or gain entry into a stable interplanetary political system, such as the Galactic Club expounded on by Bracewell. Finally, by tapping into the extraterrestrials' immense wisdom, we should seek scientific, technological, and other information that will help us advance our civilization.[97]

Although the prospect of aliens visiting Earth may seem far-fetched, it would be helpful to have some protocol in place in the event that such a situation ever comes to pass. For if representatives of an extraterrestrial civilization do ever intend to visit Earth, it could be problematic for them to land and move about. There would be the issue of the invasion of airspace, which could occasion a response from a nation's air defenses. And even if the visiting aliens were peace-loving, it would behoove them to bring with them some sort of suitable weaponry in case they need to defend themselves.[98] Consequently, there is the risk of misinterpreting signals, which could lead to a violent exchange between the two parties. By their nature, bureaucracies may be ill-suited to respond to the prospect of contact with alien civilizations in that bureaucratic decisions tend to be based on precedent and operations that follow standard procedures.[99] It may take quite a bit of forward-thinking to plan for such speculative diplomacy to deal with this type of scenario.

New hopes of life out there in the cosmos emerged in October 2010, when astronomers discovered an exoplanet in the star system Gliese 581, 20 light years away. Believed to be of similar size to Earth, this exoplanet might reside in a habitable zone in its solar system.[100] Contact with intelligent sentient beings would be of monumental significance and could have numerous scientific, political, and theological implications.[101] Understandably, based on certain episodes of human history, some people view this proposition with trepidation. Despite the potential dangers, more than likely, we need not worry. Because the Milky Way is vast, containing billions of planets and stars, even if advanced extraterrestrial intelligence is out there, they may never get around to visiting our planet. To initiate first contact, we may have to venture out into the cosmos as detailed in the next chapter.

Chapter 13

Exploring the Cosmos

Considering the multitude of stars and planets in the galaxy, SETI scientists reasonably assume that there is a strong likelihood that there could be many extraterrestrial civilizations far more advanced than ours on Earth. Perhaps this is so, but we have no concrete evidence yet for this assumption. To establish first contact with alien life forms, humans may have to journey beyond our solar system. The quest to explore might originally be taken out of curiosity and a cosmic sense of wanderlust, but someday it will be necessary to secure the existence and future of our species. As Arthur C. Clarke once exclaimed, it is dangerous for the human race to have all of its eggs in one basket—Earth.[1] As our sun becomes hotter, its habitable zone will expand, thus eventually making the Earth unsuitable for life. Consequentlty, space exploration is not just idle speculation and wishful thinking, but ultimately necessary to ensure the long-term survival of our species.

The time to start thinking about interstellar travel is now, for once a culture turns inward it is often difficult to look outward again. As the noted molecular biologist Gunther S. Stent pointed out, a particular human characteristic, what Nietzsche called the "will to power," will be necessary to impel man to explore the cosmos.[2] This Faustian trait animated civilizations to explore and colonize distant lands, but was not constant throughout their histories. For instance, the ancient Polynesians were great seafarers and explorer-settlers spreading throughout the area of the South Seas; however, by the time the European explorers arrived, the Polynesians had lost their drive to expand. Likewise, at its apogee, the Ming Dynasty in China of the fifteenth century sent out an expeditionary fleet as far as the East Coast of Africa and perhaps even to America. But after this initial burst of exploratory exuberance, the Ming Dynasty turned inward and was subsequently conquered by the Tungus in the mid-seventeenth century.[3]

There are several important reasons why we should undertake interstellar travel. As Robert Forward pointed out, first, there is the mountain climber's motive—we go to the stars and other planets because they are there. More practically, there is the survival of the species motive. Eventually, our sun will leave its main sequence and by doing so, render the Earth uninhabitable. And finally, the discovery of life beyond Earth would provide biologists and physicians with new research opportunities that would stretch the theories of life, possibly leading to new products and treatments.[4]

All of these reasons to explore are so easy to say. Even when extrapolating contemporary technology well into the future, interstellar travel would seem to be monumentally challenging. It was not until the mid-nineteenth century that scientists began to realize the magnitude of interstellar distances.[5] Even sending a probe to Proxima Centauri, the nearest star, will require a long-term commitment comparable to medieval thinking, the sort "that built cathedrals in Chartres, Salisbury, and Cologne."[6] Interstellar travel is so demanding, there will be a tendency to delay the proposition until new technology comes on line to make it more feasible.[7] Joe Haldeman colorfully commented on this conundrum: "You start out on a centuries-long project, and before you've gone a tenth of the way, someone passes you with a faster, cheaper way to travel. By the time you get to Alpha Centauri, people are commuting from Earth by wormhole, and the virgin wilderness is all suburbs and shopping malls."[8] As Robert Forward counseled, if a mission takes too long—50 years or more—it is likely to be passed by a follow-on mission launched decades later. For that reason, those voyages estimated to take 50 years or longer, should not be undertaken; instead the funding should be invested in designing a better propulsion system.[9]

Before venturing outside of our solar system, we will first have to master interplanetary travel. So what should be our first destination? Some space engineers argue that the logical first step to reach the stars is establishing a permanent base on the Moon. Writing in the early 1970s, Carl Sagan predicted that semi-permanent bases would be established on the Moon by the 1980s, conjecturing that they would initially be resupplied with material and personnel from Earth, but over time, they would become self-sustaining by utilizing lunar resources.[10] A NASA study completed in 2000, reported that a colony could be created several feet beneath the Moon's surface or covered with an existing crater to protect its residents from the constant bombardment of high-energy cosmic radiation. The study envisaged an onsite nuclear power plant, solar panel arrays, and various methods for extracting carbon, silicon, aluminum and other useful materials from the lunar surface.[11] Conceivably, the Moon, with its less intense gravity, would be a good site for constructing spacecraft for the next major journey into the solar system.[12] Harvey Wichman noted that there are advantages to returning to the Moon, as it could serve as an important step toward exploratory missions to Mars and future missions to asteroids. The Moon would provide a good base on which to test equipment to be used for a mission to Mars. Using the analogy of the settlement of the American West, Wichman noted that the frontier was initially opened up by intrepid explorers. They were followed by trappers and hunters who found ways to exploit the resources on the new land. Next, settlements were established, which ultimately turned into permanent communities.[13] Likewise, a base on the moon would serve as a springboard for the colonization of the solar system.

Some space-settlement plans call for skipping the Moon entirely. Neil de Grasse Tyson has argued that using the Moon as a stepping stone toward Mars is unnecessary insofar as it would take energy for a spacecraft to slow down, land, and then take off again from this way station.[14] Likewise, Harley Thronson, the senior scientist for advanced concepts in the Astrophysics Science Division at NASA's Goddard Space Flight Center, has counseled that making a prolonged stopover to the moon runs counter to the spirit and history of exploration. Sending astronauts to "stepping stones"—whether the moon or nearby asteroids—could significantly delay a manned Mars effort, especially in a tight fiscal environment that is likely to persist for some time.[15]

To date, the Space Age reached its apogee on July 20, 1969, when the *Apollo* 11 crew landed on the moon. Mars seemed to be the next logical destination, but the age of exploration quickly gave rise to the age of experimentation, as both the U.S. and Russian space programs focused on low-orbit missions and stints on space stations. Well into its sixth decade, people wonder what achievements await the Space Age. The popular next choice would be a manned mission to Mars. New hope for reaching Mars came in 2012, when Dutch entrepreneur Bas Lansdorp announced his intention to begin establishing a human colony on the red planet in 2023. According to the plan, the intrepid pioneers will fly on a one-way trip to Mars, where they will begin building a permanent settlement. Every two years, a new set of four astronauts would join them. Funding would come in part from a global reality television program in which candidates for the expedition would be selected.[16]

Currently, Mars is a cold barren planet. With a thin atmosphere, the planet does not have a sufficient greenhouse effect to generate enough heat for a habitable environment. This is somewhat of a mystery, because the planet's topography suggests that rivers, lakes, and perhaps even oceans once flowed on the surface 4 billion years ago. There could be some natural instability in Mars' climate that if something was released in the atmosphere, it might return the planet to its ancient clement state. Carl Sagan speculated that the introduction of certain elements—including chlorofluorocarbons, ammonia, and carbon dioxide—could warm the planet so that it might be able to sustain life.[17] For Sagan, the first voyage of humans to Mars was the key step in transforming us into a multi-planet species.[18] Possibly, Mars could be terraformed using water from ice beneath the planet's surface. Alternatively, water could be imported from an ice asteroid so that a thin ocean, and later an atmosphere, could provide breathable air and a protective shield against cosmic radiation.[19] By increasing the surface temperature, it is conceivable that someday in the future the ice caps on Mars could melt, releasing both water and carbon dioxide so that the atmosphere would become thicker. A greenhouse effect could then take hold, making the planet suitable for surface life.[20] One ambitious plan called for detonating four fusion nuclear warheads on Mars in such a way as to penetrate its surface so that the explosion would disperse enough dust to initiate a runaway greenhouse effect. As a result, this could produce an atmosphere thick enough so that liquid water would exist. Then, once things settled, space settlers could walk around the surface of Mars with respirators rather than full space suits and plants could grow.[21] Eventually, colonies would be established and over a generation or two, the settlers would begin to develop a distinct Martian culture.[22]

What would be the next step after Mars? Although planets are the likely place for life to begin, Freeman Dyson argued that they would not be the only home of a large technological society. With their plentitude of biologically useful materials—including carbon, nitrogen, oxygen, and hydrogen—he believed that comets would be the more likely habitats of advanced alien civilizations.[23] As he noted, the combined surface area of the comets in our solar system is roughly 1,000 to 10,000 times that of the Earth. With the relative propinquity of comets, spacefaring humans could hop from comet to comet before seeking to traverse interstellar expanses. He presaged that biological engineering would enable man to create oases in space in the form of habitable comets. By using self-replicating von Neumann machines, we could send out teams of advanced builders that would create and prepare habitats for humans who would settle later.[24] But to escape our solar system, we will first need to develop advanced vehicles for exploration.

When considering interstellar travel, two important factors to keep in mind are the tremendous distances between solar systems and the enormous energy requirements that would be necessary to fuel space vessels that could traverse such distances. The most challenging task of a spaceship is to escape the pull of Earth's gravity. Traditionally, spacecraft have been launched by way of multistage rockets. Various components of the rocket carry fuel used to propel the craft into space. Once a component's fuel is exhausted, it is discarded, which lightens the load of the craft. Rockets using chemical propellants require a very large percentage of the craft to be dedicated to carrying the fuel payload. Much fuel is needed for lift off, as well as for slowing down. Conventional rocketry will not be adequate for interstellar distances. For viable interstellar travel, aeronautics must find a lightweight and efficient method of propulsion whose pound-for-pound punch would vastly exceed that of conventional chemical fuels.[25]

Numerous proposals have been put forward for the next generation of spaceships. Engines using ion gas could produce a sudden dramatic blast that could propel conventional rockets. A more powerful version would use plasma to propel the craft through space.[26] The so-called fourth state of matter (the others being solid, liquid, and gas), plasma forms when a gas is heated sufficiently so that electrons will break away from the nuclei of atoms, causing the electrons and nuclei to move independently. As plasma particles are electrically charged, they can be manipulated by magnetic traps to contain them without coming into contact with the walls of a combustion chamber. Once trapped, the plasma could be directed out of the rear of the spacecraft, thus serving as a propellant.[27]

Even more ambitious would be using nuclear power to propel space vessels. The Orion rocket, first proposed by Stanislaw Ulam at the Los Alamos National Laboratory in the late 1950s, planned to use plutonium for the fuel, but the propellant could be any material that could be vaporized between the exploding bombs and a pusher plate.[28] The original goal was to send a manned spacecraft to Mars and Venus, using the detonation of mini-nuclear bombs to propel a spaceship.[29] By ejecting mini-nuclear bombs out of the back of a rocket in sequence, the spacecraft would ride on the shockwaves created by the blasts.[30] The shock wave of the explosion would impinge on a "pusher plate" that would transfer the momentum to the spacecraft.[31] The project, however, was canceled after the United States and the Soviet Union signed the Limited Nuclear Test Ban Treaty in 1963.[32] Nevertheless, the idea of using nuclear power to fuel space vehicles continued.[33]

The first nuclear rocket engine—the Kiwi 1—was tested under NASA's Project Rover in 1959. In the 1960s NASA and the US Atomic Energy Agency created the Nuclear Engine for Rocket Vehicle Applications (NERVA), which was the first nuclear rocket to be tested vertically, rather than horizontally, but in 1972 the nuclear rocket program was closed. It was feared that a runaway nuclear reaction would explode the rocket.[34] Not to be deterred, in 1973 the British Interplanetary Society began a six-year study—Project Daedalus—

which called for an unmanned spaceship using a pulsed-fusion rocket engine. The plan was to assemble the spacecraft in Earth's orbit. It would then head for Jupiter, where the Jovian atmosphere would be skimmed for helium 3, a very rare isotope of helium. With its tanks filled with helium 3, the ship would proceed for Barnard's star—a red dwarf about 5.9 light years from the sun. The ship's engine would burn deuterium, a comparatively rare isotope of hydrogen, and helium 3. According to the proposal, a one-gram pellet of a frozen mixture of deuterium and helium 3 could be compressed and heated by high-electron beams. This would induce a fusion reaction, thus creating a "micro explosion."[35] If fully operational, it was hoped that the craft could travel at about 12 percent of the speed of light.[36]

In the coming years, safe nuclear reactors could enable nuclear rocket propulsion.[37] Theoretically, a spacecraft fueled by a nuclear reactor could fly for a long time without refueling, thus requiring far less weight to carry.[38] Moreover, space probes do not have to be powered by rockets for the duration of their flights; rather, they can coast for most of the journey. Even more effective then fission would be to use fusion, a process in which two very light nuclei of hydrogen and/or helium combine to release millions of electron volts of energy per fusion event. To date, man can only produce nuclear fusion with hydrogen bombs. However, if a fusion reactor could one day be developed, it would hold great potential for propelling interstellar vehicles. Hydrogen, which presumably would be the element used in a fusion reactor, accounts for more than 90 percent of the atoms in the universe. Therefore, astronauts could find this fuel source virtually anywhere along their journey through space.[39] A ramjet fusion engine would scoop hydrogen as the craft traveled, thus providing it with an inexhaustible source of fuel as hydrogen is abundant in the universe. Theoretically, the ramjet could propel a craft indefinitely, enabling it to reach distant stars in the galaxy.[40] Moreover, the ramjet would have the potential to reach ultrarelativistic speeds, since the faster it would travel, the more fuel it would gather.[41] However, as Frank Drake noted, the scoop part of the craft would have to be some two hundred miles wide to gather sufficient fuel in the near-vacuum of space. Also problematic is the fact that the atoms that the scoop encounters would strike with great force because of the very high speed of the spacecraft. Consequently, the atoms would more than likely bounce off of the scoop rather than fold into it.[42]

Currently, the National Ignition Facility at the Lawrence Livermore Laboratory is working on the development of nuclear fusion as a source of electrical power. Scientists are pursuing an approach in which isotopes of hydrogen, deuterium, and tritium are contained in a small glass bead, about 5 mm in diameter. The temperature and pressure of those gases would be raised for the values necessary for nuclear fusion to occur, which are of the order of 100 million degrees Celsius and 1 billion times normal atmospheric pressure. The main obstacle is to get the gases hot and dense enough before the inevitable explosion causes the gas to disperse before the fusion can commence. A solution to the problem is to heat the gas so quickly that there would be a confinement by the inertia of the gas itself. To accomplish this task, a very powerful pulsed laser would be required.[43]

Perhaps the power of the sun could one day be used to propel vehicles through space. As far back as 1611, Johannes Kepler wrote about this concept in his story *Somnium*, conjecturing that that it might be possible to somehow harness the solar wind to propel a spaceship.[44] In 1958, IBM's Richard Garwin published the first scientific paper in the English language that laid down the basic principles of solar sailing.[45] In his plan, a huge solar sail would capture the solar power of the Sun and other stars to fuel a spaceship.[46] One of the advantages of the solar sail concept is that sunlight remains plentiful as long as the spacecraft is within a reasonable distance of the sun. Another advantage of using the solar sail is that the spacecraft would not have to carry any fuel. Instead, the solar sail would be impelled by photons released from the sun.[47] Although photons are massless, they nevertheless exert a force on those objects on which they impinge. NASA began taking an interest in the solar sail concept back in the 1980s. The agency considered using a solar sail as a propulsion system for a spacecraft that would rendezvous with Halley's Comet.[48] As far back as 1977, the Jet Propulsion Laboratory developed a study—the Interstellar Precursor Mission—which examined what would be needed to explore the heliopause, the theoretical boundary where the Sun's wind effectively ends and the interstellar medium begins. Robert Forward came up with the bold idea of creating a space-borne lens the size of Texas that would focus the light from thousands of solar collectors. By doing so, he estimated that the spacecraft could be accelerated to one-tenth the speed of light.[49] Another variations of the solar sail concept known as the lightsail involves firing a beam of microwaves from the ground or space stations onto the sail as a means of propulsion.[50]

Anti-matter could have great potential to propel spacecraft (which incidentally was the method used for the *Enterprise* in *Star Trek*). Whereas nuclear explosions are only about 1 percent efficient in converting mass to energy, anti-matter yields a 100 percent conversion of mass to energy.[51] It was the physicist, Paul A.M. Dirac who first conceptualized anti-matter in 1928. Combining quantum theory with special relativity, he concluded that an anti-particle should exist for the electron. The "positron" would have the same mass as an electron, but host a positive charge. The reversal of normal matter, anti-matter has an opposite charge and spin. When brought into contact, matter and anti-matter completely annihilate each other, thus producing pure energy consistent with Einstein's famous formula $E = mc^2$.[52] In 1932, Carl Anderson discovered the anti-particle in his laboratory at the California Institute of Technology. Anti-matter occurs only very rarely in nature, however, and only under the most extraordinary circumstances. To acquire anti-matter in sufficient quantity, it must be artificially created.[53] In September 2002, scientists at CERN's antiproton accelerator facility in Switzerland announced that they had created for the first time a large number of antihydrogen atoms.[54] And in 2010, CERN scientists trapped atoms of anti-matter for the first time in history. The potential implications of this achievement are significant. As a propulsion fuel, an anti-matter reaction would carry 1,000 times the energy of a nuclear fission reaction and 100 times that of fusion.[55] Two major difficulties in using anti-matter, however, are making enough of it to feasible for long-duration missions and finding ways of storing the material and then channeling the energy into a practical rocket engine.[56]

During the 1950s, the U.S. aerospace industry seriously considered the notion of anti-gravity propulsion systems. The idea behind this method is to artificially produce "a matter-attracting, gravity potential well just beyond" the spaceship's bow so that this force would tug the vehicle forward "just as if a very massive, planet-sized body had been placed ahead of it.[57] An American physicist, Thomas Townsend Brown (1905–1985), was a pioneering researcher in the field. He postulated that the most efficient shape for an anti-gravity vehicle would be a saucer. In 1948, a businessman, Robert Babson, founded the Gravity Research Foundation to develop methods to reduce the effects of gravity.[58] And during the 1950s, the aerospace industry in the United States and Canada conducted research in the area of anti-gravity, most notably, the Glen Martin Aircraft Company (which later became Martin-Marietta in 1961).[59] But with a lack of demonstrable progress, interest in anti-gravity research waned. Moreover, the Mansfield Amendment of 1973 restricted Department of Defense spending to those areas of scientific research with direct military applications approved by the Defense Advanced Research Projects Agency. As a result, U.S. government support for anti-gravity research projects were terminated.[60] But from 1996 through 2002, NASA provided funding to the Breakthrough Propulsion Physics program, which studied a number of hypothetical designs for space propulsion that were not receiving support through normal commercial and university channels. Under the name "diametric drive," various anti-gravity-like concepts were investigated. The work of the Breakthrough Physics program continues under the independent Tau Zero Foundation, which is not affiliated with NASA.[61] A prominent Princeton physicist, Raymond Y. Chiao, hopes to one day to convert gravity into both electricity and light and vice versa.[62] Conceivably, this conversion could be applied to interstellar spacecraft.

Even more exotic, zero-point energy might someday be harnessed to fuel spacecraft. Although far beyond our current capabilities, perhaps one day a spaceship could mine the quantum vacuum for energy and acquire fuel throughout its journey. According to quantum theory, a region of space devoid of matter and light will still contain some energy. Scientific tests confirmed that even in the depths of a vacuum chilled to absolute zero (-273.5°C), energy would not go away.[63] According to some speculations, zero-point energy could be extracted because in a quantum vacuum, fleeting particles continually pop in and out of existence, billions of times every second, on every conceivable frequency, and in every possible direction. If so, it might be possible for electronic circuits to extract this virtually unending supply of vacuum energy.[64]

Understandably, some scientists are highly skeptical of the prospects of using zero-point energy in any practical sense.[65]

Even if the preceding methods could be implemented, interstellar travel at high speeds could be perilous. The protons, elections, and occasional atoms that suffuse the cosmos would normally not present a hazard, but even a collision with a small object the size of a grain of sand would cause a massive explosion when traveling near the speed of light. One proposal would be to place a shield of dust in the form of a cloud ahead of the spacecraft. Any objects passing through the cloud would be heated and vaporized before they could damage the spacecraft.[66]

One way to get around these hazards would be to travel as pure energy. Stretching the bounds of physics, Isaac Asimov once speculated that it might be possible to teleport, that is, convert all the particles of mass in a ship, including the crew and passengers, into photons of different types. The photons could then be sent at the speed of light without the necessity of acceleration and without the normal expenditure required to achieve this velocity. Once the photons arrived at the desired location, they could be converted into the original particles. Although, this travel could still not exceed the speed of light, the travelers would not age insofar as they travel at this speed as consistent with Einstein's special theory of relativity. However, reconstituting the photons back to the original structures would be a daunting challenge to say the least.[67]

To successfully teleport a person, one would have to know the precise location of every atom in the living body, which would probably violate the Heisenberg uncertainty principle. According to that principle, one cannot simultaneously know the exact velocity and position of a subatomic particle. The very act of observation influences the properties of the particle. Nevertheless, research scientists have made progress in the area of teleportation. For instance, in 1993 scientists at IBM led by Charles Bennett demonstrated that it was possible to teleport objects at the atomic level. This was followed in 1997 by the first demonstration of quantum teleportation, in which photons of ultraviolet light were teleported at the University of Innsbruck. The next year, experimenters at Cal Tech conducted an even more precise experiment that involved teleporting photons. Then, in 2004, quantum teleportation was carried out with actual atoms. And in 2006, for the first time, a macroscopic object was successfully teleported. Physicists at the Niels Bohr Institute in Copenhagen and the Max Planck Institute in Germany successfully entangled a beam of light with a gas of cesium atoms, an experiment that involved trillions upon trillions of atoms. Finally, in 2007 another breakthrough was made when scientists at the Australian Research Council developed a teleportation method that does not require quantum entanglement. Although there is a long way to go before teleporting humans through space is possible, much progress has been made and, possibly within a few decades, scientists may be able to teleport the first DNA molecule and virus.[68] In fact, in September 2013, scientists at the Harvard-MIT Center for Ultracold Atoms created a special type of medium in which the photons interacted with each other so strongly that they binded together to form molecules—a state of matter![69]

In 1994, the Jet Propulsion Laboratory in Pasadena hosted an Advanced Quantum/Relativity Workshop that mulled over the prospect of faster-than-light (super-luminal) space travel. Among the approaches considered were using cosmic wormholes and interacting with another dimension to manipulate the fabric of space and time.[70] According to Einstein's Special Theory of Relativity, which he propounded in 1905, practical interstellar travel is seemingly impossible because no vessel could travel faster than the speed of light. Consequently, at sub-luminal speeds, it would take centuries or millennia to travel the distances between solar systems. However, Einstein later superseded his own theory in 1915 with the General Theory of Relativity, in which faster than the speed of light travel is theoretically possible. For instance, one could use large amounts of energy to continuously stretch space and time. The fabric of space could be compressed so that the spacecraft could travel though the medium at a sub-luminal speed, but arrive at a point and time in space comparable to having traveled at a super-luminal speed. Inasmuch as only empty space is contracting or expanding, one could exceed the speed of light by this fashion.[71] Planning for this future, Miguel Alcubierre, a theoretical physicist in Mexico, advanced the concept of using a warp drive to get around Einstein's speed of light barrier. Using the "Alcubierre drive," space-time would be distorted and expanded behind a spacecraft while space-time would be contracted in front of it. By doing so, the spacecraft could ride on an expanding fabric of space-time not unlike a surfer riding a wave. Like a wave, the motion of the underlying substance—in this instance, the fabric of space-time—would move the vehicle.[72] Working off of this concept, in June 2006, Harold "Sonny" White at NASA's Johnson Space Center revealed a model for a warp drive spacecraft that could reach Alpha Centuri in two weeks. By bending the space around it shorter, the spacecraft would technically not be traveling faster than the speed of light, but would nevertheless attain super-luminal travel.[73] Although the Alcubierre drive would not technically violate the laws of physics, the energy required for its operation would be more than what is available in the entire universe.[74]

Another way to avoid the problem of the speed of light—the cosmic speed limit—would be to use a wormhole as a shortcut across space and time. Einstein and Nathan Rosen once speculated that a so-called "Einstein-Rosen bridge" could connect two universes, thus creating a shortcut through time and space.[75] A functioning wormhole could serve as a bridge between two different regions of the universe. One could enter one end of the hole and emerge out of the other hole moments later in a place thousands of light years from the starting point.[76] Carl Sagan depicted this device as a means of interstellar travel that was featured in his novel *Contact*.[77] Likewise, in the film *Stargate*, a wormhole was used to transport an Egyptologist and his crew to another planet, which was ruled by a technologically superior tyrant. A stargate could consist of a series of wormholes connected at many points in the universe, not unlike an old-fashioned telephone switchboard used to shunt messages from one caller to another.[78] To date, no one knows for sure if wormholes exist, but if they do, then it might be possible to tear the very fabric of space and time.[79]

Theoretically, one could create a wormhole by compressing an object so that it becomes smaller than its "event horizon." Kip Thorne once theorized that traversable wormholes could be created by using negative energy that would hold open the throat of a wormhole long enough for astronauts to pass safely from one universe to another.[80] The Russian physicist, Sergei Krasnikov, once demonstrated that a certain class of wormhole could be created using positive mass-energy matter.[81] The energy requirements for such a system, though, would be, in a word, massive, and well beyond the scope of our capabilities.

According to Stephen Hawking, interstellar travel via black holes would probably not be feasible. As he points out, the mere presence of a spaceship would probably destroy the passage leading from the black hole to the white hole (a hypothetical region in which nothing can enter, but from which matter and energy can escape). He cautions that if an astronaut attempted to travel by way of this method, he would not emerge looking the same as he entered.[82] If an astronaut fell into a black hole, his mass would increase, but eventually the energy equivalent of his mass would be returned to the universe in the form of radiation. As a result, the astronaut would be recycled into energy.[83]

For years, physicists assumed that anyone unfortunate to enter a black hole would be obliterated. However, the New Zealand mathematician Roy Kerr conjectured that the centrifugal forces of spinning black holes could counteract the intense gravity of the event horizon. Theoretically, if the ring at the center of the black hole was large enough, someone could journey through it and enter a parallel world or another area of the universe safely.[84] But even if a space traveler could avoid being ripped apart by the strong gravitational forces, how, Hawking asks, could one determine the time and location of the universe after the exit from the black hole?[85] For travel by way of wormholes to be viable, there are many details that need to be worked out.

Perhaps new breakthroughs in the area of particle physics will provide the requisite energy to make interstellar travel feasible someday. The Higgs boson, also referred to as the "God particle," interacts with other subatomic particles is such a way to slow them down, thus creating inertia.[86] In 1964, Peter Higgs of the University of Edinburgh theorized that that the Higgs boson was responsible for providing the mass that interacts with electromagnetic and nuclear forces. The much-heralded discovery of the Higgs boson could revolutionize space travel in the future.[87] Conceivably, objects could be "un-massed" or huge items could be launched into space by "switching off" the Higgs, thus making light-speed travel possible.[88]

Besides thinking through the tremendous requirements necessary to carry vehicles interstellar distances, the human component must be considered as well. A mission to Mars and beyond will test the physical and mental endurance of astronauts. Furthermore, humans are difficult and expensive to send into space. For that reason, the first spacefarers were not humans, but test animals. Animal subjects played a vital role in the early space program. For the initial space flights, piloting duties were light to nonexistent. The first space capsules were more akin to bullets than rockets, so nonhumans could be used for these missions. Whereas NASA launched monkeys and chimpanzees on its rockets because primates were physiologically similar to humans, the Soviet program preferred to send dogs into space, believing that canines were less excitable. A dog named "Laika" became the first living creature to orbit the Earth, but alas, there was no plan or means to bring her safely back home. The first of the great apes to be launched into outer space—Ham the Astrochimp—was more fortunate. On his successful return to Earth, he was feted as a national hero, appearing on the cover of *Life* magazine in his flight suit. Some of his fans even sent their copies of *Life* with requests for Ham's "autograph," which consisted of an ink print of his hand. On a practical level, his lever-pushing performance demonstrated that such tasks could be carried out in space. To beat the Soviets to the Moon, NASA once considered sending a chimp on a one-way mission; however, even if successful, the agency would have had to deal with the public relations fiasco of a dead chimpanzee hero.[89]

Everything we take for granted on Earth must be rethought, relearned, and rehearsed when venturing into space. Humans have evolved for life on Earth, and they are not accustomed to living in space. For prolonged periods, weightlessness is problematical for astronauts. Initially, weightlessness can be an exhilarating experience; however, the novelty quickly wears off and floating astronauts soon dream of walking.[90] Astronauts often experience motion sickness, which can lead to vomiting—a problematic situation in a gravity-free environment. In fact, NASA space helmets are built with air channels that directly flow down over the face a six cubic feet per minute, so that vomit will blow down away from the face and into the body of the suit.[91]

Living in a near-zero gravity situation for a protracted period can have serious health consequences. Currently, the practical limit on weightlessness is about six months. Weightlessness affects circulation resulting in increased blood volume in the upper half of the body, which also happens to be where the body's blood sensors are located. When astronauts are in space, the sensors misinterpret this effect as a surplus of blood, which causes the body to reduce its production. As a result, astronauts in space make do with 10 to 15 percent less blood than they do on Earth.[92] The lack of the usual gravitational force in space can also affect different brain mechanisms.[93]

The most serious consequence of weightlessness, however, is that bones thin and muscles atrophy. Although muscle mass is regained in a matter of weeks, bones take three to six months to recover. In fact, some studies suggest that the skeletons of astronauts of long-duration missions never quite recover.[94] Even with a rigorous exercise regimen, after a year on the space station, Russian cosmonauts experienced significant atrophy in their bones and muscles. When they first returned to Earth, they could barely crawl like babies.[95] According to the research of Dennis Carter of Stanford University, astronauts on a two-year mission to Mars could expect to eventually lose from one-third to one-half of their bone mass in their lower bodies—a situation not unlike a paraplegic. An astronaut returning from a mission to Mars could very well risk snapping a bone when stepping out of the capsule into Earth's gravity.[96] Not only must astronauts acclimate to weightlessness, they must also deal with high g-forces of acceleration associated during launch and deceleration during reentry. One method to facilitate long-duration space travel would be to produce centrifuge gravity in the space vessel. This method was used in the rotating spaceship that was featured in the film *2001: A Space Odyssey*.[97]

Provisioning space crews has always been challenging. The first space food was not unlike baby food; astronauts had to suckle their meals from tubes. Carbonated drinks are not feasible in space because the bubbles rise to the surface; instead, there is only a foamy froth. In recent years, the quality of the space food has improved, as meals no longer have to be compressed or dehydrated.[98] Maintaining good body hygiene is difficult in space. Showers are unavailable; instead, astronauts wipe themselves with moistened towels and rinseless shampoo. Although clothes are normally good absorbers of sebum (the oil produced by glands) and sweat, when astronauts are unable to change garments for long durations, the garments begin to decompose.[99] Excretion can be difficult because in a weightless environment, fecal material never becomes heavy enough to break away and drop down on its own.[100]

The psychological aspects of space travel will loom larger as astronauts spend more time with one another with greater autonomy from Mission Control. From its inception, psychological screening played an important role in the American space program. Early astronauts were chosen for their "right stuff"—namely, their temerity and charisma. Initially, NASA intended to recruit astronauts from a wide variety of backgrounds, including military and commercial aviators, mountain climbers, polar explorers, and deep sea submersible operators, but with strong pressure from the White House, the candidate pool was limited to military test pilots who had already demonstrated their skills and temerity.[101] Intelligence, motivation, physical fitness, good decision-making skills, high tolerance for stress, and emotional maturity were the necessary traits for the demanding job of an astronaut.

The early space psychologists did not worry much about astronauts getting along with each other on flights, as their missions lasted only a few hours, or at most, a few days. Furthermore, the first astronauts flew solo. The overriding concern at the time was what effect the silent, black vacuum of space would have on the spacefarers. Claustrophobia and solitude were the two most salient conditions on the minds of space psychologists.[102] But there were salutogenic (enhancing mental and physical wellbeing) effects as well. Some astronauts felt the experience in space euphoric, commenting that they felt nearer to God. NASA was concerned not so much about euphoria itself, but whether this elation could overtake good sense.[103] On a long mission to a distant destination it could be difficult to anticipate how much astronauts will miss the Earth when they are deprived of it for an extended period of time. The "Earth-out-of-view" problem could arise with astronauts on a trip to Mars, in which the home planet will be reduced to an insignificant-looking dot in space.[104]

From the earliest days of the space program, the astronauts were looked upon as exemplars of their respective societies. Assumed to be rugged individualists, the early astronauts could not depend on anyone but themselves and a small crew. As the U.S. space program progressed, interest in psychology waned. After five Mercury flights without serious performance deficits, NASA officials decided that there was no pressing need to continue exhaustive psychological testing procedures. Inasmuch as NASA managers were preoccupied with the scientific and engineering facets of spaceflight, behavioral health received short shrift. Therefore, psychiatrists and psychologists played only a minimal role in the selection process of astronauts from Gemini until well into the early Shuttle missions.[105] The faintest possibility that a space mission would be compromised by psychological factors could be a public relations disaster for NASA. The stereotype of the "right stuff" deterred snooping and prying that might suggest a real or imagined blemish that could lead to mission disqualification—a most dreaded result for an astronaut. Even the American press was loath to report negatively on any aspects of the lives of astronauts. Rather, the media sought confirmation that the astronauts embodied America's deepest virtues.[106]

But by the start of the twenty-first century, cracks began to appear in this image. Research suggested that astronauts suffered from a number of maladies as a result of their missions. There was concern that NASA, which has historically downplayed psychological problems, was unlikely to spend much time and resources to investigate those issues.[107] Albert A. Harrison and Edna Fiedler found that when adjusting to their lives back on Earth, some astronauts reported depression, substance abuse problems, marital discord, and jealousy.[108] Psychological issues could no longer be ignored. Not only could unresolved mental problems jeopardize space missions, they could result in a public relations disaster. As NASA funding depends in large part on public perceptions, the space agency could not risk such an eventuality.

Several developments rekindled interest in space psychology. The first missions into space were of short duration, lasting only hours or days, and the crews were small. But since the end of the Apollo program, the trend has been in the direction of larger crew sizes, greater crew diversity, and longer mission duration. Two particular factors in particular that will compound an astronaut's stress are living in a confined environment for a protracted period of time and the inability to escape it. Current estimates suggest that a round trip mission to Mars along with a scientific expedition would take roughly two and a half years. To date, astronauts have not experienced missions of such long duration. Using the analogy of a runner, the same rapid pace that can be sustained for brief sprints cannot be sustained for marathons.

In the contemporary Space Age, crews are larger, more diverse, and spend more time in space. This has changed the group dynamics on board spaceflights. What was the "right stuff" for the Mercury era could now very well be the "wrong stuff" for long duration missions. Increasingly, astronauts have to be persons who get along well with others.[109] In the early years of spaceflight, the most pressing issues concerned life support, the human-machine interface, and the optimization of human performance to ensure mission success. But gradually, a shift from a purely "displays and knobs" orientation to a more holistic approach, which involved project managers, engineers, and behavioral researchers, began to take hold.[110]

The Space Age commenced as a rivalry between the two superpowers, but since the end of the Cold War, crewed flights have transitioned from fiercely competitive national space programs to collaborative efforts with international crews. Over time, space exploration took on a more multinational character, as the United States and Russia included crewmembers from their respective allies, and eventually, from each other, in their missions. Despite national rivalries, the quest for knowledge at times motivated enemies to work together in space exploration.[111] For example, the two countries collaborated on the Apollo-Soyuz Test Project. On July 17, 1975, an Apollo Command and Service Module docked with a Soviet Soyuz spacecraft.[112] As the Cold War was winding down, Russian cosmonauts rode in Space Shuttles and American astronauts made trips to the Salyut space station.[113] At first, the inclusion of international crewmembers was primarily a propaganda move. Both superpowers offered room and board in the space capsules to citizens of their respective blocs. In the early international missions, a landlord-tenant relationship existed between the Soviets/Russians and Americans who controlled access to space. The foreign spacefarers were granted only limited access to their host's spacecraft. But gradually, this arrangement shifted to a more collaborative enterprise. In 1984, then-President Ronald Reagan approved another U.S. space station to replace *Skylab*, but its construction was delayed for almost 15 years. In 1998, then-President Bill Clinton finally decided to cast the project as a truly worldwide venture, which became the International Space Station.[114] This change in orientation resulted in more inclusive crews representing a greater number of nations. International crewmembers bring with them different cultures that can impact interpersonal relations. For example, studies suggest that whereas Russian culture values collectivism, hierarchy, and paternalism, American culture favors individualism, egalitarianism, and autonomy.[115]

Despite potential friction, Juris G. Draguns and Albert A. Harrison make the case that international flights make good sense for a variety of reason. First, drawing from an international pool allows space managers to select astronauts from a broad range of interests and skills. Second, inasmuch as space missions are overarching and superordinate endeavors that encourage nations to work together, they could serve as a prototype for other collaborative ventures. Third, the more nations that participate in a mission, the greater the number of people can identify (if only vicariously) with the challenge and triumphs of spaceflight. Finally, international missions can reduce duplication of effort, thus defraying the enormous costs of spaceflight.[116]

The experiences of cosmonauts and astronauts on both the *Mir* and the International Space Station have demonstrated that crews from a variety of different national and ethnic groups can work effectively together. But as Harvey Wichman notes, to date, these missions have taken place in quasi-military structures. We are now on the cusp on a new age of space flight that will allow tourists to visit space. In fact, Bigelow Aerospace in Las Vegas has successfully launched two inflatable habitats that orbit the Earth.[117] The firm intends to someday lease the station for research, industrial testing, and space tourism.[118] Space tourism will vastly broaden the spectrum of participants in the space program. This historic shift in space exploration will occasion a reconsideration in the way engineers, designers, and flight managers approach their tasks.[119]

For better or for worse, as spacefarers travel farther and farther away from Earth, they will work under conditions of greater autonomy from Mission Control. Communications between the crew and ground control will become increasingly delayed. For example, depending on where they are in their respective orbits, the time required for round-trip electronic transmissions between Earth and Mars would range from six to forty-four minutes.[120] Astronauts en route to Mars will not be able to rely upon automated life-support systems or short-term rescue possibilities in the case of emergencies. Consequently, autonomy will surely increase.[121]

Greater autonomy could lead to possible breakdowns in interpersonal interactions both within the crew and with Mission Control. There is a danger that crewmembers may increasingly view ground personnel as an out-group, leading to mutual tension and misunderstanding. On occasion, conflicts have emerged between the flight crew and the ground control. Such was the case in November 1973, when after complaining of being pushed too hard, crewmembers of the *Skylab 4* staged a protest by refusing to work for a day. After this episode, the workload schedule was modified, and by the end of the mission, the crew had actually completed even more work that what had originally been planned before launch.[122] To mitigate tensions, Nick Kanas recommends periodic "bull sessions" whereby crewmembers and Mission Control can address issues before they begin to fester. He counsels that it is important to deal effectively with tensions that are sure to arise in the stressful environment of space. Group sensitivity training involving both crew and ground personnel could reduce the influence of personal, cultural, national, and other peculiarities during the mission.[123]

It is worth mentioning that increased autonomy can have some positive effects as well. Exploring pilot studies from numerous space-analogous settings,[124] Kanas and his research team found that greater autonomy was associated with increased creativity, improved performance, and higher mood and morale.[125] Likewise, Peter Roma and his colleagues concluded that greater autonomy was associated with enhanced performance and fewer negative emotional states.[126]

New breakthroughs in nanotechnology and computers could someday enable long-range space missions. But astronauts who embark on long duration missions will face many perils. For instance, astronauts would encounter increased level of cosmic radiation. They could not rely on the Earth's Van Allen Radiation Belts for protection against solar flares and cosmic rays. Furthermore, the Earth's ozone layer protects us from ultraviolet rays.[127] Using current technology, a round trip to Mars would require 24 months. With 6 months on station, the journey would take 30 months. A round trip to Saturn without any time on a station would take 14 years.[128] This will create a situation in which crewmembers experience an acute sense of isolation and separation from

Earth. The crew will have only infrequent supplies, and it will be impossible to evacuate to Earth for emergencies.[129] For such long trips, candidates might be reluctant to apply because of the terrestrial opportunities they would have to forgo. For example, a 40-year-old astronaut on a mission to the planet Saturn would be well into his or her 50s when he or she returned. Children would have grown up in the interval. For interstellar travel, an astronaut might be kept in some form of suspended animation. But if he would ever return to Earth, all of his family and friends would be dead. How could the person be reintegrated after such a long time away from Earth?[130]

On board a spacecraft traveling at velocities close to the speed of light, time dilation will kick in, thus significantly shortening the time for the crew.[131] In accordance with Einstein's Special Theory of Relativity, time would pass by more slowly for astronauts as their spacecraft approached the speed of light. However, for extremely long flights, the time elapsed on spaceship could still exceed a human lifetime. Conceivably, space travelers could be held in some form of suspended animation. One method proposed would be cryogenics, in which the body temperatures of space travelers could be lowered so that their bodily functions would nearly cease. At the present time, cryonic suspension is only theoretical, not practical. As animals, including humans, are composed primarily of water, serious damage is done to the cells both during freezing and thawing. During freezing, water expands, thus the volumes of cells increase. As a consequence, they encroach on one another and their internal structure is disrupted. Conversely, during thawing, comparable contractions occur. Conceivably, an anti-freezing chemical could someday be used; however, it would be challenging to saturate a human with such chemicals without killing the person first. Some organisms, including certain fish and frogs, create a natural "anti-freeze" that prevents the formation of ice crystals in their cells. Humans do not have this capability, but recent progress has been made in the suspended animation in mammals that do not naturally hibernate, such as mice and dogs. For instance, in 2005 scientists at the University of Pittsburgh were able to reanimate dogs after their blood had been drained and replaced with a special ice-cold solution. Although the dogs were clinically dead for three hours, their hearts were restarted. Most of the dogs remained healthy after the procedure, but a few suffered brain damage.[132] Someday, innovative possibilities might be used for humans. For example, the Swedish biologist Carl-Gören Hedén once noted that at high pressures there are other kinds of ice that form with different crystal structures and different densities from those of ordinary ice. Possibly, a person might be safely brought to, and maintained at, an ambient pressure comparable to several thousand atmospheres and then quickly and carefully frozen to a very low temperature.[133] Alternatively, cell damage could be minimized by very fast freezing followed by very slow thawing.[134]

One way out of this conundrum would be for whole communities to travel together in space in self-sustaining vessels. Several scientists have speculated that giant self-contained generation ships might be used someday to travel across the galaxy. The Russian space pioneer Konstantin Tsiolkovsky (1857–1935) is credited with being the first person to conceive of a multi-generation star ship, which he wrote about in a 1928 essay. It was not long before the theme caught on in science fiction. In 1941, Robert Heinlein published a series of short stories that involved a multi-generation star ship. His 1957 book, *Citizen of the Galaxy*, involved a crew of 80 "Free Traders" on board the spaceship *Sisu*, which is organized by sex, seniority, family, and clan, each of which has specific responsibilities. To sustain and reproduce their communities, the Free Traders occasionally rendezvoused with other space colonies to court spouses for their young crewmembers.[135] Brian Aldiss published a novel on a similar theme entitled *Non-stop* in 1958.[136] Christened "freedom ships" by Isaac Asimov, an ecological balance could be maintained indefinitely on these vessels given a secure energy source and the replacement of minimal material.[137] So excited by the idea, Timothy Leary once approached Frank Drake and asked him to design a space-faring ark that could be used to save a remnant of humanity from the global nuclear Armageddon that he feared was inevitable.[138]

It was the American physicist and space activist, Gerard Kitchen O'Neill (1927–1992), who put together the most elaborate plan for a multi-generation spacecraft. He speculated that space colonies would travel the cosmos in spherical, cylindrical, and doughnut-shaped objects that would be large enough to hold from 10,000 to 10 million people. Humans cold live on the inner surface of these objects that would spin at a rate that would produce a centrifugal effect that would simulate Earth's gravity.[139] (Incidentally, the 2013 film *Elysium* featured a luxurious space habitat based on this model.) As there would be no gravity in outer space, the habitats could remain structurally sound even at sizes large enough to accommodate thousands, even millions of residents.[140] Theoretically, the ship could last indefinitely, as it would be equipped with renewable supplies. Someday these spaceships might use advanced propulsion systems to break out of the sun's orbit and venture outward in the cosmos. For the most part, these vessels would be coasting so they would not require that much energy to travel once they attained a high speed. To insure the crew's safety, an unmanned ship companion ship carrying provisions and traveling nearby the main ship could be converted to serve as a habitat in the event that the first ship became unsuitable for human habitation.[141] After a lapse of many generations, a free ship might approach another star. If the star is sun-like, the crew might investigate its planets and search for other civilizations.[142] Occasionally, the crew might stop on these planets for exploration and to stock up on new supplies.[143]

There would seem to be psychological advantages for crews aboard generation ships that would be absent for small crews. According to the research of Jason P. Kring and Megan A Kaminski, evidence from spaceflights and analogous settings such as Antarctica and submarines suggest that the size of the crew has a major impact on its members. Large crews possess several advantages over smaller ones, for instance, a greater range of skills and abilities, as well as presenting opportunities for forming friendships and creating a more interesting social experience. Moreover, members of larger crews appear to get along better, exhibit less hostility, are more stable, and make better and more efficient decisions.[144] Despite all of the associated pitfalls, living in isolated and confined environments can also have growth enhancing effects.[145]

The issue of procreation will be important on multi-generation missions. There is much uncertainty how traveling in space would influence conception and pregnancy. What biological perils could await an embryo conceived in space? Beyond the protection of the Earth's atmosphere, cosmic and solar radiation levels rise significantly. Dividing cells are extremely sensitive to radiation, which could increase the risk of mutation and miscarriage.[146] A generation ship could board numerous families in a voyage across the galaxy. As the parents retired, their children would take over, and so would their children, and so on. John H. Moore developed a demographic plan for multigenerational space crews. He posited that a starting population of 150–180 persons could sustain itself indefinitely by following a certain social compact that he proposed. The starting population would consist of young, childless married couples, not unlike the Polynesian seafaring colonists who set out in canoe flotillas to settle unoccupied Pacific islands. To maintain genetic variation, members of the crew would be advised to postpone parenthood until late in a woman's reproductive period. Two desirable consequences would follow. First, it would lengthen generations so that fewer chromosomes would be sloughed off per century in the population. Second, there would be smaller sibships (groups of children to the same parents) because women who postpone childbirth until age 35–40 would be unlikely to have more than two or three children. This would result in discrete demographic echelons. Inasmuch as humans tend to choose mates of a similar age, the echelons tend to remain bounded and distinct through time. Organized in such a structure, Moore submits that life onboard a multigenerational ship could be pleasurable, rather than regarded as an ordeal.[147]

Over time, the crew on the multigenerational spaceship would experience genetic drift from the human population on Earth. The smaller size of the ship's population and the greater number of generations removed from Earth, the more dramatic this effect.[148] The psychological impact of a permanent divorce from the home plant is unclear. To be sure, the effects on the first generation of colonists would be profound, as they would vividly remember life on Earth. For subsequent generation, though, their total existence and reference point would be the generation ship. Images and stories of Earth would be preserved, perhaps forming the subject of future lore as the colony evolved over time. But inevitably, the emotional attachments the people on these ships had to ancestral Earth and even the sun would weaken as time went by. Each settlement would develop its own styles in clothing, music, art, literature, religion, and so on.[149]

Perhaps generation ships could carry out a policy of galactic colonization. Seemingly the most practical way to colonize the galaxy would be for colonists not to return to their home planet. The archetypical example would be the migration of the prehistoric Polynesian people who spread from island to island across the Pacific.[150] Because of the tremendous distances between solar systems, extraterrestrial civilization might have to settle for a policy of decentralized decision-making when it comes to governing interstellar colonies. In a scenario where communication is slow or inaccurate, this would be more practical insofar as the central authority would have less timely information than the local commanders about conditions in the field.[151] After all, it takes 100,000 years for a message traveling at the speed of light to cross the length of the Milky Way. A single civilization could spread colonizing tentacles in all promising directions, perhaps eventually engulfing the entire galaxy. According to a model developed by Eric Jones, a single space-faring extraterrestrial civilization could colonize the Milky Way within 5 to 60 million years, which on an astronomical time scale would not be very long.[152] Similarly, Geoffrey Landis produced a percolation model to explain how an alien civilization might spread throughout the galaxy. He rejected the notion of a galactic empire under central control. Instead, he believed that a more realistic pattern would be a patchwork quilt of diverse local cultures emerging as colonization evolves. Some colonies will prefer to remain put and consolidate, while others will prefer to expand rapidly. He assumed that violent clashes between civilizations in the style of *Star Wars* to be exceedingly unlikely.[153]

New technology will surely lead to new human-machine partnerships. In fact, with advances in artificial intelligence, it might soon be possible to send androids into space. In his 2005 book *The Singularity Is Near: When Humans Transcend Biology*, the noted futurist Ray Kurzweil predicted that by the year 2042 computers will become self-aware and greatly exceed humans in intelligence. If this scenario does indeed lie ahead in the near future, then it might be more practical to send intelligent machines on the first long-range space missions to Mars and beyond. Sending humans into deep space is an enormously expensive and dangerous proposition. From the perspective of space engineers, sending robots might seem more feasible than sending humans, as the former requires no water, oxygen, or food supply. Moreover, there is no need to worry about interpersonal issues that might arise with a human crew. With their physical limitations and short life spans, assigning human astronauts to deep space missions might seem increasingly quixotic when more able androids are up to the task.

As technology continues to improve, it becomes more practical to send robots into space in lieu of humans. Based on this reasoning, in 1997 NASA administrator Daniel S. Goldin announced that building a robotic probe that could reach another star would be a goal of the space agency.[154] There are several advantages to this approach. For one thing, it is vastly cheaper; in most cases, sending robots into space is just one-fiftieth the cost of sending people. Robots do not require life-support systems, nor are they fazed by ionizing radiation that would harm human astronauts. Robots can operate in a vast range of temperatures incapable for humans. Robots would not suffer serious side effects of space travel, including the loss of bone and muscle mass.[155] Finally, robots could endure the long periods of travel that are beyond the life expectancies of present-day humans. For these reasons, robots are the likely first candidates for interstellar travel.

Speculating on this scenario, in 2005 the Discovery Channel aired a realistic program titled *Alien Planet*, which was based on an imaginary story of exploring an exoplanet—Darwin IV—6.5 light years from Earth. The mother ship—the Von Braun—carries exploratory robotic probes to the planet. Travelling at 37,000 miles per second, the journey takes 42 years. The two robots—Ike and Leo—are sent to the surface of the planet. They are programmed to see, hear, move, and manipulate objects, and to communicate their experiences to the mother ship. On the planet, they find a variety of creatures, both small and large. Some even use gas-filled bladders (somewhat like balloons) as a form of biological propulsion with methane gas for fuel.[156] One of the robotic probes encounters a seemingly intelligent, but not technologically advanced species, with whom he attempts to communicate. Unfortunately, his would-be interlocutors destroy him.

Despite the advantages, robots, as currently constructed, have their limitations. Most notable, is their inability to make serendipitous discoveries that come from a lifetime of experience. By contrast, humans can draw upon a vast catalogue of experiences that they have accumulated over a lifetime to make quick decisions.[157] Furthermore, humans possess intuition, a trait that has yet to be duplicated in artificial intelligence. At the present time, robots cannot fully embrace revolutionary scientific discoveries that may be waiting to be found on distant planets—and as Neil de Grasse Tyson noted, "those are the ones you don't want to miss."[158]

In the future, however, artificial intelligence could improve so that robots would be fully capable of carrying out the functions of human astronauts. Just as humans learn from experience, robots could soon do the same. In fact, research is currently underway in the field of genetic algorithms, that is, computer software routines that adapt to circumstances by testing different solutions to a problem. Instead of a fixed sequence of steps, the genetic, or evolutionary, algorithm generates slight variations to its own code and then implements these modifications through a series of mutations to see what works best. One such effort, the Golem Project, was created at Brandeis University by two scientists, Jordan B. Pollack and Hod Lipson. Their goal is to develop a sustained machine evolution. Such research could lead to autonomous spacecraft that are able to think for themselves and adapt to changing circumstances and unseen contingencies.[159] This quality is important because on interstellar missions there will be a high level of autonomy in that ground personnel will exert less and less control over the probe.

John von Neumann and Ronald Bracewell once conjectured that extraterrestrial civilizations could explore the cosmos by way of robotic probes. As self-reproducing automatons, these probes would be programmed to extract raw materials and construct more machines that would diffuse the universe. As the von Neumann machines would be self-replicating, they would require no additional effort or expenditure from the civilization that created the original prototype. Frank Tipler once estimated that it would take only 300 million years for a galaxy to be suffused with von Neumann machines once the initial prototype was launched. By astronomical standards, some scientists reason that it would seem to follow that at least one advanced civilization would have developed von Neumann probes by now, in which case they should be ubiquitous in the galaxy. Some skeptics have cited their absence as further validation of the Fermi paradox: The fact that we have not detected them proves that we are alone in the galaxy.[160]

Regrettably, Frank Tipler found it unlikely that flesh-and-blood humans would ever practice interstellar travel.[161] Nevertheless, he predicted that in the near future every human mind or "soul" could be uploaded in a tiny quantum computer. With thousands of human minds contained in a mass of only a few grams, the tiny payload could fit into a small spaceship and travel throughout the cosmos. As a substitute to direct contact, Joe Haldeman speculated that it might be possible one day to experience "virtual colonization," in which robotic surrogates on other planets send back information back to Earth about what it is like living on their new worlds. If the information were precise and in the right form, humans could plug into an illusion of being on the far away planet.[162]

In recent missions, robotic probes have demonstrated great potential. In August of 2012 an automated motor vehicle—the *Curiosity* rover—began exploring and conducting experiments on the Gale Crater on Mars and sending back exciting results. But nothing excites the human mind more to know that flesh and blood people are exploring the cosmos. Without the public's interest and support, it is all the more difficult for NASA to secure funding. In short, for NASA there will be "no bucks without Buck Rogers."[163] But a manned mission to Mars will be enormously expensive. With an estimated price tag of anywhere between 100 billion and 1 trillion dollars, this proposition would seemingly be a hard sell. Nevertheless, according to a 2013 poll, 75 percent of Americans surveyed agreed that NASA's budget should be substantially increased to fund this endeavor.[164] Perhaps we will soon have the scientific capabilities, public support, and political will for the next major achievement of the Space Age.

On contemplating the prospects for human space travel, Paul Davies observed that futurologists are divided into two camps. One camp predicts that advances in science and continual innovations in technology will reinvigorate our push into space. Pessimists, on the other hand, see exploration as rooted in the politics of the Cold War. With the costs of human space exploration so prohibitively expensive and the commercial returns so negligible, the space program will begin to dwindle, especially in a tight fiscal environment.[165] By its nature, space exploration is a long-term project, which makes it vulnerable to the downturns in the economy and the vicissitudes of politics. Only by organizing extraordinary missions, Neil de Grasse Tyson argues, can we attract people of extraordinary talent.[166] On that note, then-President John F. Kennedy succeeded when he exhorted the American people to send astronauts to the Moon by the end of the decade. Many bright people answered his call and worked on the Apollo program, which produced many spillover benefits in other fields, including communications, computers, and medicine. Likewise, in 2004, then-President George W. Bush announced a long-term plan of space exploration that would include a return trip to the moon and thereafter a trip to Mars and beyond, but the Obama administration retreated from these goals in the face of a deteriorating fiscal position.[167]

Success is vital to building public support for space exploration. In 1969, less than half of Americans surveyed thought that the *Apollo* program was worth the cost. By 1994, however, that number had risen to two-thirds.[168] Traditionally, space exploration has been driven primarily by a defense imperative.[169] But since the end of the Cold War, commercial interests have loomed larger in space projects. Doug Bason argues that the broad, ambitious goal of interstellar travel could provide a focus for NASA and the U.S. economy. It would allow the nation to invest in its technological and scientific base. An interstellar program would facilitate the commercialization of space and promote space-based industries, products, and services. New technologies developed for this mission would undoubtedly have some useful terrestrial applications as well.[170] At the present time, most of the progress toward space settlement is being conducted in the private sector. For example, in December 2011 Elon Musk's SpaceX completed a successful test flight of a reusable capsule capable of carrying seven astronauts. The company has a contract with NASA to shuttle cargo to the International Space station at per-pound costs far below the current rate. Other companies, including Virgin Galactic and Space Adventures, have begun offering flights into the Earth's low orbit and brief stays in space stations. And Bigelow Aerospace has plans to launch an inflatable "space hotel" by 2015. As Rick Tumlinson, the cofounder of the Space Foundation (a group of entrepreneurs aiming to harness private enterprise to explore space) explained, a compelling profit motive is necessary to achieve where an increasingly risk-averse NASA has failed. In his estimation, we will achieve permanent human settlements in the cosmos when investors begin to make money.[171]

At the present time, our technology is not sophisticated enough to realistically plan a course of interstellar migration. One somber explanation for the Fermi paradox would be that interstellar travel is impossible. Some limits, such as the speed of light, are just that—limits that perhaps will never be surmounted. If there is no way to break the speed of light barrier, interstellar travel will be impractical because of the long periods of time necessary to traverse such vast distances. But as Ray Kurzweil predicts, if the singularity is really near, intelligence could skyrocket in a relatively short period, making the prospect of interstellar travel possible sooner than we might think. As explained in Chapter 4, in the future, a superior intelligence might be able to create "smart matter," which although nominally following the laws of physics, will be able to harness these laws in such a way as to manipulate matter and energy to its will. But there are still many important questions to ponder—to wit, is anyone really out there? Interstellar travel is an enormously ambitious undertaking that will require tremendous effort, intelligence, resources, and time. It is the last requirement—time—that is most critical for the survival of our species and the likelihood that we will ever establish contact with an extraterrestrial civilization.

Conclusion

A Race Against Time

The discovery of more and more Earth-like exoplanets raises the prospect that we will someday make contact with extraterrestrial intelligence. Moreover, proponents of convergent evolution point out that insofar as environmental conditions influence the course of biological progression, there is a strong likelihood that organisms similar to humans could arise on these planets. Life emerged on Earth relatively quickly, which suggests that biology is merely a function of the right chemistry—the right ingredients available under the right conditions. In the early years of Earth's existence, life could not take hold because the planet's surface was too hot. What is more, the planet was regularly bombarded with asteroids, which would likely snuff out nascent life. Fossil evidence suggests that simple organisms came into being a mere half a billion years after the Earth was formed. Expressed as a proportion, this means that life emerged on Earth just one-ninth into its present age of 4.5 billion years. This fact would seem to augur well for the existence of intelligent life elsewhere in the cosmos.

But the progression to advanced life is far from certain, for it would take another three billion years for multi-cellular life to emerge. Moreover, on a geological scale, intelligent life, in the form of *Homo sapiens*, is a recent development. And civilizations are of even more recent vintage. Although our sun has roughly 5 billion more years remaining in its main sequence, our planet will be not habitable for the duration of this period. As the sun gets hotter, environmental conditions on Earth will change. As a consequence, by geological standards, there is not much time for intelligent life to develop, survive, and build enduring civilizations capable of communicating with one another—hence the quest to make contact with extraterrestrial life is a race against time.

Despite the multitude of stars and galaxies, intelligent life may not exist beyond our solar system. As mentioned earlier, the so-called Fermi paradox suggests that advanced extraterrestrial civilizations are uncommon—perhaps nonexistent—in our galaxy. If extraterrestrial civilizations exist and have preceded ours by eons, so the argument goes, then they should have established contact by now. Are the odds against life in the universe? Although our planet may seem relatively tranquil, there are numerous perils—both natural and man-made—that could eradicate humans on Earth. For instance, there is the threat posed by the roughly 200 known asteroids whose paths take them near Earth. About 20 percent of them, sooner or later, are bound to strike Earth, with possibly devastating consequences.[1] On average, every few hundred years, the Earth is struck by an object about 70 meters in diameter. The resulting energy is roughly to about 50 megatons—equivalent to the largest nuclear weapon ever detonated.[2] Even worse, approximately every 10,000 years, a 200-meter object collides with Earth which could precipitate serious climactic changes on the planet. And every million years, an impact by an object over 2 kilometers occurs, the impact of which produces an explosion equivalent to nearly a million megatons. Such an impact would provoke a global catastrophe, wiping out a significant fraction of the human race—perhaps up to a billion people on the planet.[3]

The good news is that the human race might someday develop the capacity to deflect potentially dangerous asteroids. Conceivably, a nuclear weapon could be used to destroy an incoming asteroid obliterating it into enough small pieces so that it would produce a harmless, though spectacular, meteor shower. Alternatively, a neutron bomb (a Cold War-era nuclear weapon that would primarily kill people but leave buildings intact) could be used to heat up one side of the asteroid which would cause material to spew forth, inducing the asteroid to recoil. Yet another solution would be to nudge an asteroid out of harm's way with slow but steady rockets that could be attached to one side. Finally, a gravitational tractor could be employed by parking a probe in space near the killer asteroid. As their mutual gravity attracts each other, an array of retro rockets would fire, causing the asteroid to draw toward the probe and off its collision course with Earth. As Neil de Grasse Tyson explained, if humans one day become extinct because of a catastrophic collision, it would not "be because we lacked brainpower to protect ourselves, but because we lacked the foresight and determination."[4]

Deflection technology, however, would have a great potential for misuse. Its dual-use nature means that it could be utilized for offensive military operations, which would make arms control efforts problematic, as Carl Sagan explained:

> If we're too quick in developing the technology to move worlds around, we may destroy ourselves; if we're too slow, we will surely destroy ourselves. The reliability of world political organizations and the confidence they inspire will have to make significant strides before they can be trusted to deal with a problem of this seriousness. At the same time, there seems to be no acceptable national solution. Who would feel comfortable with the means of world destruction in the hands of some dedicated (or even potential) enemy nation, whether or not our nation had comparable powers? The existence of interplanetary collision hazards, when widely understood, works to bring our species together. When facing a common danger, we humans have sometimes reached heights widely thought impossible; we have set aside our differences—at least until the danger passed.[5]

Currently, our sun is about half way into its main sequence—the period during which it emits a roughly steady amount of heat. An optimistic figure sets the sun's remaining life at 5 billion more years.[6] However, a star is not an adequate incubator of life throughout its existence. As the sun grows older and brighter, its habitable zone will shift farther and farther out from the center of the solar system. As the sun expands and increases its luminosity, the top of Earth's atmosphere will become exceedingly hot. The oceans will boil away and the atmosphere will evaporate into space, leaving no sustenance for life.[7] A conservative estimate is that the Earth will be uninhabitable within a billion years.[8] As mentioned in Chapter 3, fossil evidence indicates that life existed on Earth as far back as nearly 4 billion years ago. Assuming that the Earth will remain habitable for another billion years, this means that life has about a 5-billion-year tenancy on the planet. To date, life has used up 80 percent of this time. The remaining 20 percent will determine if intelligent life will develop the capabilities to become immortal.

Some scientists have speculated that advanced technology might enable us to stave off the destruction of our planet through astroengineering. By the twenty-second century, Carl Sagan predicted that we might be able to use nuclear fusion engines, or their equivalents, to move small worlds around the solar system.[9] By doing so, humans might be able to expand its orbit around the sun by commandeering asteroids from the Kuiper Belt and altering their trajectories so that they pull the Earth farther away from the Sun.[10] If successful, this project would give the Earth more time to spend in the solar system's habitable zone. Astroengineering might also be used to keep the moon's orbit close to the Earth. The gravitational influence of the moon acts as a stabilizer to Earth, which *inter alia*, is responsible for the cycle of seasons that are so vital to life on the planet.[11]

But even if humans do master astroengineering, it will only forestall the inevitable. After a star condenses to the point at which nuclear fires begin at its core, it begins to radiate light. When the condensation reaches a stable state and the radiation reaches some maximum state, the star is then said to have entered its "main sequence." The inward pull of gravity and the outward push of temperature remain in balance. But at some critical point, when a large amount of the hydrogen has been used up, the star is destabilized. The star begins to expand and leaves the main sequence. As it grows to an enormous size, the star cools and becomes merely red hot, though the total radiation emitted from its vast surface is much greater than it had been before.[12] In about 5 billion years, the sun will expand and completely consume the orbits of Mercury, Venus, and Earth.[13] After the sun leaves its main sequence, it will remain a red giant for several million years—a relatively short period of time on the astronomical scale.[14] As the hydrogen is consumed, the core will grow hotter and hotter. Finally, the sun will collapse when the energy created by the nuclear fusion and its core fails as the nuclear fuel is depleted. As a result, the sun will no longer remain distended against its own gravity. The more massive the star, the greater is the explosion. What is left of the star shrinks to a very tiny and dense ball.[15]

The human race also faces natural threats on the planet. As the Earth's planetary core cools over time, volcanic activity will slow down, and in doing so reduce the carbon dioxide from the atmosphere. Consequently, the Earth's atmosphere will not be suitable for life.[16] Furthermore, in about 4 billion years, the entire core of the Earth will have solidified. As a result, the Earth will lose its magnetic field, which provides protection to the planet from harmful cosmic rays.[17]

In addition to natural calamities, the human race will also have to overcome social and political obstacles. Some SETI scientists have warned that for humans to survive, they will have to make it through a "technological adolescence" in which the threat of nuclear annihilation, global warming, environmental despoliation, and overpopulation loomed large. According to J. Richard Gott III, an astrophysicist at Princeton University, to endure over the long haul, life forms must pass through periodic filters. Based on his statistical analysis, he concluded with a 97.5 percent confidence level that our species will survive for no more than 8 million more years.[18] If the eerie silence we hear is real, then something filters out most civilizations, either by preventing their formation in the first place or by annihilating them soon after they become established. Paul Davies concludes that if life is a cosmic imperative, the great silence is indeed eerie, and in fact, is positively sinister as far as the fate of humanity is concerned.[19] Although Carl Sagan found some merit in Gott's analysis, he countered that if we could survive the short term, our long-term chances would be even brighter than Gott calculates.[20] Similarly, Seth Shostak argued that rather than being on the road to civilizational self-destruction, we may merely be passing through a bottleneck—that is, we have the technological capacity to destroy ourselves, but have not yet purchased the anti-extinction insurance policy of colonizing other worlds.[21]

Initially, hopes were raised after the end of the Cold War that the world would enter a period of unprecedented tranquility and economic opportunity. Francis Fukuyama's 1989 article "The End of History?" presaged the coming *Zeitgeist.* According to his analysis, governments around the world were converging on a model consisting of democracy and free markets. As a consequence, grand ideological conflicts—the kind that brought so much destruction in the twentieth century—would disappear. The new era heralded peace and increased multilateralism in global politics. And, to a certain extent, empirical evidence seemed to bear this prediction out. According to many indices—as Stephen Pinker noted in *The Better Angels of Our Nature: Why Violence has Declined*—the world is actually becoming a safer place.[22] Both the United States and Russia have drastically reduced their nuclear arsenals since the end of the Cold War.[23] More and more, nations and governments reject the appropriateness of waging war.

Sill, conflict has not completely disappeared. In fact, some trends in global politics could reverse the peace trend. As the human population rises, the world is becoming a more crowded place. In 2013 the world's population was estimated to be 7.1 billion, but this figure is projected to reach close to 10 billion by mid-century.[24] Rapidly growing countries with a high proportion of young people are especially prone to civil unrest and are less able to create or maintain democratic institutions.[25] As Michael T. Klare noted, three forces could raise the specter of resource competition on a global scale. First is the increase in worldwide demand for resources, which is fueled in large part by the increase in the global population. Also problematic in this regard is that those countries in which resource competition will become most acute are also those undergoing the most rapid growth in population. What is more, the spread of industrialization throughout more and more areas of the world engenders greater per capita demand for resources, as wealth tends to follow economic development. The second factor contributing to resource competition is the prospect of looming shortages of vital commodities such as oil, natural gas, timber, and minerals. Moreover, in upcoming years, it will become increasingly difficult to keep extracting additional resources from the known deposits of reserves. Wells will have to be dug deeper and deeper as deposits are depleted. Finally, the third factor is the increasing proclivity for nations to contest ownership rights over resources. Many key sources and deposits of these resources are shared by two or more nations, or lie in contested border areas or offshore economic zones. And to make things worse, often this competition for resources occurs against a backdrop of preexisting hostilities, such as areas of the Middle East and parts of Africa.[26]

Even in the absence of conflict, population pressures could impact the world in profound ways. For example, food shortages could force a large segment of the world's population into vegetarianism.[27] One of the most pressing challenges for contemporary states is delivering the basic services that citizens demand from their governments. Persistent poverty breeds a variety of social dysfunctions, including criminal gangs, narcotrafficking, terrorism, and a general feeling of insecurity on the large part of ordinary people.[28] Faced with increasing populations, governments might be strapped with huge burdens to meet the day-to-day necessities of their constituents. As a consequence, there will be fewer resources for grand projects, such as planning for interstellar travel.

By the early 1990s, other voices began to express a less sanguine view of the process of globalization and saw disturbing trends antagonistic to democratization, *viz.*, the return of tribalism and violent religious fundamentalism. In 1993 Patrick Moynihan saw the specter of violent ethnic conflict afflicting nations and leading to "pandemonium" in various parts of the world.[29] That same year, the eminent Harvard professor, Samuel Huntington, introduced his "clash of civilizations" model of international relations, in which he predicted that conflict in the twenty-first century would be most pronounced along "civilizational fault lines." Consequently, atavistic hostilities were reemerging around the globe, as religious and ethnic identities hardened. Ominously, Huntington warned of possible civilizational conflicts of truly apocalyptic proportions. According to his analysis, the forces of integration in the contemporary world were real, but warned that they generated "counterforces of cultural assertion and civilizational consciousness."[30] Especially worrisome to Huntington was the potential for violence emanating from the Islamic world. In fact, for several years, Huntington contended that Islam and the West had been engaged in a "quasi-war" and were headed toward a collision course.[31]

Some observers fear that societal fragmentation could undermine the security of countries in the West, including the United States. In 2006, British

Rear Admiral Chris Parry, a senior military strategist, raised eyebrows when he announced at a conference that the nations of the West faced peril from the hordes of the Third World—not unlike ancient Rome faced centuries earlier from barbarian invasions. In an era of globalization, Parry reasoned that large immigrant communities in the West eschew assimilation seeing it as "redundant and old-fashioned." Moreover, porous borders render some areas beyond governmental control.[32] In a similar analysis made over a decade ago, Robert Kaplan prognosticated in his influential article, "The Coming Anarchy," that it was not entirely clear that the United States will be able to survive exactly in its present form because it is a multi-ethnic society, where the concept of the nation state has always been more fragile than it is in homogeneous countries. As he pointed out, during the 1960s, "America began a slow but unmistakable process of transformation" resulting in a more fragmented country.[33] This theme was taken up in 2004, when Samuel Huntington released his book, *Who Are We: The Challenges to America's National Identity*, in which he argued that the rise of various group identities based on race and ethnicity threaten to diminish the larger American national identity, which he believes is essential for the long-run survival of the country as a unified political entity.[34]

Two contemporaneous, but contradictory, trends—integration and fragmentation—are shaping the contours of contemporary global politics. One the one hand, there is greater integration of the nations of the world. On the other hand, there are indications that fragmentation is taking place in many countries, not only in the developing world, but in the advanced countries of the west as well. In fact, the general trend in the post-World War II era has been the expansion in the number of states as more and more countries have been created by seceding from other states. As of 2013, the United Nations included 193 member nations.[35] The encroaching process of globalization will undoubtedly engender a certain amount of opposition from those who feel threatened with a loss of identity and culture.

Other observers, though, are more sanguine about the process of globalization and believe that trends are on the right course to global peace and prosperity. In his major study, *The Pentagon's New Map*, a former defense analyst, Thomas P.M. Barnett, divided the world into two broad regions. Countries in the "functioning core" are integrated into a world system and operate under "rule sets." As a result, they arbitrate their differences through international bodies, such as the UN, or the World Trade Organization, and are less likely to go to war against other nations in the core.[36] By contrast, countries in the "Gap" do not follow these rule sets and are the areas from which most of the problems that bedevil the world today emanate. The real enemy, according to Barnett, is disconnectedness insofar as it separates people from the rest of the world. An unabashed economic determinist, he believes that when people are integrated into the economy, they are far less likely to succumb to radicalism. In short, Barnett's long-term strategic goal is to shrink the gap, and by doing so, dry up the reservoir for international conflict.[37]

It appears that the human race is at a critical juncture in its history. The theoretical physicist Michio Kaku believes that the existence of extraterrestrial life is practically a certainty.[38] Yet, he finds it puzzling that despite over 50 years of effort, the SETI project has not detected any signs of alien intelligence in the cosmos.[39] Using the Kardashev scale as his framework, Kaku argues that the transition from a Type 0 to Type I civilization carries a strong risk of self-destruction. Consequently, a short-lived civilization would be unlikely to establish contact. Perhaps, he wonders, the galaxy could be riddled with planets that never made the perilous transition from a type 0 to a Type I civilization. To Kaku, this transition is akin to a period of tribulation which he depicts in a millennial tone:

> This transition is perhaps the greatest in human history. In fact, the people living today are the most important ever to walk the surface of the planet, since they will determine whether we attain this goal [planetary civilization] or descend into chaos. Perhaps 5,000 generations of humans have walked the surface of the earth since we first emerged in Africa about 100,000 years ago, and of them, the ones living in this century will ultimately determine our fate.[40]

Despite the remarkable scientific progress made over the past century, in the background, lurks the specter of nuclear war, global warming, and the outbreak of deadly pandemics.[41] As Kaku explains, advanced civilizations will eventually have to deal with the power of the atom. Although he sees great potential in nuclear technology for fulfilling the planet's energy needs, Kaku warns that the pitfalls must be seriously considered as well.[42] For example, following the March 2011 Tōhoku earthquake and tsunami, there were reactor containment failures that resulted in the release of radioactive materials at the Fukushima Nuclear Power Plant in Japan. Furthermore, today there is much consternation over the increasing availability of weapons of mass destruction and the prospect of proliferation.[43] In addition to the nuclear peril, Kaku has also warned on the danger of global warming. If these twin global disasters can be averted, then he believes that science will inevitably pave the way for our society to rise to the level of a planetary civilization.

By the close of the twenty-first century, Kaku predicts that the people of our planet will be forced to cooperate on a scale never before seen in history. From his perspective, we are on the cusp of a planetary civilization as evidenced by several important trends. He describes the Internet as a Type I planetary phone system, connecting people all over the world. Trade blocs, such as the European Union and NAFTA, are the prototypes of an emerging planetary economy. Concomitant with that development is the decline of nation states. Higher standards of living are giving rise to a global middle class for whom political stability and consumer goods, not wars, religion, or strict moral codes, are the primary goals. English is rapidly emerging as the future Type I language. Enhanced communications and networking are forging a planetary youth culture based on the popularity of rock and roll music, Hollywood movies, fashion, and franchise food chains. Kaku is sanguine that the new global culture will erase long-standing cultural and national barriers that often led to war.[44] Moreover, cooperation and coordination on a global scale will enable the planet to protect the environment.[45] Increasing levels of education will lead to greater technological advancement. And greater transparency will strengthen the trend toward democratization around the world.[46]

The path to a planetary civilization, though, is not without its hazards. There is a dark side to human nature where the forces of fundamentalism, sectarianism, terrorism, racism, and intolerance are still at work. Some parties—such as Islamic extremists and dictatorships—seek to resist this wave of history because instinctively they know that it threatens their cherished beliefs and powers.[47] As Kaku explains, despite evolution, human nature has not changed much in 100,000 years, except that we now have nuclear, biological, and chemical weapons to settle old scores.[48] If we can navigate through these perils, Kaku predicts that we will attain type I civilizational status in roughly a hundred years. Advancing to Type II status could take approximately 800 years. Finally, to achieve Type III status would require mastering the physics of interstellar travel and could take 10,000 years or more.[49] Such a civilization, Kaku mused, would be immortal.

If we are indeed alone in the universe, that makes the human mission all the more vital. If humans become extinct, it is unlikely that another species could evolve and take over the reins of civilization. During man's tenure as master of the Earth, he has depleted natural resources that took eons to develop. As Fred Hoyle and John Gribbin both noted, even if some advanced species were to emerge on Earth after man, it would not have the necessary raw materials to rebuild a civilization. Consequently, there would be no second chances for intelligent life on the planet. Man's rapid consumption of natural resources as an energy source would preclude the reestablishment of civilization for any species that would attain preeminence after man's reign on Earth had ended.[50]

Arthur C. Clarke once mused, "Either we are alone in the universe, or we are not. Either thought is frightening." But will we ever know which answer for sure? As Paul Davies noted, there are three possibilities with respect to life in the cosmos. First, the universe is teeming with intelligent life. This prospect would bode well for humanity, for it would suggest a bright future. The second is that Earth is a unique oasis for life, which would place an awesome burden of responsibility on our shoulders. Finally, life could be widespread in the universe but we are the only ones intelligent enough to be able to celebrate it. Although life could be common in the cosmos, intelligent life could be extremely rare.[51] As a scientist, owing to the myriad of contingencies necessary to give rise to intelligent life, Davies believes that we are probably the only intelligent beings in the observable universe. But as a philosopher, the prospect of a universe devoid of intelligent life makes him feel uneasy insofar as only we, "lowly *Homo sapiens*," are able to appreciate the cosmos. Finally, as a human being, he would like very much to believe that the universe is intrinsically friendly to life in the universe. But at the end of the day, he concedes that "we just don't know" which makes the SETI enterprise so tantalizing.[52]

The discovery of a civilization beyond Earth, Carl Sagan once noted, would help us calibrate our place in the universe. Although confirming contact with an extraterrestrial civilization might not end war here on Earth, it could make the common bond of humanity more apparent. Likewise, Donald E. Tarter sees the SETI enterprise as an opportunity for us to regain our sense of human unity, for if we are able to detect evidence of other civilizations in the cosmos, we may find it easier to envisage ourselves as one civilization among many.[53]

But if we are alone, that means that the universe, or at least the Milky Way, is ours for the taking. Moreover, if intelligent life is able to transcend biology, then our long-term chances of survival would be greatly enhanced. According to Ray Kurzweil's 2005 book, *The Singularity is Near: When Humans Transcend Biology*, somewhere around the year 2042, humans will begin entering a post-biological era. If that is the case, then this post-biological intelligence would no longer require a habitable planet, but could survive anywhere in the cosmos where computers could operate.

But even on grand timescales, our galaxy may not be a safe place to live. In hundreds of millions to billions of years, the centers of galaxies will explode.[54] On this somber note, the mathematician and philosopher Bertrand Russell once lamented:

> that no fire, no heroism, no intensity of thought or feeling, can preserve a life beyond the grave; that all the labors of the ages, all the devotion, all the inspiration, all the noonday brightness of human genius, are destined to extinction in the vast death of the solar system; and the whole temple of Man's achievements must inevitably be buried beneath the debris of a universe in ruins.[55]

After astronomers determined that the universe was expanding, the logical question to ask was, would this expansion continue? If this expansion was not greater than the escape velocity of the Big Bang, then the universe would eventually contract.[56] What Einstein once called his greatest blunder—the cosmological constant—actually accounts for the largest source of matter and energy in the universe and is driving its expansion.[57] Ultimately, the property of dark energy and matter will determine the fate of the universe.[58] If the property of this dark energy is repulsive enough, the universe will expand forever, resulting in a "Big Freeze." Data from the WMAP satellite suggest that the expansion of the universe, rather than slowing, is actually accelerating.[59] Taken to its logical conclusion, eventually, entropy will continue to increase in the universe until its temperature approaches absolute zero, thereby extinguishing all life. Michio Kaku, however, sees a loophole in this seemingly somber future.

According to Kaku, string theory is no mere academic exercise; rather it could be used as a way to escape the death of our cosmos. To avoid this fate, an advanced civilization may decide to make the ultimate journey to another universe.[60] Perhaps someday, Kaku speculates, we could learn how to harness the power of dark matter and dark energy. With this capability, we might be able to expand the microscopic black holes that exist at the quantum level so that we could pass into other universes and thereby escape the Big Freeze. Alternatively, if an advanced civilization could miniaturize its total information content to the molecular level, it could inject it through one of these microscopic gateways and then have it self-assemble on the other side.[61] Through this method, an advanced civilization might be able to send its "seed" through a wormhole. Using nanotechnology, this seed could copy the important properties of the civilization. Once a suitable environment was found, these nanobots could construct factories that would produce replicas of themselves and make a large cloning laboratory with the mission of regenerating whole organisms and eventually the entire species.[62] Finally, one more possible escape plan would be for an advanced civilization to use a wormhole as a time machine and return to the universe of an earlier era when life was still possible.[63] Although such escape scenarios may seem preposterous, as Kaku points out, an advanced civilization that endures for billions of years might be able to meet the challenges of surmounting any of these hurdles.[64]

To date, scientists have not really had much time to conduct an adequate search for life in the universe. Skeptics counter with the question, with all of its multitude of stars and planets, why does the cosmos seem so silent? Should we have not heard something by now? The short answer is that the universe is really not that old. Although 13.7 billion years may seem like a long time, the first two generations of stars could probably not support life, as they were composed almost entirely of hydrogen and helium. The heavier elements that are necessary for life—carbon, oxygen, and iron—are created when massive stars blow up as supernovae. Presumably, it is only the third-generation stars like our sun that host planets that contain the necessary heavier elements in sufficient abundance so that intelligent life can evolve. By astronomical standards, extraterrestrial life still has not had much time to spread throughout the cosmos. Galactic colonization would be an extremely arduous and time-consuming enterprise. Based on our current knowledge of physics, even if a species were able to travel at very high speeds and were dedicated to nothing else, it would not be possible to explore even a fraction of the Milky Way because of its vastness.[65] As Martin Rees once suggested, the unfolding of intelligence is probably near its cosmic beginnings.[66]

If our civilization survives, it is reasonable to assume that our technology and knowledge of the cosmos will continue to advance. In the near future, a concatenation of new technologies and scientific discoveries—nanotechnology, quantum computing, fusion, and string theory—implies that progress will proceed very rapidly—exponentially, rather than linearly. With these capabilities, we may someday be ready to journey outside of our solar system. If that is the case, then one day we might actually come face to face with civilizations beyond Earth.

But to reach this threshold, our species must surmount the difficult problems that lay ahead in this century, *inter alia*, overpopulation, global warming, resource depletion, economic uncertainty, proliferation of weapons of mass destruction, and ethnic and sectarian conflict. In the event of a major civilizational setback, there may not be adequate time and resources to recover. We have only a limited window of opportunity. As mentioned in Chapter 4, Ray Kurzweil predicts that by the year 2042, we will reach the "singularity" in which human minds will be able to merge with artificial intelligence, thus achieving immortality. He will be 96 years of age if he lives long enough to reach that milestone. To increase his chances, he practices a strict regimen that entails consuming roughly 150 pills and supplements a day. If effect, he hopes to live long enough so that he can live forever.[67] Likewise, collectively as a civilization, if we can navigate through our current problems, then someday we may develop the capacity for ambitious projects, including astro-engineering, interstellar travel, and exotic ways to harness enormous amounts of energy. By doing so, we can reach our cosmic destiny of colonizing the universe, either alone, or with the help of extraterrestrial intelligence.

Suggested Reading

Alexander, John B., *UFOs: Myths, Conspiracies, and Realities*. New York: St. Martin's Press, 2011.

Allman, Toney, *Are Extraterrestrials a Threat to Humankind?* San Diego, CA: Reference Point Press, 2012.

Asimov, Isaac, *Extraterrestrial Civilizations*. New York: Crown Publishers, Inc., 1979.

Basalla, George, *Civilized Life in the Universe: Scientists on Intelligent Extraterrestrials*. Oxford and New York: Oxford University Press, 2006.

Birnes, William J., *The Everything UFO Book: An investigation of sightings, cover-ups, and the quest for extraterrestrial life*. Avon, MA: Adams Media, 2012.

Bracewell, Ronald N., *The Galactic Club: Intelligent Life in Outer Space*. San Francisco, CA: W.W. Freeman and Company, 1975.

Bullard, Thomas E., *The Myth and Mystery of UFOs*. Lawrence, KS: University Press of Kansas, 2010.

Carey, Thomas J. and Donald R. Schmitt, *Witness to Roswell: Unmasking the Government's Biggest Cover-Up* (Revised and Expanded Edition). Franklin Lakes, NJ: New Page Books 2009.

Cook, Nick, *The Hunt for Zero Point: Inside the Classified World of Antigravity Technology*. New York: Broadway Books, 2003.

Coppens, Philip, *The Ancient Alien Question: A New Inquiry Into the Existence, Evidence, and Influence of Ancient Visitors*. Pompton Plains, NJ: New Page Books.

Corso, Philip J. with William J. Birnes, *The Day After Roswell.* New York: Pocket Books, 1997.

Crowe, Michael J., *The Extraterrestrial Life Debate: Antiquity to 1915.* Notre Dame, IN: University of Notre Dame Press, 2008.

Davenport, Marc, *Visitors From Time: The Secret of the UFOs.* Tuscaloosa, AL: Greenleaf Publications, 1994.

Davies, Paul, *The Eerie Silence: Renewing Our Search for Alien Intelligence.* Boston and New York: Houghton Mifflin Harcourt, 2010.

Dick, Steven J., *Life on Other World: The 20th-Century Extraterrestrial Life Debate.* New York: Cambridge University Press, 1998.

Drake, Frank and Dava Sobel, *Is Anyone Out There? The Scientific Search for Extraterrestrial Intelligence.* New York: Delacorte Press, 1992.

Friedman, Stanton T., *Flying Saucers and Science: A Scientist Investigates the Mysteries of UFOs.* The Career Press: Pompton Plains, NJ, 2008.

Fukuyama, Francis, *The Origins of Political Order: From Prehuman Times to the French Revolution.* New York: Farrar, Straus, and Girous, 2011.

Gilster, Paul, *Centauri Dreams: Imagining and Planning Interstellar Exploration.* New York: Copernicus Books, 2004.

Goldberg, Dave and Jeff Blomquist, *A User's Guide to the Universe: Surviving the Perils of Black Holes, Time Paradoxes, and Quantum Uncertainty.* Hoboken, NJ: Wiley, 2010.

Gribbin, John, *Alone in the Universe: Why Our Planet is Unique.* Hoboken, NJ: Wiley, 2011.

Greene, Brian, *The Elegant Universe: Superstrings, Hidden Dimensions, and the Quest for the Ultimate Reality.* New York and London: W.W. Norton & Company, 2003.

Greene, Brian, *The Hidden Reality: Parallel Universes and the Deep Laws of the Cosmos.* New York: Alfred A. Knopf, 2011.

Hamilton III, William F., *Cosmic Top Secret: America's Secret UFO Program.* New Brunswick, NJ: Inner Light Publications, 1991.

Harrison, Albert A., *After Contact: The Human Response to Extraterrestrial Life.* New York: Plenum Press, 1997.

Hawking, Stephen, *Black Holes and Baby Universes and Other Essays.* New York: Bantam Books, 1993.

Hawking, Stephen, *A Brief History of Time: From the Big Bang to Black Holes.* New York: Bantam Books, 1998.

Hawking, Stephen and Leonard Mlodinow, *The Grand Design*. New York: Bantam Books, 2010.

Huntington, Samuel P., *The Clash of Civilizations: Remaking of World Order*. New York: Touchstone, 1996.

Jessup, Morris K., *The Allende Letters and the Varo Edition of The Case for the UFOs*. New Brunswick, NJ: Conspiracy Journal/Global Communications, 2007.

Kaku, Michio, *Einstein's Cosmos: How Albert Einstein's Vision Transformed Our Understanding of Space and Time*. New York and London: W.W. Norton & Company, 2004.

Kaku, Michio, *Hyperspace: A Scientific Odyssey Through Parallel Universes, Time Warps, and the 10th Dimension*. New York: Doubleday, 1994.

Kaku, Michio, *Parallel Worlds: A Journey through Creation, Higher Dimensions, and the Future of the Cosmos*. New York: Anchor Books, 2005.

Kaku, Michio, *Visions: How Science Will Revolutionize the 21st Century*. New York: Anchor Books, 1997.

Kaku, Michio, *Physics of the Impossible: A Scientific Exploration into the World of Phasers, Force Fields, Teleportation, and Time Travel*. New York: Anchor Books, 2008.

Kasten, Len, *Secret Journey to Planet Serpro*. Rochester, VT and Toronto, Canada: Bear & Company, 2013.

Kean, Leslie, *UFOs: Generals, Pilots, and Government Officials Go on the Record*. New York: Harmony Books, 2010.

Kondo, Yoji, Frederick Bruhweiler, John Moore, and Charles Sheffield (eds.), *Interstellar Travel and Multi-Generation Space Ships*. Burlingon, Ontario, Canada: Apogee Books, 2003.

Krauss, Lawrence M., *A Universe From Nothing: Why There Is Something Rather than Nothing*. Free Press: New York, 2012.

Kurland, Michael, *The Complete Idiot's Guide to Extraterrestrial Intelligence*. New York: alpha books, 1999.

Kurzweil, Ray, *The Singularity is Near: When Humans Transcend Biology*. New York: Penguin Books, 2005.

LaViolette, Paul A., *Secrets of Antigravity Propulsion: Tesla, UFOs, and Classified Aerospace Technology*. Rochester, VT: Bear & Company, 2008.

Maloney, Mack, *UFOs in Wartime: What They Don't Want You to Know*. New York: Berkley Books, 2011.

Marrs, Jim, *Alien Agenda: Investigating the Extraterrestrial Presence Among Us*. New York: Harper Paperbacks, 1997.

Michaud, Michael A.G., *Contact with Alien Civilizations: Our Hopes and Fears about Encountering Extraterrestrials*. New York: Copernicus Books, 2007.

McTaggart, Lynne, *The Field: The Quest for the Secret Force of the Universe*. New York: Harper, 2008.

Panek, Richard, *The 4 Percent Universe: Dark Matter, Dark Energy, and the Race to Discover the Rest of Reality*. Boston and New York: Houghton Mifflin Harcourt, 2011.

Pinker, Stephen, *The Better Angels of Our Nature: Why Violence has Declined*. New York: Penguin Books, 2012.

Plaxco, Kevin and Michael Gross, *Astrobiology: A Brief Introduction*. Second Edition. Baltimore, MD: The Johns Hopkins University Press, 2011.

Pope, Nick, *Open Skies, Closed Minds: For the First Time A Government UFO Expert Speaks Out*. London and New York: Simon & Schuster, 1996.

Pye, Michael and Kirsten Dalley (eds.), *UFOs & Aliens: Is There Anybody Out There?* Pompton Plains, NJ: The Career Press, 2011.

Randle, Kevin D., *Case MJ-12: The True Story Behind the Government's UFO Conspiracies*. New York: HarperTorch, 2002.

Randles, Jenny, *Breaking the Time Barrier: The Race to Build the First Time Machine*. New York: Paraview Pocket Books, 2005.

Randles, Jenny, *The Truth Behind Men in Black: Government Agents—Or Visitor From Beyond*. New York: St. Martin's Press, 1997.

Redfern, Nick, *Contactees: A History of Alien-Human Interaction*. Pompton Plains, NJ: New Page Books, 2010.

Redfern, Nick, *The Real Men in Black: Evidence, Famous Cases, and True Stories of these Mysterious Men and Their Connection to UFO Phenomena*. Pompton Plains, NJ: New Page Books, 2011.

Roach, Mary, *Packing for Mars: The Curious Science of Life in the Void*. New York and London: W.W. Norton & Company, 2013.

Sagan, Carl and Thornton Page (eds.), *UFOs A Scientific Debate*. New York: W.W. Norton & Company, 1972.

Sagan, Carl (ed.), *Communication with Extraterrestrial Intelligence*. Cambridge, MA: MIT Press, 1973.

Sagan, Carl, *Cosmos*, New York: Ballantine Books, 1985.

Sagan, Carl, *Pale Blue Dot: A Vision of the Human Future in Space*. New York: Ballantine Books, 1994.

Sagan, Carl, *Carl Sagan's Cosmic Connection: An Extraterrestrial Perspective*. Cambridge: Cambridge University Press, 2000.

Salla, Michael E., *Exopolitics: Political Implications of the Extraterrestrial Presence*. Temple, AZ: Danelion Books, 2004.

Sanderson, Ivan T., *Invisible Residents: The Reality of Underwater UFOs*. Kempton, IL: Adventures Unlimited Press, 2005.

Schmidt, Stanley, *Aliens and Alien Societies: A writer's guide to creating extraterrestrial life-forms*. Cincinnati, OH: Writer's Digest Books, 1995.

Shklovskii, I.S. and Carl Sagan, *Intelligent Life in the Universe*. San Francisco, London, and Amsterdam: Holden-Day, Inc., 1966.

Shostak, Seth, *Confessions of an Alien Hunter: A Scientist's Search for Extraterrestrial Intelligence*. Washington, D.C.: National Geographic, 2009.

Sitchin, Zecharia, *The 12th Planet: Book I of the Earth Chronicles*. New York: Harper-Collins, 2007.

Steiger, Brad, *The Philadelphia Experiment & Other UFO Conspiracies*. New Brunswick, NJ: Inner Light Publications, 1990.

Steiger, Brad and Sherry Hansen Steiger, *Real Aliens, Space Beings, and Creatures from Other Worlds*. Canton, MI: Visible Ink Press, 2011.

Strassman, Rick, *DMT: The Spirit Molecule: A Doctor's Revolutionary Research into the Biology of Near-Death and Mystical Experiences*. Rochester, VT: Part Street Press, 2000.

Strassman, Rick, Slawek Wojtowicz, Luis Eduardo Luna, and Ede Frecska, *Innter Paths to Outer Space: Journeys to Alien Worlds though Psychadelics and Other Spritual Technologies*. Rochester, VT: Park Street Press, 2008.

Strieber, Whitley, *Communion: A True Story*. New York: Avon, 1987.

Swords, Michael and Robert Powell, *UFOs and Government: A Historical Inquiry*. San Antonio and Charlottesville: Anomalist Books, 2012.

Talbot, Michael, *The Holographic Universe: The Revolutionary Theory of Reality*. New York: Harper/Perennial, 2001.

Taylor, Travis S. and Bob Boan, *Alien Invasion: The Ultimate Survival Guide for the Ultimate Attack*. Riverdale, NY: Baen Publishing Enterprises, 2011.

The UFO Phenomenon. Alexandria, VA: Time Life Books, 1987.

Tyson, Neil deGrasse, *Space Chronicles: Facing the Ultimate Frontier*. New York: W.W. Norton & Company Ltd., 2012.

United States Air Force, *The Roswell Report: Case Closed*. New York: Skyhorse Publishing, 2013.

Vakoch, Douglas A. (ed.), *Communication with Extraterrestrial Intelligence*. Albany: State University Press of New York, 2011.

Vakoch, Douglas A. and Albert A. Harrison (eds.), *Civilizations Beyond Earth: Extraterrestrial Life and Society*. New York and Oxford: Berghan Books 2011.

Vakoch, Douglas A. (ed.), *On Orbit and Beyond: Psychological Perspectives on Human Spaceflight*. Heidelberg, New York, Dordrecht, and London: Springer 2013.

Vakoch, Douglas A. (ed.), *Extraterrestrial Altruim: Evolution and Ethics in the Cosmos*. Heidelberg, New York, Dordrecht, and London: Springer, 2013.

Von Däniken, Erich, *Chariots of the Gods*. New York: Berkley Books, 1999.

Endnotes

Introduction

1 John Gribbin, *Alone in the Universe: Why Our Planet is Unique*. Hoboken, NJ: Wiley, 2011.

2 P. C.W. Davis, "Searching for a shadow biosphere on Earth as a test of the 'cosmic imperative,'" *Philosophical Transactions of the Royal Society*, no. 369 (2011), pp. 625–626.

3 Paul Davies, *The Eerie Silence: Renewing Our Search for Alien Intelligence*. Boston and New York: Houghton Mifflin Harcourt, 2010, pp. 24–27.

4 Ibid., p. 33.

5 Davis, "Searching for a shadow biosphere on Earth as a test of the 'cosmic imperative,'" pp. 625–626.

6 "Cosmic census finds crowd of planets in galaxy." Associated Press, February 20, 2011.

7 Seth Shostak, "Are We Alone? Estimating the Prevalence of Extraterrestrial Intelligence," in Douglas A. Vakoch and Albert A. Harrison (eds.), *Civilizations Beyond Earth: Extraterrestrial Life and Society*. New York and Oxford: Berghan Books, 2011, p. 35.

8 George Michael, "Extraterrestrial Aliens: Friends, Foes, or Just Curious?" *Skeptic Magazine*, Vol. 16, No. 3, (2011), pp. 46-53.

9 Cosmos is a Greek word for the order of the universe. Carl Sagan, *Cosmos*. New York: Ballantine Books, 1985, p. 10.

10 Dennis Overbye, "Far-Off Planets Like the Earth Dot the Galaxy," *The New York Times*,

November 4, 2013, http://www.nytimes.com/2013/11/05/science/cosmic-census-finds-billions-of-planets-that-could-be-like-earth.html?_r=0.

11 Michael A.G. Michaud, *Contact with Alien Civilizations: Our Hopes and Fears about Encountering Extraterrestrials*. New York: Copernicus Books, 2007, p. 59.

12 Ibid., p. 63.

13 Albert A. Harrison, "The Science and Politics of SETI: How to Succeed in an Era of Make-Believe History and Pseudoscience," in Douglas A. Vakoch and Albert A. Harrison (eds.), *Civilizations Beyond Earth: Extraterrestrial Life and Society*. New York and Oxford: Berghan Books 2011, p. 144.

14 Rick Strassman, *DMT: The Spirit Molecule: A Doctor's Revolutionary Research into the Biology of Near-Death and Mystical Experiences*. Rochester, VT: Part Street Press, 2000.

15 Michio Kaku, *Physics of the Impossible: A Scientific Exploration into the World of Phasers, Force Fields, Teleportation, and Time Travel*. New York: Anchor Books, 2008, p. 142.

16 Michael Swords and Robert Powell (eds.), "Chapter 13: Battle in the Desert," in *UFOs and Government: A Historical Inquiry*. San Antonio and Charlottesville: Anomalist Books, 2012, p. 297.

17 Stanton T. Friedman, *Flying Saucers and Science: A Scientist Investigates the Mysteries of UFOs*. The Career Press: Pompton Plains, NJ, 2008, p. 136.

18 Harrison, "The Science and Politics of SETI", pp. 141–155.

chapter 1

1 I.S. Shklovskii and Carl Sagan, *Intelligent Life in the Universe*. San Francisco, London, and Amsterdam: Holden-Day, Inc., 1966, p. 3.

2 Michael Kurland, *The Complete Idiot's Guide to Extraterrestrial Intelligence*. New York: alpha books, 1999, p. 6.

3 Isaac Asimov, *Extraterrestrial Civilizations*. New York: Crown Publishers, Inc., 1979, p. 24 and I.S. Shklovskii and Carl Sagan, *Intelligent Life in the Universe*, p. 3.

4 Michael J. Crowe, *The Extraterrestrial Life Debate: Antiquity to 1915*. Notre Dame, IN: University of Notre Dame Press, 2008, p. 13.

5 Steven J. Dick, *Life on Other Worlds: The 20th-Century Extraterrestrial Life Debate*. New York: Cambridge University Press, 1998, p. 8.

6 Quoted in Shklovskii and Sagan, *Intelligent Life in the Universe*, p. 3.

7 Crowe, *The Extraterrestrial Life Debate*, pp. 8–13.

8 George Basalla, *Civilized Life in the Universe: Scientists on Intelligent Extraterrestrials*. Oxford and New York: Oxford University Press, 2006, p. 4.

9 Crowe, *The Extraterrestrial Life Debate*, p. 13.

10 Michael A.G. Michaud, *Contact with Alien Civilizations: Our Hopes and Fears about Encountering Extraterrestrials*. New York: Copernicus Books, 2007, p. 11.

11 Ockham's razor is often invoked in science when considering numerous hypotheses. Also known as the law of parsimony, economy, or succinctness, it posits that the one which makes the fewest assumptions should be selected. In essence, Ockham's razor favors the most straight-forward explanation to a question.

12 Dick, *Life on Other Worlds*, pp. 8–9.

13 Michaud, *Contact with Alien Civilizations*, p. 13.

14 Basalla, *Civilized Life in the Universe*, p. 11.

15 Dick, *Life on Other Worlds*, p. 10.

16 Crowe, *The Extraterrestrial Life Debate*, p. 518.

17 Ibid., p. 36.

18 Basalla, *Civilized Life in the Universe*, pp. 4–5.

19 Crowe, *The Extraterrestrial Life Debate*, p. 35.

20 Michaud, *Contact with Alien Civilizations*, p. 13.

21 Basalla, *Civilized Life in the Universe*, p. 8.

22 Dick, *Life on Other Worlds*, p. 10.

23 Basalla, *Civilized Life in the Universe*, p. 4.

24 Crowe, *The Extraterrestrial Life Debate*, pp. 18–19.

25 Ibid., p. 4.

26 Basalla, *Civilized Life in the Universe*, p. 36.

27 Asimov, *Extraterrestrial Civilizations*, p. 24.

28 Basalla, *Civilized Life in the Universe*, p. 3.

29 Kepler demonstrated that the orbits of the planets around the sun were not circles, but ellipses.

30 Basalla, *Civilized Life in the Universe*, p. 22.

31 Seth Shostak, *Confessions of an Alien Hunter: A Scientist's Search for Extraterrestrial Intelligence*. Washington, D.C.: National Geographic, 2009, pp. 26–27.

32 Basalla, *Civilized Life in the Universe*, pp. 23–25.

33 Michaud, *Contact with Alien Civilizations*, p. 15.

34 Basalla, *Civilized Life in the Universe*, p. 52.
35 Shklovskii and Sagan, *Intelligent Life in the Universe*, p. 6.
36 Dick, *Life on Other Worlds*, p. 14.
37 Basalla, *Civilized Life in the Universe*, p. 50.
38 Crowe, *The Extraterrestrial Life Debate*, pp. 116–119.
39 Dick, *Life on Other Worlds*, p. 12.
40 Basalla, *Civilized Life in the Universe*, p. 35.
41 Shostak, *Confessions of an Alien Hunter*, p. 25.
42 Crowe, *The Extraterrestrial Life Debate*, p. 105.
43 Basalla, *Civilized Life in the Universe*, p. 41–42.
44 Ibid., pp. 36–40.
45 Dick, *Life on Other Worlds*, p. 15.
46 Ibid., p. 16.
47 Ibid., p. 16.
48 Ibid., p. 17.
49 Ibid., p. 24.
50 Crowe, *The Extraterrestrial Life Debate*, p. 333.
51 Ibid., pp. 333–354.
52 Dick, *Life on Other Worlds*, p. 107.
53 Ibid., p. 18. It is worth mentioning that Alfred Russell Wallace was a co-discoverer of evolution by natural selection. Carl Sagan, *Cosmos*. New York: Ballantine Books, 1985, p. 88.
54 Crowe, *The Extraterrestrial Life Debate*, p. 383.
55 Dick, *Life on Other Worlds*, pp. 18–19.
56 Ibid., pp. 19–20.
57 Crowe, *The Extraterrestrial Life Debate*, pp. 404–427.
58 Shklovskii and Sagan, *Intelligent Life in the Universe*, p. 7.
59 Dick, *Life on Other Worlds*, p. 31.
60 Crowe, *The Extraterrestrial Life Debate*, p. 481.
61 Basalla, *Civilized Life in the Universe*, pp. 56–67.
62 Shostak, *Confessions of an Alien Hunter*, pp. 29–30.
63 Paul Davies, *The Eerie Silence: Renewing Our Search for Alien Intelligence*. Boston and New York: Houghton Mifflin Harcourt, 2010, p. 16.

64 Asimov, *Extraterrestrial Civilizations*, p. 50.

65 Dick, *Life on Other Worlds*, p. 34.

66 Basalla, *Civilized Life in the Universe*, p. 75.

67 Ibid., p. 82.

68 Ibid., p. 82.

69 Ibid., pp. 86–87.

70 Shostak, *Confessions of an Alien Hunter*, pp. 30–32.

71 Crowe, *The Extraterrestrial Life Debate*, pp. 492–507.

72 Dick, *Life on Other Worlds*, pp. 38–39.

73 Ibid., p. 53.

74 Shklovskii and Sagan, *Intelligent Life in the Universe*, p. 6.

75 Crowe, *The Extraterrestrial Life Debate*, pp. 427–437.

chapter 2

1 Michael J. Crowe, *The Extraterrestrial Life Debate: Antiquity to 1915*. Notre Dame, IN: University of Notre Dame Press, 2008, pp. 130–138.

2 George Basalla, *Civilized Life in the Universe: Scientists on Intelligent Extraterrestrials*. Oxford and New York: Oxford University Press, 2006, p. 129.

3 Michio Kaku, *Physics of the Impossible: A Scientific Exploration into the World of Phasers, Force Fields, Teleportation, and Time Travel*. New York: Anchor Books, 2008, p. 292. In 1781, Immanuel Kant's *Critique of Pure Reason* presented the first claim that the Milky Way was a disk-shaped structure. Crowe, *The Extraterrestrial Life Debate*, p. 138.

4 Michio Kaku, *Einstein's Cosmos: How Albert Einstein's Vision Transformed Our Understanding of Space and Time*. New York and London: W.W. Norton & Company, 2004, p. 133.

5 Frank Drake and Dava Sobel, *Is Anyone Out There? The Scientific Search for Extraterrestrial Intelligence*. New York: Delacorte Press, 1992, p. 73.

6 Richard Panek, *The 4 Percent Universe: Dark Matter, Dark Energy, and the Race to Discover the Rest of Reality*. Boston and New York: Houghton Mifflin Harcourt, 2011, pp, 4–5.

7 According to Michio Kaku, superstring theory can predict what happened before the Big Bang. String theory posits that at the moment of the Big Bang, the four forces (gravity, electromagnetism, and the strong and weak nuclear forces) were all united. Moreover, the four dimensions of which we are aware—three spatial dimensions and

time—were once combined with six other dimensions. However, together these ten dimensions were unstable and "cracked" into two pieces, leaving us with the four we are familiar with and six others that we cannot detect. String theory views the Big Bang as merely an aftershock of a much larger and much greater cataclysm—the cracking of space and time itself. Michio Kaku, *Hyperspace: A Scientific Odyssey Through Parallel Universes, Time Warps, and the 10th Dimension*. New York: Doubleday, 1994, p. 27.

8 Paul Davies, *The Eerie Silence: Renewing Our Search for Alien Intelligence*. Boston and New York: Houghton Mifflin Harcourt, 2010, p. 198.

9 Stanley Schmidt, *Aliens and Alien Societies: A writer's guide to creating extraterrestrial life-forms*. Cincinnati, OH: Writer's Digest Books, 1995, p. 23.

10 Michael Kurland, *The Complete Idiot's Guide to Extraterrestrial Intelligence*. New York: alpha books, 1999, p. 227.

11 Isaac Asimov, *Extraterrestrial Civilizations*. New York: Crown Publishers, Inc., 1979, p. 130.

12 John Gribbin, *Alone in the Universe: Why Our Planet is Unique*. Hoboken, NJ: Wiley, 2011, p. 74.

13 Neil de Grasse Tyson, *Space Chronicles: Facing the Ultimate Frontier*. New York: W.W. Norton & Company Ltd., 2012, p. 258.

14 Michael A.G. Michaud, *Contact with Alien Civilizations: Our Hopes and Fears about Encountering Extraterrestrials*. New York: Copernicus Books, 2007, pp. 58–59.

15 Crowe, *The Extraterrestrial Life Debate*, p. 370.

16 Pascale Ehrenfreund, Marco Spaans, and Nils G. Holm, "The evolution of organic matter in space," *Philosophical Transactions of the Royal Society*, no. 369 (2011), pp. 538–554.

17 Albert A. Harrison, *After Contact: The Human Response to Extraterrestrial Life*. New York: Plenum Press, 1997, p. 13.

18 de Grasse Tyson, *Space Chronicles*, p. 240.

19 Ibid., p. 36.

20 Freeman J. Dyson, "Looking for Life in Unlikely Places: Reasons Why Planets May Not Be the Best Places to Look for Life," in Yoji Kondo, Frederick Bruhweiler, John Moore, and Charles Sheffield (eds.), *Interstellar Travel and Multi-Generation Space Ships*. Burlington, Ontario, Canada: Apogee Books, 2003, p. 107.

21 Harrison, *After Contact*, p. 10. (It was actually Rev. William Whewell who first coined the term "nebular hypothesis." Crowe, *The Extraterrestrial Life Debate*, p. 302.)

22 I.S. Shklovskii and Carl Sagan, *Intelligent Life in the Universe*. San Francisco, London, and Amsterdam: Holden-Day, Inc., 1966, p. 162.

23 Steven J. Dick, *Life on Other Worlds: The 20th-Century Extraterrestrial Life Debate*. New York: Cambridge University Press, 1998, p. 72.

24 Shklovskii and Sagan, *Intelligent Life in the Universe*, p. 165.

25 Ibid., pp. 165–175.

26 Robert Holz, "Scientists Uncover Evidence of New Planets Orbiting Star," *Los Angeles Times*, April 22, 1992.

27 Dick, *Life on Other Worlds*, p. 71.

28 Harrison, *After Contact*, p. 11.

29 Charles Sheffield, "Fly Me to the Stars: Interstellar Travel in Fact and Fiction," in Yoji Kondo, Frederick Bruhweiler, John Moore, and Charles Sheffield (eds.), *Interstellar Travel and Multi-Generation Space Ships*. Burlington, Ontario, Canada: Apogee Books, 2003, p. 24.

30 These estimates vary depending on the source. This figure is from the NASA Exoplanet Archive, which can be found at http://exoplanetarchive.ipac.caltech.edu/, accessed June 14, 2014.

31 Jill Tarter, "Exoplanets, Extremophiles, and the Search for Extraterrestrial Intelligence," in Douglas A. Vakoch (ed.), *Communication with Extraterrestrial Intelligence*. Albany: State University Press of New York, 2011, pp. 4–5.

32 Seth Shostak, "Are We Alone? Estimating the Prevalence of Extraterrestrial Intelligence," in Douglas A. Vakoch and Albert A. Harrison (eds.), *Civilizations Beyond Earth: Extraterrestrial Life and Society*. New York and Oxford: Berghan Books, 2011, p. 36.

33 Erik A. Petigura, Andrew W. Howard, and Geoffrey W. Marcy, "Prevalence of Earth-size planets orbiting Sun-like starts," *Proceeding of the National Academy of Sciences*, vol. 110, no. 48 (November 26, 2013) pp. 1–6, http://www.pnas.org/content/early/2013/10/31/1319909110.full.pdf+html.

34 "Cosmic census finds crowd of planets in galaxy," Associated Press, February 20, 2011.

35 Dennis Overbye, "Far-Off Planets Like the Earth Dot the Galaxy," *The New York Times*, November 4, 2013. http://www.nytimes.com/2013/11/05/science/cosmic-census-finds-billions-of-planets-that-could-be-like-earth.html?_r=0.

36 See for example, Gribbin, *Alone in the Universe*.

37 Gribbin, *Alone in the Universe*, p. 72.

38 Shklovskii and Sagan, *Intelligent Life in the Universe*, p. 89.

39 Ibid., p. 108.

40 Kevin Plaxco and Michael Gross, *Astrobiology: A Brief Introduction*. Second Edition. Baltimore, MD: The Johns Hopkins University Press, 2011, p. 48.

41 Gribbin, *Alone in the Universe*, p. 74.

42 Ibid., p. 77.

43 Ibid., p. 79.

44 Ibid., p. 9.

45 Stephen Hawking and Leonard Mlodinow, *The Grand Design*. New York: Bantam Books, 2010, p. 152.

46 Hawking and Mlodinow, *The Grand Design*, p. 152 and Seth Shostak, *Confessions of an Alien Hunter: A Scientist's Search for Extraterrestrial Intelligence*. Washington, D.C.: National Geographic, 2009, p. 6.

47 Gribbin, *Alone in the Universe*, p. 85.

48 Asimov, *Extraterrestrial Civilizations*, p. 123.

49 Ibid., p. 136.

50 Gribbin, *Alone in the Universe*, p. 89.

51 Stephen Webb, *Where is Everybody? Fifty Solutions to the Fermi Paradox and the Problem of Extraterrestrial Life*. New York, Copernicus Books, 2000, p. 158.

52 Charles Q. Choi, "Newfound 'Tatooine' Alien Planet Bodes Well for E.T. Search," September 4, 2012. http://www.space.com/17416-tatooine-alien-planet-habitable-zone-life.html.

53 Davies, *The Eerie Silence*, p. 103.

54 Gribbin, *Alone in the Universe*, pp. 72–73.

55 de Grasse Tyson, *Space Chronicles*, p. 49.

56 Michaud, *Contact with Alien Civilizations*, p. 107.

57 Gribbin, *Alone in the Universe*, p. 76.

58 Davies, *The Eerie Silence*, p. 18.

59 Ibid., p. 24.

60 Michael D. Lemonick, "Anyone Out There? A New Way to Look for Alien Life," *Time*, November 30, 2011.

61 Lemonick, "Anyone Out There? A New Way to Look for Alien Life."

62 Dave Goldberg and Jeff Blomquist, *A User's Guide to the Universe: Surviving the Perils*

of Black Holes, Time Paradoxes, and Quantum Uncertainty. Hoboken, NJ: Wiley, 2010, pp. 237–239.

63 Gribbin, *Alone in the Universe*, p. 135.

64 Ibid., p. 32.

65 Plaxco and Gross, *Astrobiology*, p. 65.

66 Gribbin, *Alone in the Universe*, pp. 126–150.

67 Ibid., 2011).

68 Email to author from John Gribbin, November 14, 2012.

69 Smolin suggested that the laws of physics in our universe emerged from a process not unlike Darwinian evolution that included mutation and natural selection. As he explained, a black hole in one universe could give birth to another, different expanding universe. Thus a universe that permits the formation of black holes has descendants that will in turn produce more black holes. Conversely, a universe operating under parameters that lead to little or no black hole formation will produce little or no offspring (Webb, *Where is Everybody?* pp. 57–58). Similarly, Carl Sagan once explained that if our universe is closed and even light cannot escape from it, then it is not unreasonable to describe the universe as the interior of a black hole. As he famously put it, "If you wish to know what it is like inside a black hole, look around you." Carl Sagan, *Cosmos*, New York: Ballantine Books, 1985, p. 221.

70 Quoted in Paul Davies, "Life, the Universe and Everything," *Cosmos Magazine*, July 19, 2007. http://cosmosmagazine.com/features/life-universe-and-everything/.

71 Michaud, *Contact with Alien Civilizations*, pp. 197–198.

72 If the total energy of the universe must be zero, and it takes energy to create mass, Hawking and Mlodinow ask, how can a universe be created from nothing? As they explain, it is because of gravity, which is attractive and gravitational energy, which is negative. Consequently, this negative energy can balance the positive energy needed to create matter (Hawking and Mlodinow, *The Grand Design*, pp. 8–9, 180). Likewise, the theoretical physicist Lawrence M. Krauss argued that the universe effectively arose from nothing. According to the quantum theory of electromagnetism, particles can pop out of empty space at will disappear again on a time frame determined by Heisenberg's Uncertainty Principle. As Krauss explains, the our universe originated with a "quantum fluctuation" in the density of seemingly empty space, but rapidly expanded in a process called inflation, which accounts for all of the matter and energy in the

stars and galaxies. In that sense, the universe is, as Alan Guth put it, the ultimate "free lunch." However, as Krauss lamented, insofar as positive energy resides in empty space, our universe cannot be stable. As a consequence, our universe could disappear abruptly as it began (Lawrence M. Krauss, *A Universe From Nothing: Why There Is Something Rather than Nothing*. Free Press: New York, 2012, pp. 103 and 180).

73 Carl Sagan, *Pale Blue Dot: A Vision of the Human Future in Space*. New York: Ballantine Books, 1994, p. 35.

74 Michio Kaku, "The Paradox of Multiple Goldilocks Zones or 'Did the Universe Know We Were Coming?'" bigthink.com, May 20, 2011. http://bigthink.com/dr-kakus-universe/the-paradox-of-multiple-goldilocks-zones-or-did-the-universe-know-we-were-coming.

75 Shostak, "Are We Alone? Estimating the Prevalence of Extraterrestrial Intelligence," p. 32.

76 Brian Greene, *The Hidden Reality: Parallel Universes and the Deep Laws of the Cosmos*. New York: Alfred A. Knopf, 2011.

77 Stephen H. Dole once attempted to estimate the number of planets that could support human beings without life-support systems such as domes or spacesuits. He concluded that there could be 600 million human-habitable planets in our galaxy alone. Schmidt, *Aliens and Alien Societies*, p. 66.

78 Author interview with Steven J. Dick July 14, 2012.

79 Author interview with Michael G. Michaud, July 8, 2012.

80 William Sims Bainbridge, "Direct Contact with Extraterrestrials via Computer Emulation," in Douglas A. Vakoch and Albert A. Harrison (eds.), *Civilizations Beyond Earth: Extraterrestrial Life and Society*. New York and Oxford: Berghan Books 2011, p. 200.

81 Seth Borenstein, "NASA finds planet that could sustain life," Associated Press, December 5, 2011.

82 Paul Gilster, *Centauri Dreams: Imagining and Planning Interstellar Exploration*. New York: Copernicus Books, 2004, p. 45.

chapter 3

1 According to a 2005 poll commissioned by the National Geographic Channel and conducted by the University of Connecticut's Center for Survey Research and Analysis, 60 percent of those Americans surveyed believe that there was intelligent life elsewhere in the universe. Seth Shostak, *Confessions of an Alien Hunter: A Scientist's Search for*

Extraterrestrial Intelligence. Washington, D.C.: National Geographic, 2009, p. 3. An Internet poll conducted on 1 July 2000 by CNN in which of 6,399 people participated arrived at a figure of 82 percent. Stephen Webb, *Where is Everybody? Fifty Solutions to the Fermi Paradox and the Problem of Extraterrestrial Life*. New York, Copernicus Books, 2000, p. 4.

2 Carl Sagan, in Jerome Agel (ed.). *Cosmic Connection: An extraterrestrial Perspective*. Cambridge: Cambridge University Press, 1973, 189.

3 As outlined in Michael A.G. Michaud, *Contact with Alien Civilizations: Our Hopes and Fears about Encountering Extraterrestrials*. New York: Copernicus Books, 2007, p. 68.

4 Shostak, *Confessions of an Alien Hunter*, pp. 65–72.

5 Ibid., pp. 97–98.

6 Ibid., p. 80.

7 Toney Allman, *Are Extraterrestrials a Threat to Humankind?* San Diego, CA: Reference Point Press, 2012, p. 18.

8 Baruch S. Blumberg, "Astrobiology, space and the future age of discovery," *Philosophical Transactions of the Royal Society*, no. 369 (2011), p. 508.

9 Albert A. Harrison, "The Science and Politics of SETI: How to Succeed in an Era of Make-Believe History and Pseudoscience," in Douglas A. Vakoch and Albert A. Harrison (eds.), *Civilizations Beyond Earth: Extraterrestrial Life and Society*. New York and Oxford: Berghan Books 2011, p. 143.

10 Definition of "life" from Merriam-Webster Dictionary, http://www.merriam-webster.com/dictionary/life, accessed August, 11, 2013.

11 Kevin Plaxco and Michael Gross, *Astrobiology: A Brief Introduction*. Second Edition. Baltimore, MD: The Johns Hopkins University Press, 2011, p. 3.

12 Albert A. Harrison and Douglas A. Vakoch, "Introduction: The Search for Extraterrestrial Intelligence as an Interdisciplinary Effort," in Douglas A. Vakoch and Albert A. Harrison (eds.), *Civilizations Beyond Earth: Extraterrestrial Life and Society*. New York and Oxford: Berghan Books, 2011, p. 5.

13 Michael Kurland, *The Complete Idiot's Guide to Extraterrestrial Intelligence*. New York: alpha books, 1999, p. 227.

14 Shostak, *Confessions of an Alien Hunter*, p. 259.

15 I.S. Shklovskii and Carl Sagan, *Intelligent Life in the Universe*. San Francisco, London, and Amsterdam: Holden-Day, Inc., 1966, p. 229.

16 Carl Sagan, in Jerome Agel (ed.). *Cosmic Connection*, pp. 46–47.

17 Paul Davies, *The Eerie Silence: Renewing Our Search for Alien Intelligence*. Boston and New York: Houghton Mifflin Harcourt, 2010, p. 55.

18 Stanley Schmidt, *Aliens and Alien Societies: A Writer's Guide to Creating Extraterrestrial Life-forms*. Cincinnati, OH: Writer's Digest Books, 1995, p. 60.

19 Plaxco and Gross, *Astrobiology*, p. 13.

20 Webb, *Where is Everybody?* pp. 147–149.

21 Davies, *The Eerie Silence*, p. 198.

22 Neil de Grasse Tyson, *Space Chronicles: Facing the Ultimate Frontier*. New York: W.W. Norton & Company Ltd., 2012, p. 48.

23 Isaac Asimov, *Extraterrestrial Civilizations*. New York: Crown Publishers, Inc., 1979, p. 31.

24 Michaud, *Contact with Alien Civilizations*, p. 67.

25 Seth Shostak, "Are We Alone? Estimating the Prevalence of Extraterrestrial Intelligence," in Douglas A. Vakoch and Albert A. Harrison (eds.), *Civilizations Beyond Earth: Extraterrestrial Life and Society*. New York and Oxford: Berghan Books, 2011, p. 34.

26 Michael D. Lemonick, "Anyone Out There? A New Way to Look for Alien Life," *Time*, November 30, 2011.

27 Plaxco and Gross, *Astrobiology*, p. 14.

28 John Gribbin, *Alone in the Universe: Why Our Planet is Unique*. Hoboken, NJ: Wiley, 2011, pp. 80–81.

29 Water is believed to be second only to hydrogen as the most common molecule in the universe. Plaxco and Gross, *Astrobiology*, pp. 16–17.

30 Seth Shostak, "Are We Alone? Estimating the Prevalence of Extraterrestrial Intelligence," p. 34.

31 Plaxco and Gross, *Astrobiology*, pp. 15–16.

32 P.C.W. Davis, "Searching for a shadow biosphere on Earth as a test of the 'cosmic imperative,'" *Philosophical Transactions of the Royal Society*, no. 369 (2011), p. 625.

33 Ronald N. Bracewell, *The Galactic Club: Intelligent Life in Outer Space*. San Francisco, CA: W.W. Freeman and Company, 1975, pp. 8–10.

34 Frank Drake and Dava Sobel, *Is Anyone Out There? The Scientific Search for Extraterrestrial Intelligence*. New York: Delacorte Press, 1992, p. 50.

35 Plaxco and Gross, *Astrobiology*, p. 200.

36 Lovelock advanced the "Gaia" theory, which suggests that the planet Earth can be conceptualized as a living organism. Gribbin, *Alone in the Universe*, p. 21.

37 Carl Sagan, in Jerome Agel (ed.), *Cosmic Connection: An extraterrestrial Perspective*, p. 148.

38 Many organisms on Earth survive without oxygen; in fact, many organisms are poisoned by it. Some living organisms on Earth use a variety of methods to avoid contact with oxygen altogether or to repair the oxidation damage that occurs as a result of exposure. Carl Sagan, in Jerome Agel (ed.), *Cosmic Connection: An extraterrestrial Perspective*, p. 44.

39 Shklovskii and Sagan, *Intelligent Life in the Universe*, p. 224.

40 Plaxco and Gross, *Astrobiology*, p. 188.

41 Carl Sagan once speculated that on other planets there could be more energetic foodstuffs available. Or perhaps the metabolic process could operate more slowly, which would obviate the advantage of oxygen. Shklovskii and Sagan, *Intelligent Life in the Universe*, p. 216.

42 As Kevin Plaxco and Michael Gross, explain, "On Earth, at least, it took a reasonable fraction of the age of the Universe before an obscure, mammalian branch of the eukaryotic domain of the tree of life evolved to a point where it could spend endless hours arguing over whether Archaea [a domain of single-celled microorganisms] deserve their own domain." Plaxco and Gross, *Astrobiology*, p. 3.

43 Carl Sagan, *Cosmos*, New York: Ballantine Books, 1985, p. 22.

44 Michaud, *Contact with Alien Civilizations*, p. 69.

45 Schmidt, *Aliens and Alien Societies*, p. 58.

46 Gribbin, *Alone in the Universe*, p. 152.

47 Webb, *Where is Everybody?*, pp. 210–211.

48 Ibid., pp. 147–149.

49 The sun is type G on the scale of spectral types. On a scale ranging from hottest to coolest, the spectral types are designated by the letters O, B, A, F, G, K, and M. Only about 3 percent of the stars in the Milky Way are type G, as is the sun. Seventy percent of the stars in the galaxy are type M, while another 15 percent are type K. Paul Gilster, *Centauri Dreams: Imagining and Planning Interstellar Exploration*. New York: Copernicus Books, 2004, pp. 33–34.

50 Michio Kaku, *Physics of the Impossible: A Scientific Exploration into the World of Phas-*

ers, Force Fields, Teleportation, and Time Travel. (New York: Anchor Books, 2008), p. 140.

51 "Discussion" in Carl Sagan (ed.), *Communication with Extraterrestrial Intelligence.* Cambridge, MA: MIT Press, 1973, p. 120.

52 Albert A. Harrison, *After Contact: The Human Response to Extraterrestrial Life*. New York: Plenum Press, 1997, p. 11.

53 Shostak, *Confessions of an Alien Hunter*, p. 104.

54 Shklovskii and Sagan, *Intelligent Life in the Universe*, p. 205.

55 Ibid., p. 206.

56 Philip Coppens, *The Ancient Alien Question: A New Inquiry Into the Existence, Evidence, and Influence of Ancient Visitors*. Pompton Plains, NJ: New Page Books, p. 226.

57 Bracewell, *The Galactic Club*, p. 12.

58 Shostak, *Confessions of an Alien Hunter*, pp. 94–96.

59 Davies, *The Eerie Silence*, p. 26.

60 Asimov, *Extraterrestrial Civilizations*, p. 168.

61 Steven J. Dick, *Life on Other Worlds: The 20th-Century Extraterrestrial Life Debate*. New York: Cambridge University Press, 1998, pp. 170–171.

62 Plaxco and Gross, *Astrobiology*, p. 77.

63 Drake and Sobel, *Is Anyone Out There? The Scientific Search for Extraterrestrial Intelligence*, pp. 49–50.

64 Plaxco and Gross, *Astrobiology*, p. 77.

65 Ibid., pp. 78–79.

66 Kaku, *Physics of the Impossible*, p. 129.

67 Plaxco and Gross, *Astrobiology*, p. 79.

68 Shostak, *Confessions of an Alien Hunter*, p. 84.

69 Kaku, *Physics of the Impossible*, p. 129.

70 Plaxco and Gross, *Astrobiology*, pp. 1–2.

71 Davies, *The Eerie Silence*, p. 46.

72 Francis Fukuyama, *The Origins of Political Order: From Prehuman Times to the French Revolution*. New York: Farrar, Straus, and Giroux, 2011, p. 22.

73 Schmidt, *Aliens and Alien Societies*, pp. 52–56.

74 Dick, *Life on Other Worlds*, p. 188.

75 Gribbin, *Alone in the Universe*, p. 26.

76 "Life's Building Blocks Found on Surprising Meteorite," December 15, 2010. http://www.space.com/10498-life-building-blocks-surprising-meteorite.html.

77 Plaxco and Gross, *Astrobiology*, pp. 116–117.

78 Coppens, *The Ancient Alien Question*, p. 24.

79 Michaud, *Contact with Alien Civilizations*, p. 30.

80 Allman, *Are Extraterrestrials a Threat to Humankind?*, pp. 27–29.

81 Harrison, *After Contact*, p. 96.

82 Plaxco and Gross, *Astrobiology*, p. 117.

83 Shklovskii and Sagan, *Intelligent Life in the Universe*, p. 7.

84 Sagan, *Cosmos*, p. 25.

85 Davies, *The Eerie Silence*, p. 30.

86 F. H.C. Crick and L.E. Orgel, "Directed Panspermia," *Icarus* No. 19 (1973), pp. 341–346.

87 Quoted in Davis, "Searching for a Shadow Biosphere on Earth as a Test of the 'Cosmic Imperative,'" p. 625.

88 George Basalla, *Civilized Life in the Universe: Scientists on Intelligent Extraterrestrials.* (Oxford and New York: Oxford University Press, 2006), p. 132.

89 Kurland, *The Complete Idiot's Guide to Extraterrestrial Intelligence*, p. 210.

90 F.H.C. Crick and L.E. Orgel, "Directed Panspermia," *Icarus* No. 19 (1973), pp. 341–346.

91 Plaxco and Gross, *Astrobiology*, pp. 145–146.

92 Dick, *Life on Other Worlds*, p. 196.

93 Basalla, *Civilized Life in the Universe*, p. 179. According to the research of Brandon Carter and Robin Hanson, a "Great Filter" consisting of six crucial, but highly improbable hurdles must be surmounted en route to intelligent life. Each hurdle is roughly 800 million years apart. Their model corresponds well with Earth's fossil record. Brandon Carter argues that Earth is a very rare exception, and that the emergence of intelligent beings like humans was a freak event. Davies, *The Eerie Silence*, pp. 86–89.

94 Webb, *Where is Everybody?* p. 169.

95 Davies, *The Eerie Silence*, p. 68.

96 Kaku, *Physics of the Impossible*, p. 141.

97 Fukuyama, *The Origins of Political Order*, p. 34.

98 Shostak, *Confessions of an Alien Hunter*, p. 101.

99 Alan Penny, "The Lifetimes of Scientific Civilizations and the Genetic Evolution of the Brain," in Douglas A. Vakoch and Albert A. Harrison (eds.), *Civilizations Beyond*

Earth: Extraterrestrial Life and Society. New York and Oxford: Berghan Books, 2011, pp. 60–73.

100 "The Evolution of Technical Civilizations" in Carl Sagan (ed.), *Communication with Extraterrestrial Intelligence*. Cambridge, MA: MIT Press, 1973, p. 107.

101 Shostak, *Confessions of an Alien Hunter*, p. 103.

102 Carl Sagan, "The Number of Advanced Galactic Civilizations" in Carl Sagan (ed.), *Communication with Extraterrestrial Intelligence*. Cambridge, MA: MIT Press, 1973, p. 169.

103 Basalla, *Civilized Life in the Universe*, p. 181.

104 Davies, *The Eerie Silence*, p. 71.

105 Shostak, *Confessions of an Alien Hunter*, pp. 99–100.

106 William F. Hamilton III, *Cosmic Top Secret: America's Secret UFO Program*. New Brunswick, NJ: Inner Light Publications, 1991, p. 36.

107 Drake and Sobel, *Is Anyone Out There? The Scientific Search for Extraterrestrial Intelligence*, p. 209.

108 Michaud, *Contact with Alien Civilizations*, p. 87.

109 Asimov, *Extraterrestrial Civilizations*, p. 198.

110 Carl Sagan, in Jerome Agel (ed.), *Cosmic Connection: An extraterrestrial Perspective*, p. 41.

111 Drake and Sobel, *Is Anyone Out There? The Scientific Search for Extraterrestrial Intelligence*, p. 57.

112 Carl Sagan, in Jerome Agel (ed.), *Cosmic Connection: An extraterrestrial Perspective*, p. 179.

113 Asimov, *Extraterrestrial Civilizations*, p. 17.

114 Shostak, *Confessions of an Alien Hunter*, p. 100.

115 Michaud, *Contact with Alien Civilizations*, p. 214.

116 Christian de Duve, "Life as a Cosmic imperative?" *Philosophical Transactions of the Royal Society*, no. 369 (2011), pp. 620–623.

117 Jennifer Abbasi, "The Search is on," *Popular Science* (October 2011), p. 44.

118 Ibid.

119 Seth Shostak, "Are We Alone? Estimating the Prevalence of Extraterrestrial Intelligence," in Douglas A. Vakoch and Albert A. Harrison (eds.), *Civilizations Beyond Earth: Extraterrestrial Life and Society*. New York and Oxford: Berghan Books, 2011, pp. 31–42.

120 Seth Shostak, "Are We Alone? Estimating the Prevalence of Extraterrestrial Intelligence," p. 37.

121 Seth Shostak, "Are We Alone? Estimating the Prevalence of Extraterrestrial Intelligence," p. 36.

122 Quoted in Kathryn Coe, Craig T. Palmer, and Christina Pomianek, "ET Phone Darwin: What Can an Evolutionary Understanding of Animal Communication and Art Contribute to Our Understanding of Methods for Interstellar Communication?" in Douglas A. Vakoch and Albert A. Harrison (eds.), *Civilizations Beyond Earth: Extraterrestrial Life and Society*. New York and Oxford: Berghan Books, 2011, p. 219.

123 Basalla, *Civilized Life in the Universe*, pp. 184–185.

124 Simon Conway Morris, "Predicting what extra-terrestrials will be like: and preparing for the worst," *Philosophical Transactions of the Royal Society*, no. 369 (2011), pp. 555–571.

125 Harrison, *After Contact*, p. 14.

126 Davies, *The Eerie Silence*, p. 32.

127 Christopher P. McKay, "The Search for life in our Solar System and the implications for science and society," *Philosophical Transactions of the Royal Society*, no. 369 (2011), pp. 594–606.

128 Basalla, *Civilized Life in the Universe*, p. 95.

129 Davies, *The Eerie Silence*, p. 40.

130 Katia Moskovitch, "Water on Mars: Curiosity rover uncovers a flood of evidence," CBS News, September 23, 2013. http://www.cbsnews.com/8301-205_162-57604273/water-on-mars-curiosity-rover-uncovers-a-flood-of-evidence/.

131 Charles Q. Choi, "Ingredient for Life More Plentiful on Ancient Mars Than Earth," September 1, 2013. http://news.yahoo.com/ingredient-life-more-plentiful-ancient-mars-earth-170443455.html.

132 Basalla, *Civilized Life in the Universe*, p. 95.

133 Asimov, *Extraterrestrial Civilizations*, p. 57.

134 Davies, *The Eerie Silence*, p. 27.

135 de Grasse Tyson, *Space Chronicles*, pp. 48–49.

136 Dick, *Life on Other Worlds*, p. 65.

137 Ibid., pp. 1–2 and Davies, *The Eerie Silence*, pp. 61–62.

138 Davis, "Searching for a shadow biosphere on Earth as a test of the 'cosmic imperative,'" pp. 628–630.

139 Jill Tarter, "Exoplanets, Extremophiles, and the Search for Extraterrestrial Intelligence," in Douglas Z. Vakoch (e.d.), *Communication with Extraterrestrial Intelligence*. Albany: State University Press of New York, 2011, pp. 3–18.

140 Abbasi, "The Search is on," p. 41.

141 Schmidt, *Aliens and Alien Societies*, p. 52.

142 Abbasi, "The Search is on," p. 38.

143 Kaku, *Physics of the Impossible*, pp. 133–134.

144 Carl Sagan, in Jerome Agel (ed.), *Cosmic Connection: An extraterrestrial Perspective*, p. 45.

145 Ibid., p. 45.

146 Freeman J. Dyson, "Looking for Life in Unlikely Places: Reasons Why Planets May Not Be the Best Places to Look for Life," in Yoji Kondo, Frederick Bruhweiler, John Moore, and Charles Sheffield (eds), *Interstellar Travel and Multi-Generation Space Ships*. Burlington, Ontario, Canada: Apogee Books, 2003, p. 107.

147 Dyson, "Looking for Life in Unlikely Places," pp. 112–116.

148 Ibid., p. 118.

149 Ibid., p. 106.

150 Abbasi, "The Search is on," p. 38.

151 Basalla, *Civilized Life in the Universe*, p. 108.

152 Dick, *Life on Other Worlds*, pp. 56–57.

153 Allman, *Are Extraterrestrials a Threat to Humankind?*, pp. 4–8.

154 Dyson, "Looking for Life in Unlikely Places," p. 119.

155 Ibid., p. 108.

156 Plaxco and Gross, *Astrobiology*, p. 269.

157 Nick Pope, *Open Skies, Closed Minds: For the First Time A Government UFO Expert Speaks Out*. London and New York: Simon & Schuster, 1996, p. 204.

chapter 4

1 Michael A.G. Michaud, *Contact with Alien Civilizations: Our Hopes and Fears about Encountering Extraterrestrials*. New York: Copernicus Books, 2007, p. 5.

2 Albert A. Harison, "Cosmic Evolution, Reciprocity, and Interstellar Tit for Tat," in Douglas A. Vakoch (ed.), *Extraterrestrial Altruim: Evolution and Ethics in the Cosmos*. Heidelberg, New York, Dordrecht, and London: Springer 2013, p. 3.

3 Francis Fukuyama, *The Origins of Political Order: From Prehuman Times to the French Revolution*. New York: Farrar, Straus, and Giroux, 2011, pp. 29–46.

4 Isaac Asimov, *Extraterrestrial Civilizations*. New York: Crown Publishers, Inc., 1979, p. 19.

5 Alan Penny, "The Lifetimes of Scientific Civilizations and the Genetic Evolution of the Brain," in Douglas A. Vakoch and Albert A. Harrison (eds.), *Civilizations Beyond Earth: Extraterrestrial Life and Society*. New York and Oxford: Berghan Books, 2011, p. 61.

6 Davies, *The Eerie Silence*, p. 73.

7 Ibid., p. 73.

8 Ibid, p. 75.

9 Ibid, p. 75.

10 Asimov, *Extraterrestrial Civilizations*, p. 16.

11 Nikolai S. Kardashev, "On the Inevitability and the Possible Structures of Supercivilizations," International Astronomical Union, 1985, and Michio Kaku, *Hyperspace: A Scientific Odyssey Through Parallel Universes, Time Warps, and the 10th Dimension*. New York: Doubleday, 1994, pp. 276–280.

12 Paul Davies, *The Eerie Silence: Renewing Our Search for Alien Intelligence*. Boston and New York: Houghton Mifflin Harcourt, 2010, p. 141.

13 Kardashev, "Cosmology and Civilizations," *Astrophysics and Space Science*, (1997), 252: pp. 25–40.

14 Michio Kaku, *Visions: How Science Will Revolutionize the 21st Century*. New York: Anchor Books, 1997, p. 18.

15 Stephen Webb, *Where is Everybody? Fifty Solutions to the Fermi Paradox and the Problem of Extraterrestrial Life*. New York: Copernicus Books, 2000, p. 71.

16 Freeman J. Dyson, "Search for Artificial Stellar Sources of Infra-Red Radiation," *Science* 1311 (3414): 1667–1668.

17 Davies, *The Eerie Silence*, p. 141.

18 George Basalla, *Civilized Life in the Universe: Scientists on Intelligent Extraterrestrials*. Oxford and New York: Oxford University Press, 2006, p. 146.

19 Dyson, "Search for Artificial Stellar Sources of Infra-Red Radiation," pp. 1667–1668.

20 Ronald N. Bracewell, *The Galactic Club: Intelligent Life in Outer Space*. San Francisco, CA: W.W. Freeman and Company, 1975, p. 116. Astronomers have looked for signs of

extraterrestrial construction in distant solar systems. One 25 year study found many observations of interest, but nothing that could be interpreted as construction, such as a Dyson's sphere, undertaken by an advanced extraterrestrial civilization. Albert A. Harrison, "The Science and Politics of SETI: How to Succeed in an Era of Make-Believe History and Pseudoscience," in Douglas A. Vakoch and Albert A. Harrison (eds.), *Civilizations Beyond Earth: Extraterrestrial Life and Society*. New York and Oxford: Berghan Books 2011, p. 145.

21 Webb, *Where is Everybody?* p. 68.

22 Davies, *The Eerie Silence*, pp. 141–142.

23 Nick Pope, "What to do if we find extraterrestrial life," MSNBC, October 18, 2010. http://www.msnbc.msn.com/id/39675346/ns/technology_and_science/t/what-do-if-we-find-extraterrestrial-life/.

24 Albert A. Harrison, *After Contact: The Human Response to Extraterrestrial Life*. (New York: Plenum Press, 1997), p. 162.

25 Carl Sagan, in Jerome Agel (ed.). *Cosmic Connection: An extraterrestrial Perspective*. (Cambridge: Cambridge University Press, 1973), p. 234.

26 I.S. Shklovskii and Carl Sagan, *Intelligent Life in the Universe*. San Francisco, London, and Amsterdam: Holden-Day, Inc., 1966, p. 360.

27 Carl Sagan, in Jerome Agel (ed.), *Cosmic Connection: An extraterrestrial Perspective*, pp. 233–238. As far as SETI is concerned, Sagan once averred that a Type II civilization can communicate with its galactic neighbors while a Type III civilization can communicate across the entire universe. "On the Detectivity of Advanced Galactic Civilizations," in Carl Sagan (ed.), *Communication with Extraterrestrial Intelligence*. Cambridge, MA: MIT Press, 1973, p. 369.

28 Carl Sagan, in Jerome Agel (ed.), *Cosmic Connection: An extraterrestrial Perspective*, p. 169.

29 Frank Drake and Dava Sobel, *Is Anyone Out There? The Scientific Search for Extraterrestrial Intelligence*. New York: Delacorte Press, 1992, pp. 51–52.

30 John Gribbin, *Alone in the Universe: Why Our Planet is Unique*. Hoboken, NJ: Wiley, 2011, p. 35.

31 Seth Shostak, "Are We Alone? Estimating the Prevalence of Extraterrestrial Intelligence," in Douglas A. Vakoch and Albert A. Harrison (eds.), *Civilizations Beyond Earth: Extraterrestrial Life and Society*. New York and Oxford: Berghan Books, 2011, p. 40.

32 Drake and Sobel, *Is Anyone Out There? The Scientific Search for Extraterrestrial Intelligence*, p. 62.

33 Frank Drake, "The Search for Extra-terrestrial Intelligence," *Philosophical Transactions of the Royal Society*, no. 369 (2011), p. 633.

34 Drake and Sobel, *Is Anyone Out There? The Scientific Search for Extraterrestrial Intelligence*, p. 52.

35 Carl Sagan, "The Number of Advanced Galactic Civilizations" in Carl Sagan (ed.), *Communication with Extraterrestrial Intelligence*. Cambridge, MA: MIT Press, 1973, p. 165.

36 Carl Sagan, *Cosmos*. New York: Ballantine Books, 1985, pp. 251–257.

37 Charles Krauthammer, "Are we alone in the universe?," *The Washington Post*, December 29, 2011. http://www.washingtonpost.com/opinions/are-we-alone-in-the-universe/2011/12/29/gIQA2wSOPP_story.html.

38 Fukuyama, *The Origins of Political Order*, p. 446.

39 Ibid., pp. 446–447.

40 Ibid., p. 452.

41 Sagan, *Cosmos*, p. 251.

42 Carl Sagan, "The Lifetimes of Technical Civilizations" in Carl Sagan (ed.), *Communication with Extraterrestrial Intelligence*. Cambridge, MA: MIT Press, 1973, pp. 151–152.

43 Toney Allman, *Are Extraterrestrials a Threat to Humankind?* San Diego, CA: Reference Point Press, 2012, p. 43. In 1908, nearly a decade before the Bolshevik Revolution, the Russian author Alexander Bogdanov published a novel *Red Star*, in which he described Mars as a socialist utopia (Alexander Bogdanov, *Red Star*, Bloomington, IN: Indiana University Press, 1984). Less optimistic was Karl Marx's collaborator, Frederick Engels, who concluded in his book *Dialectics of Nature*, that the lifetime for intelligent life on any particular planet was finite (Frederick Engels, *Dialectics of Nature*, New York: Wellred Publications, 2012).

44 Shklovskii and Sagan, *Intelligent Life in the Universe*, p. 437.

45 Drake and Sobel, *Is Anyone Out There? The Scientific Search for Extraterrestrial Intelligence*, pp. 210–211.

46 Penny, "The Lifetimes of Scientific Civilizations and the Genetic Evolution of the Brain," p. 60.

47 Basalla, *Civilized Life in the Universe*, pp. 190–193.

48 Harrison, *After Contact*, p. 130.

49 Kathryn Denning, "'L' on Earth," in Douglas A. Vakoch and Albert A. Harrison (eds.), *Civilizations Beyond Earth: Extraterrestrial Life and Society*. New York and Oxford: Berghan Books, 2011, pp. 74–83.

50 Asimov, *Extraterrestrial Civilizations*, p. 191.

51 Michaud, *Contact with Alien Civilizations*, p. 103.

52 Davies, *The Eerie Silence*, p. 82.

53 Albert A. Harrison and Douglas A. Vakoch, "Introduction: The Search for Extraterrestrial Intelligence as an Interdisciplinary Effort," in Douglas A. Vakoch and Albert A. Harrison (eds.), *Civilizations Beyond Earth: Extraterrestrial Life and Society*. New York and Oxford: Berghan Books, 2011, p. 2.

54 Michaud, *Contact with Alien Civilizations*, p. 87.

55 Ray Kurzweil, *The Singularity is Near: When Humans Transcend Biology*. (New York: Penguin Books, 2005), p. 7.

56 Kurzweil, *The Singularity is Near*, pp. 14–21.

57 Ibid., p. 30.

58 Ibid., p. 7.

59 Ibid., p. 364.

60 Ibid., p. 348.

61 Ibid., pp. 342–358. Similarly, in a 1975 article published in the *Quarterly Journal of the Royal Astronomical Society*, Michael Hart argued that Earth's civilization was the first in the Milky Way. Webb, *Where is Everybody?* p. 23.

62 Shklovskii and Sagan, *Intelligent Life in the Universe*, p. 487.

63 Davies, *The Eerie Silence*, p. 161.

64 Ibid., p. 160.

65 Ibid., p. 155.

66 Ibid., p. 161.

67 Ibid., p. 162.

68 Ibid., p. 166.

69 Jennifer Abbasi, "The Search is on," *Popular Science* (October 2011), p. 44. Shostak wonders, though, that without the peer pressure to "get a job and establish a decent ranking in society," can machines be expected "to develop the complexity of interaction that we find in living things?" Seth Shostak, *Confessions of an Alien Hunter: A Scientist's*

Search for Extraterrestrial Intelligence. Washington, D.C.: National Geographic, 2009, p. 270.

chapter 5

1 Ronald N. Bracewell, *The Galactic Club: Intelligent Life in Outer Space*. San Francisco, CA: W.W. Freeman and Company, 1975, p. 38.

2 Steven J. Dick, *Life on Other Worlds: The 20th-Century Extraterrestrial Life Debate*. New York: Cambridge University Press, 1998, p. 201.

3 Giuseppe Cocconi and Philip Morrison, "Searching for Interstellar Communications," *Nature*, vol. 184 No. 4690 (September 19, 1959), pp. 844–846. Even before Cocconi and Morrison's article, pioneering engineers, including Nikola Tesla and Guglielmo Marconi recommend radio for possible interstellar communication. Seth Shostak, *Confessions of an Alien Hunter: A Scientist's Search for Extraterrestrial Intelligence*. Washington, D.C.: National Geographic, 2009, p. 190.

4 Frank Drake and Dava Sobel, *Is Anyone Out There? The Scientific Search for Extraterrestrial Intelligence*. New York: Delacorte Press, 1992, p. 31.

5 Shostak, *Confessions of an Alien Hunter*, pp. 47–48.

6 Drake and Sobel, *Is Anyone Out There? The Scientific Search for Extraterrestrial Intelligence*, pp. xi–xii.

7 Ibid., p. 27.

8 Ibid., pp. 2–5.

9 Ibid., pp. 9–12.

10 Sturve's research focused on the evolution and structure of stars which he studied by analyzing their spectra (light waves). Drake and Sobel, *Is Anyone Out There? The Scientific Search for Extraterrestrial Intelligence*, pp. 9–28.

11 Drake and Sobel, *Is Anyone Out There? The Scientific Search for Extraterrestrial Intelligence*, p. 154.

12 Ibid., pp. 215–216.

13 Eric J. Korpela, Jeff Cobb, Matt Lebofsky, Andrew Siemon, Joshua Von Korff, Robert C. Bankay, Dan Werthimer, and David Anderson, "Candidate Identification and Interference Removal in SETI@home," in Douglas Z. Vakoch (ed.), *Communication with Extraterrestrial Intelligence*. Albany: State University Press of New York, 2011, pp. 37–43.

14 Drake and Sobel, *Is Anyone Out There? The Scientific Search for Extraterrestrial Intelligence*, pp. 107–108.

15 Albert A. Harrison, *After Contact: The Human Response to Extraterrestrial Life*. New York: Plenum Press, 1997, p. 43.

16 Seth D. Baum, Jacob D. Haqq-Misra, Shawn D. Domagal-Goldman, "Would Contact with Extraterrestrials Benefit for Harm Humanity? A Scenario Analysis," Acta Astronutica, vol 68 no. 11–12, p. 5. http://sethbaum.com/ac/2011_ET-Scenarios.pdf.

17 Drake and Sobel, *Is Anyone Out There? The Scientific Search for Extraterrestrial Intelligence*, pp. 107–108.

18 Shostak, *Confessions of an Alien Hunter*, p. 161.

19 Ibid., p. 186.

20 Bracewell, *The Galactic Club*, pp. 41–50.

21 Drake and Sobel, *Is Anyone Out There? The Scientific Search for Extraterrestrial Intelligence*, p. 139.

22 Shostak, *Confessions of an Alien Hunter*, p. 179.

23 Ray Kurzweil, *The Singularity is Near: When Humans Transcend Biology*. New York: Penguin Books, 2005, p. 343.

24 "Jodie Foster Helps Revive SETI Search for Aliens," Associated Press, August 10, 2011.

25 Frank Drake, "The Search for Extra-terrestrial Intelligence," *Philosophical Transactions of the Royal Society*, no. 369 (2011), p. 640.

26 Shostak, *Confessions of an Alien Hunter*, pp. 186–188.

27 Paul Davies, *The Eerie Silence: Renewing Our Search for Alien Intelligence*. Boston and New York: Houghton Mifflin Harcourt, 2010, p. 197.

28 Shostak, *Confessions of an Alien Hunter*, p. 49.

29 Dave Goldberg and Jeff Blomquist, *A User's Guide to the Universe: Surviving the Perils of Black Holes, Time Paradoxes, and Quantum Uncertainty*. Hoboken, NJ: Wiley, 2010, pp. 237–239.

30 Webb, *Where is Everybody?* p. 108.

31 Neil de Grasse Tyson, *Space Chronicles: Facing the Ultimate Frontier*. New York: W.W. Norton & Company Ltd., 2012, p. 44.

32 Drake and Sobel, *Is Anyone Out There? The Scientific Search for Extraterrestrial Intelligence*, p. 87.

33 Shostak, *Confessions of an Alien Hunter*, p. 45.

34 Philip Coppens, *The Ancient Alien Question: A New Inquiry Into the Existence, Evidence, and Influence of Ancient Visitors.* Pompton Plains, NJ: New Page Books, p. 35. In 1992, a planetary system was found circling a pulsar (a class of extremely dense, rapidly rotating neutron stars). However, the area around a pulsar would be extremely inhospitable to life. Dick, *Life on Other Worlds*, p. 61.

35 Bracewell, *The Galactic Club*, p. 53.

36 Frank Drake, "Foreword," in Douglas A. Vakoch (ed.), *Extraterrestrial Altruim: Evolution and Ethics in the Cosmos.* Heidelberg, New York, Dordrecht, and London: Springer, 2013, p. ix.

37 Drake, "The search for extra-terrestrial intelligence," p. 642.

38 Michaud, *Contact with Alien Civilizations*, p. 264.

39 Davies, *The Eerie Silence*, p. 82.

40 The German astrophysicist Sebastian von Hoerner (1919–2003) also endorsed this method, speculating that extraterrestrials might communicate among themselves with laser systems, which might be detectable by using an optical SETI strategy. Drake and Sobel, *Is Anyone Out There? The Scientific Search for Extraterrestrial Intelligence*, p. 69.

41 Andrew Siemion, Henry Chen, Jeff Cobb, et al., "Current and Nascent SETI Instruments in the Radio and Optical," in Douglas Z. Vakoch (ed.), *Communication with Extraterrestrial Intelligence.* Albany: State University Press of New York, 2011, pp. 19–35.

42 Webb, *Where is Everybody?* p. 110.

43 Curtis Mead and Paul Horowitz, "Harvard's Advanced All-sky Optical SETI," in Douglas Z. Vakoch (ed.), *Communication with Extraterrestrial Intelligence.* Albany: State University Press of New York, 2011, pp. 125–135.

44 Paul Gilster, *Centauri Dreams: Imagining and Planning Interstellar Exploration.* New York: Copernicus Books, 2004, p. 195. Interestingly, the final method could produce patterns not unlike the crop circles which some observers attribute to extraterrestrial aliens.

45 Drake, "The search for extra-terrestrial intelligence," p. 635.

46 Drake and Sobel, *Is Anyone Out There? The Scientific Search for Extraterrestrial Intelligence*, pp. 101–102.

47 Kardashev, "On the Inevitability and the Possible Structures of Supercivilizations."

48 Louis K. Scheffer, "Large-Scale Use of Solar Power May Be Visible across Interstellar Distances," in Douglas Z. Vakoch (ed.), *Communication with Extraterrestrial Intelligence.* Albany: State University Press of New York, 2011, pp. 161–175.

49 Michio Kaku, *Hyperspace: A Scientific Odyssey Through Parallel Universes, Time Warps, and the 10th Dimension.* New York: Doubleday, 1994, p. 282.

50 Davies, *The Eerie Silence*, p. 141.

51 Carl Sagan, "Astroengineering Activity: The Possibility of ETI in Present Astrophysical Phenomena" in Carl Sagan (ed.), *Communication with Extraterrestrial Intelligence.* Cambridge, MA: MIT Press, 1973, pp. 188–189.

52 Milan Ćirković, "Macroengineering in the Galactic Context," in Viorel Badescu, Richard B. Cathcart, and Roelof D. Schuiling (eds.), *Macro-Engineering: A Challenge for the Future*, London: Springer, 2008. http://arxiv.org/ftp/astro-ph/papers/0606/0606102.pdf.

53 George Dvorsky, "NASA Scientist to Scour Kepler Data in Search of Alien Technologies," July 26, 2013. http://io9.com/nasa-scientist-to-scour-kepler-data-in-search-of-alien-923393162.

54 Isaac Asimov, *Extraterrestrial Civilizations.* New York: Crown Publishers, Inc., 1979, p. 262.

55 One astronomical unit is 93 million miles—the distance between the Earth and the sun.

56 Drake and Sobel, *Is Anyone Out There? The Scientific Search for Extraterrestrial Intelligence*, pp. 230–232.

57 Interview with Michael G. Michaud, July 8, 2012.

58 Davies, *The Eerie Silence*, p. 167.

59 Nick Pope, "What to do if we find extraterrestrial life," MSNBC, October 18, 2010. http://www.msnbc.msn.com/id/39675346/ns/technology_and_science/t/what-do-if-we-find-extraterrestrial-life/.

60 Shostak, *Confessions of an Alien Hunter*, pp. 274.

61 Michaud, *Contact with Alien Civilizations*, p. 46.

62 John Gribbin, *Alone in the Universe: Why Our Planet is Unique.* (Hoboken, NJ: Wiley, 2011), p. 41.

63 Michaud, *Contact with Alien Civilizations*, pp. 135–136.

64 Shostak, *Confessions of an Alien Hunter*, p. 199.

65 Coppens, *The Ancient Alien Question*, pp. 33–34 and George Basalla, *Civilized Life in the Universe: Scientists on Intelligent Extraterrestrials.* Oxford and New York: Oxford University Press, 2006, p. 117.

66 Carl Sagan, *Carl Sagan's Cosmic Connection: An Extraterrestrial Perspective.* Cambridge: Cambridge University Press, 2000, pp.17–33.

67 Harrison, *After Contact*, p. 163.

68 Asimov, *Extraterrestrial Civilizations*, p. 259.

69 I.S. Shklovskii and Carl Sagan, *Intelligent Life in the Universe.* San Francisco, London, and Amsterdam: Holden-Day, Inc., 1966, p. 97.

70 Davies, *The Eerie Silence*, p. 97.

71 Sagan, "Astroengineering Activity: The Possibility of ETI in Present Astrophysical Phenomena," p. 207.

72 Sagan, "Astroengineering Activity: The Possibility of ETI in Present Astrophysical Phenomena," p. 208 and Martin Harwit, "Tachyon Bit Rates" in Carl Sagan (ed.), *Communication with Extraterrestrial Intelligence.* Cambridge, MA: MIT Press, 1973, pp. 395–397.

73 The official name of "Buckyballs" is fullerene, made up of 60 carbon atoms, which in structure is analogous to a hollow soccer ball made out of carbon. They very rarely occur in nature.

74 Amir D. Aczel, *Entanglement: The Unlikely Story of How Scientists, Mathematicians, and Philosophers Proved Einstein's Spookiest Theory*. New York and London: Plume, 2003, p. 108.

75 Ibid., pp. 235–238.

76 On that note, the computer scientist C.H. Bennett and his colleagues came up with an idea on how to teleport information by way of entanglement. Taking into account the Heisenberg Principle, which posits that observing a particle will destroy some of its information content, Bennett and his colleagues contend that this difficulty could be overcome by a clever manipulation. They aver that the state of an object could be reconstructed by dividing it into two parts—a quantum part and a classical part. In their model there would be two channels for the teleportation act: a quantum channel and a classical channel (Aczel, *Entanglement*, pp. 241–247). It is now possible to extend entanglement to more than two particles. For example, some scientists have created triples of particles in a laboratory that are 100 percent correlated with one another. Whatever happens to one particle will cause instantaneous changes in the other two (Aczel, *Entanglement*, pp. 2–3).

77 Aczel, *Entanglement*, p. 238.

78 Donald E. Tarter, "Can SETI Fulfill the Value Agenda of Cultural Anthropology?" in Douglas A. Vakoch and Albert A. Harrison (eds.), *Civilizations Beyond Earth: Extraterrestrial Life and Society*. New York and Oxford: Berghan Books, 2011, pp. 87–101.

79 Drake and Sobel, *Is Anyone Out There? The Scientific Search for Extraterrestrial Intelligence*, p. xiii.

80 Ibid., p. xiii.

81 Drake and Sobel, *Is Anyone Out There? The Scientific Search for Extraterrestrial Intelligence*, pp. 102–104.

82 Abbasi, "The Search is on," p. 44.

83 Webb, *Where is Everybody?* p.99.

84 Shostak, *Confessions of an Alien Hunter*, p. 55.

85 Toney Allman, *Are Extraterrestrials a Threat to Humankind?* (San Diego, CA: Reference Point Press, 2012), p. 4–19.

86 Shostak, *Confessions of an Alien Hunter*, pp. 62.

87 Albert A. Harrison, "The Science and Politics of SETI: How to Succeed in an Era of Make-Believe History and Pseudoscience," in Douglas A. Vakoch and Albert A. Harrison (eds.), *Civilizations Beyond Earth: Extraterrestrial Life and Society*. (New York and Oxford: Berghan Books 2011), p. 148.

88 Drake and Sobel, *Is Anyone Out There? The Scientific Search for Extraterrestrial Intelligence*, pp. 191–196.

89 Shostak, *Confessions of an Alien Hunter*, p. 155–157.

90 Kevin Plaxco and Michael Gross, *Astrobiology: A Brief Introduction*. Second Edition. (Baltimore, MD: The Johns Hopkins University Press, 2011), p. 304.

91 Drake and Sobel, *Is Anyone Out There? The Scientific Search for Extraterrestrial Intelligence*, p. xii.

92 de Grasse Tyson, *Space Chronicles*, p. 35.

93 Harrison, "The Science and Politics of SETI: How to Succeed in an Era of Make-Believe History and Pseudoscience," p. 148.

94 Quoted in Allman, *Are Extraterrestrials a Threat to Humankind?*, pp. 59–60.

95 Harrison, *After Contact*, p. 63.

96 Shostak, *Confessions of an Alien Hunter*, p. 178.

97 Stan Schroeder, "SETI Shuts Down Search for Alien Life Due to Lack of Funding," *Mashable*, April 27, 2011.

98 "Jodie Foster Helps Revive SETI Search for Aliens," Associated Press, August 10, 2011.

99 Author interview with Michael G. Michaud, July 8, 2012.

100 In essence, crowdsourcing entails the pooling of efforts of a large group of people especially from an online community. Crowdsourcing has been applied to numerous endeavors including fundraising, science projects, and even terrorism and insurgency. For more on the concept of crowdsourcing, see Jeff Howe, *Crowdsourcing: Why the Power of the Crowd is Driving the Future of Business.* (New York: Crown Business, 2008).

101 Shostak, *Confessions of an Alien Hunter*, p. 216.

102 Plaxco and Gross, *Astrobiology*, p. 304.

103 Michaud, *Contact with Alien Civilizations*, p. 42.

104 Travis S. Taylor and Bon Boan, *Alien Invasion: The Ultimate Survival Guide for the Ultimate Attack.* Riverdale, NY: Baen Publishing Enterprises, 2011, p. 9.

105 Kaku, *Hyperspace*, p. 285.

106 Kaku, *Physics of the Impossible*, p. 147.

107 Shostak, *Confessions of an Alien Hunter*, p. 202.

108 Jill Tarter, "Exoplanets, Extremophiles, and the Search for Extraterrestrial Intelligence," in Douglas Z. Vakoch (ed.), *Communication with Extraterrestrial Intelligence.* Albany: State University Press of New York, 2011, p. 12. In what became known as "Moore's law," back in 1960, Gordon Moore, a cofounder of Intel, observed that the number of transistors on a computer chip (integrated circuit) tends to double every 18 months. John Robb, *Brave New War: The Next Stage of Terrorism and the End of Globalization.* Hoboken, NJ: John Wiley & Sons, Inc., 2007, p. 10.

chapter 6

1 Carl Sagan, "Prospect" in Carl Sagan (ed.), *Communication with Extraterrestrial Intelligence.* Cambridge, MA: MIT Press, 1973, p. 5.

2 Basalla, *Civilized Life in the Universe*, p. 13.

3 Seth Shostak, *Confessions of an Alien Hunter: A Scientist's Search for Extraterrestrial Intelligence*. Washington, D.C.: National Geographic, 2009, p. 229.

4 Michael A.G. Michaud, *Contact with Alien Civilizations: Our Hopes and Fears about Encountering Extraterrestrials.* New York: Copernicus Books, 2007, p. 51.

5 Nick Pope, "What to do if we find extraterrestrial life," MSNBC, October 18, 2010.

http://www.msnbc.msn.com/id/39675346/ns/technology_and_science/t/what-do-if-we-find-extraterrestrial-life/.

6 Allman, *Are Extraterrestrials a Threat to Humankind?*, p. 65.

7 Michaud, *Contact with Alien Civilizations*, p. 51.

8 Seth D. Baum, Jacob D. Haqq-Misra, Shawn D. Domagal-Goldman, "Would Contact with Extraterrestrials Benefit for Harm Humanity? A Scenario Analysis," *Acta Astronutica*, vol 68 no. 11–12, p. 25. http://sethbaum.com/ac/2011_ET-Scenarios.pdf.

9 Carl Sagan, in Jerome Agel (ed.). *Cosmic Connection: An extraterrestrial Perspective.* Cambridge: Cambridge University Press, 1973, p. 219.

10 Albert A. Harrison, *After Contact: The Human Response to Extraterrestrial Life.* New York: Plenum Press, 1997, p. 286.

11 Mark C. Langston, "The Accidental Altruist: Inferring Atruism from and Extraterrestrial Signal," in Douglas A. Vakoch (ed.), *Extraterrestrial Altruim: Evolution and Ethics in the Cosmos.* Heidelberg, New York, Dordrecht, and London: Springer, 2013, pp. 129–138.

12 Kathryn Denning, "Unpacking the Great Transmission Debate," in Douglas Z. Vakoch (ed.), *Communication with Extraterrestrial Intelligence.* Albany: State University Press of New York, 2011, pp. 237–252.

13 George Dvorsky, "New Project to Message Aliens is Both Useless and Potentially Reckless," June 12, 2012. http://io9.com/new-project-to-message-aliens-is-both-useless-and-poten-512863567.

14 Douglas A. Vakoch, "Integrating Active and Passive SETI Programs," in Douglas Z. Vakoch (ed.), *Communication with Extraterrestrial Intelligence.* Albany: State University Press of New York, 2011, pp. 253–278.

15 James Benford, Dominic Benford, and Gregory Benford, "Building and Searching for Cost-Optimized Insterstellar Beacons," in Douglas Z. Vakoch (ed.), *Communication with Extraterrestrial Intelligence.* Albany: State University Press of New York, 2011, pp. 279–306.

16 Ronald N. Bracewell, *The Galactic Club: Intelligent Life in Outer Space.* San Francisco, CA: W.W. Freeman and Company, 1975, p. 43.

17 Seth Shostak, "Limits on Interstellar Messages," in Douglas Z. Vakoch (ed.), *Communication with Extraterrestrial Intelligence.* Albany: State University Press of New York, 2011, pp. 357–369.

18 Bracewell, *The Galactic Club*, pp. 52–53.

19 Shostak, "Limits on Interstellar Messages," pp. 357–369.

20 Carl L. DeVito, "Cultural Aspects of Interstellar Communication," in Douglas A. Vakoch and Albert A. Harrison (eds.), *Civilizations Beyond Earth: Extraterrestrial Life and Society*. New York and Oxford: Berghan Books, 2011, pp. 159–169.

21 David Dunér, "Cognitive Foundations of Interstellar Communication," in Douglas Z. Vakoch (ed.), *Communication with Extraterrestrial Intelligence*. Albany: State University Press of New York, 2011, pp. 461–462.

22 Ibid., pp. 449–467.

23 Albert A. Harrison and Douglas A. Vakoch, "Introduction: The Search for Extraterrestrial Intelligence as an Interdisciplinary Effort," in Douglas A. Vakoch and Albert A. Harrison (eds.), *Civilizations Beyond Earth: Extraterrestrial Life and Society*. New York and Oxford: Berghan Books, 2011, p. 7.

24 Paul K. Wason, "Encountering Alternative Intelligences: Cognitive Archaeology and SETI," in Douglas A. Vakoch and Albert A. Harrison (eds.), *Civilizations Beyond Earth: Extraterrestrial Life and Society*. New York and Oxford: Berghan Books, 2011, pp. 43–59.

25 Jason T. Kuznicki, "The Inscrutable Names of God: The Jesuit Missions of New France as a Model for SETI-Related Spiritual Questions," in Douglas A. Vakoch and Albert A. Harrison (eds.), *Civilizations Beyond Earth: Extraterrestrial Life and Society*. New York and Oxford: Berghan Books, 2011, pp. 203–213.

26 Paul Davies, *The Eerie Silence: Renewing Our Search for Alien Intelligence*. Boston and New York: Houghton Mifflin Harcourt, 2010, p. 204.

27 DeVito, "Cultural Aspects of Interstellar Communication," pp. 159–169.

28 Carl Sagan, *Cosmos*. New York: Ballantine Books, 1985, pp. 242–244.

29 Stéphane Dumas, "A Proposal for an Interstellar Rosetta Stone," in Douglas Z. Vakoch (ed.), *Communication with Extraterrestrial Intelligence*. Albany: State University Press of New York, 2011, pp. 403–411.

30 Harrison, *After Contact*, p. 59.

31 Alexander Ollongren, "A Logic-Based Approach to Characterizing Altruism in Interstellar Messages," in Douglas A. Vakoch (ed.), *Extraterrestrial Altruim: Evolution and Ethics in the Cosmos*. Heidelberg, New York, Dordrecht, and London: Springer, 2013, p. 251.

32 Allman, *Are Extraterrestrials a Threat to Humankind?*, p. 60.

33 Shostak, *Confessions of an Alien Hunter*, pp. 242–243.

34 Morris Jones, "A Journalistic Perspective on SETI-Related Message Composition," in Douglas A. Vakoch and Albert A. Harrison (eds.), *Civilizations Beyond Earth: Extraterrestrial Life and Society*. New York and Oxford: Berghan Books, 2011, pp. 226–235.

35 Harry Letaw Jr., "Cosmic Storytelling: Primitive Observables as Rosetta Analogies," in Douglas A. Vakoch and Albert A. Harrison (eds.), *Civilizations Beyond Earth: Extraterrestrial Life and Society*. New York and Oxford: Berghan Books, 2011, p. 170–190.

36 Guillermo A. Lemarchand and Jon Lomberg, "Communication among Interstellar Intelligence Species," in Douglas Z. Vakoch (ed.), *Communication with Extraterrestrial Intelligence*. Albany: State University Press of New York, 2011, pp. 371–395.

37 Michael A.G. Michaud, *Contact with Alien Civilizations: Our Hopes and Fears about Encountering Extraterrestrials*. New York: Copernicus Books, 2007, pp. 241–247.

38 Harrison and Vakoch, "Introduction: The Search for Extraterrestrial Intelligence as an Interdisciplinary Effort," p. 10.

39 Ibid., p. 2.

40 Ivan Almar and Margaret S. Race, "Discovery of extra-terrestrial life: assessment by scales of its importance and associated risks," *Philosophical Transactions of the Royal Society*, no. 369 (2011), pp. 679–692.

41 George Pettinico, "American Attitudes About Life Beyond Earth: Beliefs, Concerns, and the Role of Education and Religion in Shaping Public Perceptions," in Douglas A. Vakoch and Albert A. Harrison (eds.), *Civilizations Beyond Earth: Extraterrestrial Life and Society*. New York and Oxford: Berghan Books 2011, p. 102.

42 Ibid., pp. 102–117.

43 Harrison, *After Contact*, p. 229.

44 Ibid., pp. 231–233.

45 I.S. Shklovskii and Carl Sagan, *Intelligent Life in the Universe*. San Francisco, London, and Amsterdam: Holden-Day, Inc., 1966, p. 395.

46 Harrison, *After Contact*, p. 249.

47 Ibid., p. 214.

48 George Basalla, *Civilized Life in the Universe: Scientists on Intelligent Extraterrestrials*. Oxford and New York: Oxford University Press, 2006, pp. 120–121.

49 Steven J. Dick, *Life on Other Worlds: The 20th-Century Extraterrestrial Life Debate*. New York: Cambridge University Press, 1998, p. 126. When asked on his 90th birthday what

he wished for before he died, Clarke answered first that he would like to find evidence of extraterrestrial life (Stanton T. Friedman, *Flying Saucers and Science: A Scientist Investigates the Mysteries of UFOs*. Pompton Plains, NJ: The Career Press, 2008, p. 197).

50 In *The Sentinel*, a geologist working on the Moon discovers a crystal pyramidal structure roughly twice as high as a man. After close inspection, he determines that it is an ancient edifice probably created before humans emerged on Earth. He concludes that its purpose was to serve as a type of an alarm. He reasoned that its builders "were not concerned with races still struggling up from savagery." Rather, they would be interested in Earth's civilization if it demonstrated its fitness to survive by crossing space and so escaping its cradle, Earth. Once this challenge was passed, it was only a matter of time before the sentinel was discovered. The story closes with the geologist wondering from which part of the galaxy the emissaries would be coming. Arthur C. Clarke, *The Sentinel*. New York: Avon Periodicals Inc, 1951.

51 Shklovskii and Sagan, *Intelligent Life in the Universe*, p. 462.

52 Natalie Wolchover, "'Monolith' Object on Mars? You Could Call It That," LiveScience.com, April 11, 2012. http://news.yahoo.com/monolith-object-mars-could-call-214004772.html.

53 Ted Peters, "The implications of the discovery of extra-terrestrial life for religion," *Philosophical Transactions of the Royal Society*, no. 369 (2011), p. 649.

54 Harrison, *After Contact*, pp. 311–314.

55 Peters, "The implications of the discovery of extra-terrestrial life for religion," p. 645.

56 Albert Harrison and Vakoch, "Introduction: The Search for Extraterrestrial Intelligence as an Interdisciplinary Effort," pp. 10–11.

57 Albert Harrison and Vakoch, "Introduction: The Search for Extraterrestrial Intelligence as an Interdisciplinary Effort," p. 10.

58 Basalla, *Civilized Life in the Universe*, p. 8.

59 Davies, *The Eerie Silence*, p. 189.

60 Nick Pope, *Open Skies, Closed Minds: For the First Time a Government UFO Expert Speaks Out*. London and New York: Simon & Schuster, 1996, p. 185.

61 His books eventually formed the scriptures of the Swedenborgians, as his followers came to be known. More formally, their organization was referred to as the Church of the New Jerusalem. Michael J. Crowe, *The Extraterrestrial Life Debate: Antiquity to 1915*. Notre Dame, IN: University of Notre Dame Press, 2008, pp. 215–221.

62 Basalla, *Civilized Life in the Universe*, pp. 128–129.

63 Davies, *The Eerie Silence*, p. 189.

64 Peters, "The implications of the discovery of extra-terrestrial life for religion," pp. 652–653.

65 Albert Harrison and Vakoch, "Introduction: The Search for Extraterrestrial Intelligence as an Interdisciplinary Effort," p. 10. Cargo cults can be found in the Melanesian region of Oceania. The most widely known period in the formation of cargo cults occurred during and after World War II. They arose in the wake of contact between the indigenous people on the islands and representatives of the commercial networks of colonizing societies. The natives wondered how the visitors were provisioned with the vast amount of supplies. The natives reasoned that if they would ritualistically mimic the actions of the visitors, then perhaps the gods would send provisions to them as well. Ian Robertson, *Sociology* (Third Edition). New York: Worth Publising, Inc., 1987, p. 553.

66 Ted Peters, "The Implications of the Discovery of Extra-terrestrial Life for Religion," *Philosophical Transactions of the Royal Society*, no. 369 (2011), p. 653.

67 Albert Harrison and Vakoch, "Introduction: The Search for Extraterrestrial Intelligence as an Interdisciplinary Effort," p. 11.

68 Peters, "The implications of the discovery of extra-terrestrial life for religion," p. 653.

69 Allman, *Are Extraterrestrials a Threat to Humankind?*, p. 54.

70 "Protocols for an ETI Signal Detection," SETI Institute, http://www.seti.org/post-detection.html, accessed June 15, 2014.

71 Shostak, *Confessions of an Alien Hunter*, p. 232.

72 Paul Davies is the current chairman of the Task Group. Davies, *The Eerie Silence*, p. 169.

73 Nick Pope, "What to do if we find extraterrestrial life," MSNBC, October 18, 2010, http://www.msnbc.msn.com/id/39675346/ns/technology_and_science/t/what-do-ifwe-find-extraterrestrial-life/.

74 Allman, *Are Extraterrestrials a Threat to Humankind?*, p. 58.

75 Nick Pope, "What to do if we find extraterrestrial life," MSNBC, October 18, 2010, http://www.msnbc.msn.com/id/39675346/ns/technology_and_science/t/what-do-ifwe-find-extraterrestrial-life/.

76 William Sims Bainbridge, "Cultural Beliefs about Extraterrestrials: A Questionnaire

Study," in Douglas A. Vakoch and Albert A. Harrison (eds.), *Civilizations Beyond Earth: Extraterrestrial Life and Society*. New York and Oxford: Berghan Books, 2011, pp. 118–140.

77 William Sims Bainbridge, "Direct Contact with Extraterrestrials via Computer Emulation," in Douglas A. Vakoch and Albert A. Harrison (eds.), *Civilizations Beyond Earth: Extraterrestrial Life and Society*. (New York and Oxford: Berghan Books 2011), pp. 191–202.

78 Shostak, *Confessions of an Alien Hunter*.

chapter 7

1 Stephen Webb, *Where is Everybody? Fifty Solutions to the Fermi Paradox and the Problem of Extraterrestrial Life*. New York, Copernicus Books, 2000, pp. 17–25.

2 Kevin Plaxco and Michael Gross, *Astrobiology: A Brief Introduction*. Second Edition. Baltimore, MD: The Johns Hopkins University Press, 2011, p. 302.

3 George Basalla, *Civilized Life in the Universe: Scientists on Intelligent Extraterrestrials*. Oxford and New York: Oxford University Press, 2006, p. 14.

4 *The UFO Phenomenon*. Alexandria, VA: Time Life Books, 1987, p. 14.

5 Michio Kaku, *Physics of the Impossible: A Scientific Exploration into the World of Phasers, Force Fields, Teleportation, and Time Travel*. New York: Anchor Books, 2008, p. 148.

6 Thomas E. Bullard, *The Myth and Mystery of UFOs*. Lawrence, KS: University Press of Kansas, 2010, p. 99.

7 Jim Marrs, *Alien Agenda: Investigating the Extraterrestrial Presence Among Us*. New York: Harper Paperbacks, 1997, p. 44.

8 Philip Coppens, *The Ancient Alien Question: A New Inquiry Into the Existence, Evidence, and Influence of Ancient Visitors*. Pompton Plains, NJ: New Page Books, pp. 112–118.

9 Ibid.

10 Albert A. Harrison and Douglas A. Vakoch, "Introduction: The Search for Extraterrestrial Intelligence as an Interdisciplinary Effort," in Douglas A. Vakoch and Albert A. Harrison (eds.), *Civilizations Beyond Earth: Extraterrestrial Life and Society*. New York and Oxford: Berghan Books, 2011, p. 10.

11 *The UFO Phenomenon*, p. 14.

12 Michael J. Crowe, *The Extraterrestrial Life Debate: Antiquity to 1915*. Notre Dame, IN: University of Notre Dame Press, 2008, p. 116.

13 I.S. Shklovskii and Carl Sagan, *Intelligent Life in the Universe*. San Francisco, London, and Amsterdam: Holden-Day, Inc., 1966, p. 454.

14 Ede Frecska, "Close Encounters of the Ancient Kind and Spontaneous DMT Release," in Rick Strassman, Slawek Wojtowicz, Luis Eduardo Luna, and Ede Frecska, *Innter Paths to Outer Space: Journeys to Alien Worlds though Psychadelics and Other Spritual Technologies*. Rochester, VT: Park Street Press, 2008, p. 228.

15 Coppens, *The Ancient Alien Question*, pp. 39–40.

16 Ibid., pp. 39–40.

17 Morris K. Jessup, *The Allende Letters and the Varo Edition of The Case for the UFO*. New Brunswick, NJ: Conspiracy Journal/Global Communications, 2007, p. x.

18 Jessup, *The Allende Letters and the Varo Edition of The Case for the UFO*, p. 124.

19 Ibid., p. 156.

20 Ibid., p. 148.

21 As Sagan explained, if the Earth was chosen for regular alien visitations, one would have to assume that somehow our planet was unique in the universe. This assumption, he continued, went exactly against the idea that there are many alien civilizations in the cosmos. If there were many alien civilizations, then Earth's would be unremarkable. But if Earth is not common, then there would not be many advanced extraterrestrial civilizations capable of sending visitors. Hence, "Sagan's paradox" suggested that extraterrestrial life most likely exists but would probably not have anything to do with UFOs. *The UFO Phenomenon*, p. 123.

22 Basalla, *Civilized Life in the Universe*, pp. 143–144.

23 Shklovskii and Sagan, *Intelligent Life in the Universe*, pp. 456–460.

24 Erich von Däniken, *Chariots of the Gods?* New York: Berkley Books, 1999, pp. viii–ix.

25 von Däniken, *Chariots of the Gods?*, p. 13.

26 Ibid., p. 64.

27 Ibid., p. 13.

28 Ibid., p. 32.

29 Ibid., p. 45.

30 Ibid., p. 53.

31 Michael Kurland, *The Complete Idiot's Guide to Extraterrestrial Intelligence*. (New York: alpha books, 1999), pp. 87–88.

32 von Däniken, *Chariots of the Gods?*, p. 114.

33 Ibid., p. 96.

34 Ronald N. Bracewell, *The Galactic Club: Intelligent Life in Outer Space*. San Francisco, CA: W.W. Freeman and Company, 1975, p. 98.

35 von Däniken, *Chariots of the Gods?*, p. 101.

36 Marrs, *Alien Agenda*, p. 68. Similarly in 1531, a French cartographer by the name of Oronce Fine produced a map that displayed Antarctica in the middle of a circular chart. It even showed the continent without an obscuring layer of ice. Kurland, *The Complete Idiot's Guide to Extraterrestrial Intelligence*, pp. 38–42.

37 Marrs, *Alien Agenda*, p. 71.

38 von Däniken, *Chariots of the Gods?*, p. 85.

39 Coppens, *The Ancient Alien Question*, pp. 139–143.

40 von Däniken, *Chariots of the Gods.*

41 Coppens, *The Ancient Alien Question.*

42 Ibid., pp. 56–59.

43 Quoted in Coppens, *The Ancient Alien Question*, p. 41.

44 Marrs, *Alien Agenda*, p. 84.

45 Shklovskii and Sagan, *Intelligent Life in the Universe*, p. 356.

46 Frecska, "Close Encounters of the Ancient Kind and Spontaneous DMT Release," p. 234.

47 Zecharia Sitchin, *The 12th Planet: Book I of the Earth Chronicles*. New York: Harper Collins, 2007, pp. 177–178.

48 Sitchin, *The 12th Planet: Book I of the Earth Chronicles*, p. 261.

49 Jim Marrs, *Rule By Secrecy: The Hidden History that Connects the Trilateral Commission, the Freemasons, and the Great Pyramids*. New York: Perennial, 2000, p. 381.

50 Sitchin, *The 12th Planet: Book I of the Earth Chronicles*, p. 285.

51 Frecska, "Close Encounters of the Ancient Kind and Spontaneous DMT Release," p. 237.

52 Marrs, *Rule By Secrecy*, p. 382.

53 Sitchin, *The 12th Planet: Book I of the Earth Chronicles*, pp. 177–341.

54 Ibid., pp. 177–344.

55 Brad Steiger and Sherry Hansen Steiger, *Real Aliens, Space Beings, and Creatures from Other Worlds*. (Canton, MI: Visible Ink Press, 2011), p. 69.

56 Frecska, "Close Encounters of the Ancient Kind and Spontaneous DMT Release," p. 237.

57 Marrs, *Rule By Secrecy*, p. 385.

58 Sitchin, *The 12th Planet: Book I of the Earth Chronicles*, pp. 177–368.

59 Marrs, *Rule By Secrecy*, p. 389.

60 Ibid., p. 390. According to some legends, members of the royalty were said to have been a hybrid of reptoid aliens and humans. Frecska, "Close Encounters of the Ancient Kind and Spontaneous DMT Release," p. 237. More fantastic theories of ancient aliens includes those advanced by David Icke who averred that a reptilian species of aliens have subverted the world's governments in an evil conspiracy to enslave the human race. Icke claims that a remnant of the Anunnaki rules the Earth. Through interbreeding with select humans, they became the aristocratic "blue bloods" of Europe and other nations (Coppens, *The Ancient Alien Question*, pp. 43–44 and Michael E. Salla, *Exopolitics: Political Implications of the Extraterrestrial Presence*. Temple, AZ: Danelion Books, 2004, p. 28).

61 Sitchin, *The 12th Planet: Book I of the Earth Chronicles*, pp. 1–3.

62 Ibid., pp. 387–411.

63 Ibid., p. 420.

64 Marrs, *Rule By Secrecy*, pp. 400–402.

65 Marrs, *Alien Agenda*, p. 43.

66 Ibid., pp. 74–78.

67 Ibid., pp. 74–75.

68 Ibid., p. 62.

69 Kaku, *Physics of the Impossible*, pp. 127–128.

70 Basalla, *Civilized Life in the Universe*, pp. 101–103.

71 Ibid., p. 101.

72 Marrs, *Alien Agenda*, pp. 14–15.

73 Ibid., pp. 1–18.

74 Seth Shostak, *Confessions of an Alien Hunter: A Scientist's Search for Extraterrestrial Intelligence*. Washington, D.C.: National Geographic, 2009, p. 130.

75 Basalla, *Civilized Life in the Universe*, p. 104.

76 Frank Drake and Dava Sobel, *Is Anyone Out There? The Scientific Search for Extraterrestrial Intelligence*. New York: Delacorte Press, 1992, p. 176.

77 Shostak, *Confessions of an Alien Hunter*, pp. 128–134.

78 Ibid., p. 133.

79 Isaac Asimov, *Extraterrestrial Civilizations*. New York: Crown Publishers, Inc., 1979, p. 196.

80 Coppens, *The Ancient Alien Question*, p. 52.

81 Ibid., p. 54.

82 Carl Sagan, in Jerome Agel (ed.). *Cosmic Connection: An extraterrestrial Perspective.* Cambridge: Cambridge University Press, 1973, pp. 205–206.

83 Carl Sagan, *Broca's Brain: Reflections on the Romance of Science.* New York: Random House, 1979, pp. 63–71.

84 Carl Sagan, *Cosmos.* New York: Ballantine Books, 1985, p. 253.

85 Michaud, *Contact with Alien Civilizations*, p. 141.

86 Bracewell, *The Galactic Club*, pp. 98–101.

87 Ibid., pp. 98–100.

88 Marrs, *Alien Agenda*, p. 41.

chapter 8

1 Seth Shostak, *Confessions of an Alien Hunter: A Scientist's Search for Extraterrestrial Intelligence.* Washington, D.C.: National Geographic, 2009, p. 112.

2 George Pettinico, "American Attitudes About Life Beyond Earth: Beliefs, Concerns, and the Role of Education and Religion in Shaping Public Perceptions," in Douglas A. Vakoch and Albert A. Harrison (eds.), *Civilizations Beyond Earth: Extraterrestrial Life and Society.* New York and Oxford: Berghan Books, 2011, pp. 111–112. Other surveys, however, suggest that only a minority of Americans believe that UFOs originate from outside of our world. William Sims Bainbridge presented the results from an online questionnaire—*Survey 2001*—which was sponsored by the National Geographic Society and the National Science Foundation. The survey questioned people on the prospects of alien life and UFOs. While a significant majority of the respondents expressed a belief that intelligent life probably exists beyond Earth, just over one-fifth of them agreed that UFOs are probably spaceships. Bainbridge found a strong correlation between believing in UFOs and astrology, but no connection between the latter and the belief in extraterrestrial life. William Sims Bainbridge, "Cultural Beliefs about Extraterrestrials: A Questionnaire Study," in Douglas A. Vakoch and Albert A. Harrison (eds.), *Civilizations Beyond Earth: Extraterrestrial Life and Society.* (New York and Oxford: Berghan Books, 2011), pp. 118–140.

3 Thomas E. Bullard, *The Myth and Mystery of UFOs.* Lawrence, KS: University Press of Kansas, 2010, p. 5.

4 William J. Birnes, *The Everything UFO Book: An investigation of sightings, cover-ups, and the quest for extraterrestrial life*. Avon, MA: Adams Media, 2012, p. 8.
5 Ibid., p. 18.
6 Ibid., p. 21.
7 Mack Maloney, *UFOs in Wartime: What They Don't Want You to Know*. New York: Berkley Books, 2011, pp. 18–22.
8 Nick Pope, *Open Skies, Closed Minds: For the First Time A Government UFO Expert Speaks Out*. London and New York: Simon & Schuster, 1996, p. 11.
9 Maloney, *UFOs in Wartime*, pp. 32–38.
10 Ibid.
11 Ibid., pp. 41–47.
12 Jim Marrs, *Alien Agenda: Investigating the Extraterrestrial Presence Among Us*. New York: Harper Paperbacks, 1997, p. 93.
13 Michael Swords and Robert Powell (eds), "Chapter 1: World War II and the Immediate Post-War Era," in *UFOs and Government: A Historical Inquiry*. San Antonio and Charlottesville: Anomalist Books, 2012, p. 3.
14 Maloney, *UFOs in Wartime*, pp. 95–96.
15 Ibid., p. 97.
16 Swords and Powell (eds.), "Chapter 1: World War II and the Immediate Post-War Era," pp. 3–7.
17 Travis S. Taylor and Bon Boan, *Alien Invasion: The Ultimate Survival Guide for the Ultimate Attack*. Riverdale, NY: Baen Publishing Enterprises, 2011, pp. 143–144.
18 Michael Swords and Robert Powell (eds.), "Chapter 2: Ghost Rockets," in *UFOs and Government: A Historical Inquiry*. San Antonio and Charlottesville: Anomalist Books, 2012, p. 12.
19 Ibid., p. 14.
20 Ibid., p. 16.
21 Maloney, *UFOs in Wartime*, pp. 105–110.
22 Swords and Powell (eds.), "Chapter 2: Ghost Rockets," p. 26.
23 Edwin K. Wright to President Harry Truman, memorandum, subject: "Ghost Rockets over Scandinavia," 1 August 1946, FOIA Request to the USAF in Michael Swords and Robert Powell (eds.), "Chapter 2: Ghost Rockets," in *UFOs and Government: A Historical Inquiry*. San Antonio and Charlottesville: Anomalist Books, 2012, p. 28.

24 "Doolittle skall bara prata bensin – har hört om spökbomber," August 21, 1946 in Michael Swords and Robert Powell (eds.), "Chapter 2: Ghost Rockets," in *UFOs and Government: A Historical Inquiry*. San Antonio and Charlottesville: Anomalist Books, 2012, p. 28.

25 Swords and Powell (eds.), "Chapter 2: Ghost Rockets," p. 21.

26 *The UFO Phenomenon*. Alexandria, VA: Time Life Books, 1987, p. 37.

27 Quoted in Marc Davenport, *Visitors From Time: The Secret of the UFOs*. (Tuscaloosa, AL: Greenleaf Publications, 1994), p. 83.

28 Birnes, *The Everything UFO Book*, p. 34.

29 Steven J. Dick, *Life on Other World: The 20th-Century Extraterrestrial Life Debate*. New York: Cambridge University Press, 1998, p. 141.

30 *The UFO Phenomenon*, p. 37.

31 Michael Swords and Robert Powell (eds.), "Chapter 3: The Flying Disks and the United States," in *UFOs and Government: A Historical Inquiry*. San Antonio and Charlottesville: Anomalist Books, 2012, pp. 33–35.

32 *The UFO Phenomenon*, p. 419.

33 Kevin D. Randle, *Case MJ-12: The True Story Behind the Government's UFO Conspiracies*. New York: HarperTorch, 2002, p. 79.

34 Michael Swords and Robert Powell (eds.), "Chapter 4: A Formalized UFO Project," in *UFOs and Government: A Historical Inquiry*. San Antonio and Charlottesville: Anomalist Books, 2012, p. 51.

35 Bullard, *The Myth and Mystery of UFOs*, pp. 8–9; Randle, *Case MJ-12*, pp. 80–81. Interestingly, some personnel involved in the Skyhook Project have reported UFO activities. For example, in 1950, *True* magazine published a story written by Commander Robert J. McLaughlin, a naval officer and guided-missile expert who was stationed at White Sands Proving Grounds in New Mexico. As he explained, on April 24, 1949, he and a group of engineers were preparing to launch a Project Skyhook high-altitude research balloon. As a preliminary, they launched a small weather balloon to establish the wind patterns. When the theodolite operator positioned his instrument to track the balloon, a strange object crossed its path. McLaughlin described the object as elliptical and close to 105 feet in diameter and flying at an extremely high altitude of about 56 miles. Engineers calculated that it moved through space at roughly 18,000 miles per hour. *The UFO Phenomenon*, pp. 46–47.

36 Pope, *Open Skies, Closed Minds*, p. 192.

37 Bullard, *The Myth and Mystery of UFOs*, pp. 62–63.

38 Michael Swords and Robert Powell (eds.), "Chapter 15: After the Close of Blue Book," in *UFOs and Government: A Historical Inquiry*. San Antonio and Charlottesville: Anomalist Books, 2012, p. 340.

39 Leslie Kean, *UFOs: Generals, Pilots, and Government Officials Go on the Record*. New York: Harmony Books, 2010, pp. 85–92.

40 Birnes, *The Everything UFO Book*, p. 210.

41 Ibid., p. 209.

42 Kean, *UFOs: Generals, Pilots, and Government Officials Go on the Record*, pp. 93–98.

43 Ibid., p. 149.

44 Ibid., p. 139.

45 Ibid., p. 139.

46 Maloney, *UFOs in Wartime*, pp. 220–221.

47 Quoted in Kean, *UFOs: Generals, Pilots, and Government Officials Go on the Record*, p. 144.

48 John B. Alexander, *UFOs: Myths, Conspiracies, and Realities*. New York: St. Martin's Press, 2011, p. 169.

49 Kean, *UFOs: Generals, Pilots, and Government Officials Go on the Record*, pp. 144–145.

50 Alexander, *UFOs: Myths, Conspiracies, and Realities*, pp. 171–172.

51 Billy Cox, "The Buck stops where?" HeraldTribune.com, July 3, 2012, http://devoid.blogs.heraldtribune.com/13107/the-buck-stops-where/.

52 Marc Ambinder, "Failure Shuts Down Squadron of Nuclear Missiles," *The Atlantic*, October 26, 2010. http://www.theatlantic.com/politics/archive/2010/10/failure-shuts-down-squadron-of-nuclear-missiles/65207/.

53 "Moseley and Wynne forced out," Air Force Times, June 5, 2008, http://www.airforcetimes.com/article/20080605/NEWS/806050301/Moseley-Wynne-forced-out.

54 Maloney, *UFOs in Wartime*, pp. 260–261.

55 Ibid., p. 262.

56 Alexander, *UFOs: Myths, Conspiracies, and Realities*, p. 156.

57 Birnes, *The Everything UFO Book*, pp. 98–99.

58 Ibid., p. 99.

59 Alexander, *UFOs: Myths, Conspiracies, and Realities*, p. 156.

60 Maloney, *UFOs in Wartime*, p. 263.

61 Nick Pope, "The Real X-Files," In Michael Pye and Kirsten Dalley (eds.), *UFOs & Aliens: Is There Anybody Out There?* Pompton Plains, NJ: The Career Press, 2011, p. 36.

62 Maloney, *UFOs in Wartime*, p. 263.

63 Ibid., p. 269.

64 Pope, "The Real X-Files," p. 35.

65 Kean, *UFOs: Generals, Pilots, and Government Officials Go on the Record*, pp. 179–188.

66 Maloney, *UFOs in Wartime*, p. 263.

67 Pope, "The Real X-Files,"p. 36.

68 Alexander, *UFOs: Myths, Conspiracies, and Realities*, p. 159.

69 Ibid., p. 158.

70 Quoted in Pope, *Open Skies, Closed Minds*, p. 149.

71 Birnes, *The Everything UFO Book*, p. 100.

72 Alexander, *UFOs: Myths, Conspiracies, and Realities*, pp. 159–160.

73 Bullard, *The Myth and Mystery of UFOs*, p. 83.

74 Quoted in Birnes, *The Everything UFO Book*, p. 102.

75 *The UFO Phenomenon*, p. 8.

76 Pope, *Open Skies, Closed Minds*, p. 202.

77 Kean, *UFOs: Generals, Pilots, and Government Officials Go on the Record*, pp. 153–155.

78 Ibid., pp. 65–68.

79 The report was co-authored by Haines; William Pucket, a meteorologist who formerly worked with the Environment Protection Agency; Laurence Lemke, an aerospace engineer who worked with NASA on advanced space missions; Donald Ledger, a Canadian pilot and aviation professional; and five other specialists. Bullard, *The Myth and Mystery of UFOs*, pp. 2–3.

80 Bullard, *The Myth and Mystery of UFOs*, p. 3.

81 Kean, *UFOs: Generals, Pilots, and Government Officials Go on the Record*, pp. 247–261.

82 Alexander, *UFOs: Myths, Conspiracies, and Realities*, pp. 165–166.

83 Bullard, *The Myth and Mystery of UFOs*, pp. 92–93.

84 Fife Symington III, "Setting the Record Straight," in Kean, *UFOs: Generals, Pilots, and Government Officials Go on the Record*, pp. 262–264.

85 Marrs, *Alien Agenda*, p. 371.

86 Pope, *Open Skies, Closed Minds*, p. 109.

87 Marrs, *Alien Agenda*, p. 372.

88 Bullard, *The Myth and Mystery of UFOs*, p. 82.

89 Marrs, *Alien Agenda*, pp. 385–387.

90 Birnes, *The Everything UFO Book*, pp. 195–201.

91 Philip J. Corso with William J. Birnes, *The Day After Roswell*. New York: Pocket Books, 1997, p. 197.

92 Pope, *Open Skies, Closed Minds*, pp. 91–92.

93 Michael Kurland, *The Complete Idiot's Guide to Extraterrestrial Intelligence*. New York: alpha books, 1999, p. 138.

94 Birnes, *The Everything UFO Book*, p. 187.

95 Pope, *Open Skies, Closed Minds*, pp. 100–101.

96 Birnes, *The Everything UFO Book*, p. 189.

97 Pope, *Open Skies, Closed Minds*, p. 102.

98 Kurland, *The Complete Idiot's Guide to Extraterrestrial Intelligence*, p. 139.

99 Jim Marrs, *Alien Agenda: Investigating the Extraterrestrial Presence Among Us*. New York: Harper Paperbacks, 1997, p. 416.

100 Birnes, *The Everything UFO Book*, p. 190.

101 Kurland, *The Complete Idiot's Guide to Extraterrestrial Intelligence*, p. 138.

102 Ibid., p. 139.

103 Marrs, *Alien Agenda*, pp. 416–417.

104 Seth Shostak, *Confessions of an Alien Hunter: A Scientist's Search for Extraterrestrial Intelligence*. Washington, D.C.: National Geographic, 2009, pp. 124–125.

105 Jim Marrs, *Alien Agenda: Investigating the Extraterrestrial Presence Among Us*. New York: Harper Paperbacks, 1997, pp. 434–435.

106 Marrs, *Alien Agenda*, pp. 434–435.

107 Seth Shostak, *Confessions of an Alien Hunter: A Scientist's Search for Extraterrestrial Intelligence*. Washington, D.C.: National Geographic, 2009, p. 143.

108 Shostak, *Confessions of an Alien Hunter*, p. 142.

109 Michio Kaku, *Physics of the Impossible: A Scientific Exploration into the World of Phasers, Force Fields, Teleportation, and Time Travel*. New York: Anchor Books, 2008, pp. 148–149.

110 Kaku, *Physics of the Impossible*, p. 150.

111 Ibid., p. 149.

112 John White, "The UFO Problem: Toward a Theory of Everything?" In Michael Pye and

Kirsten Dalley (eds.), *UFOs & Aliens: Is There Anybody Out There?* Pompton Plains, NJ: The Career Press, 2011, p. 178.

113 Donald H. Menzel, "UFOs—The Modern Myth," in Carl Sagan and Thornton Page (eds.), *UFOs A Scientific Debate*. New York: W.W. Norton & Company, 1972, p. 144.

114 Bullard, *The Myth and Mystery of UFOs*, p. 5.

115 Albert A. Harrison, *After Contact: The Human Response to Extraterrestrial Life*. New York: Plenum Press, 1997, p. 81.

116 Ibid., p. 84.

117 James E. McDonald, "Twenty-two Years of Inadequate UFO Investigations," in Carl Sagan and Thornton Page (eds.), *UFO's A Scientific Debate*. New York: W.W. Norton & Company, 1972, p. 53.

118 As Hall explained: "I would find it puzzling and behaviorally anomalous if witnesses to a dramatic, ambiguous event promptly interpreted it in a way that lay outside their previous beliefs and contrary to the beliefs of others around them unless, indeed, their observations seemed quite unequivocal. This would be an extraordinary suspension of the usual laws of human behavior." Robert L. Hall, "Sociological Perspectives on UFO Reports," in Carl Sagan and Thornton Page (eds.), *UFOs A Scientific Debate*. New York: W.W. Norton & Company, 1972, p. 216.

119 Bruce Maccabee, "Forewords" in Stanton T. Friedman, *Flying Saucers and Science: A Scientist Investigates the Mysteries of UFOs*. The Career Press: Pompton Plains, NJ, 2008, p. 18.

120 William F. Hamilton III, *Cosmic Top Secret: America's Secret UFO Program*. New Brunswick, NJ: Inner Light Publications, 1991, p. 48.

121 Marc Davenport, *Visitors From Time: The Secret of the UFOs*. Tuscaloosa, AL: Greenleaf Publications, 1994.

122 Ibid., pp. 171–172.

123 Ivan T. Sanderson, *Invisible Residents: The Reality of Underwater UFOs*. Kempton, IL: Adventures Unlimited Press, 2005.

124 Davenport, *Visitors From Time*, p. 206.

125 Paul Davies, *The Eerie Silence: Renewing Our Search for Alien Intelligence*. Boston and New York: Houghton Mifflin Harcourt, 2010, pp. 121–122.

126 Hamilton, *Cosmic Top Secret: America's Secret UFO Program*, p. 97.

127 Ivan T. Sanderson, *Invisible Residents: The Reality of Underwater UFOs*. Kempton, IL: Adventures Unlimited Press, 2005.

128 Brad Steiger and Sherry Hansen Steiger, *Real Aliens, Space Beings, and Creatures from Other Worlds*. Canton, MI: Visible Ink Press, 2011, p. 257.

129 Dick, *Life on Other Worlds*, p. 159.

130 Shostak, *Confessions of an Alien Hunter*, p. 143.

131 Philip Coppens, *The Ancient Alien Question: A New Inquiry Into the Existence, Evidence, and Influence of Ancient Visitors*. Pompton Plains, NJ: New Page Books, p. 31.

132 Stephen Hawking, *A Brief History of Time: From the Big Bang to Black Holes*. New York: Bantam Books, 1998, p. 166.

133 For example, in an interview with this author, Steven Dick commented: "It all comes down to good evidence—not recollections, fuzzy photos, etc., but real unambiguous data. The extraterrestrial hypothesis should be the last resort—"extraordinary claims require extraordinary evidence." I am open to extraordinary evidence." Interview with Steven J. Dick, July 14, 2012.

134 Shostak, *Confessions of an Alien Hunter*, pp. 122–123.

135 Ibid., pp. 134–135.

136 Ibid., p. 144.

137 Leslie Kean, *UFOs: Generals, Pilots, and Government Officials Go on the Record*. New York: Harmony Books, 2010.

138 Albert A. Harrison, "The Science and Politics of SETI: How to Succeed in an Era of Make-Believe History and Pseudoscience," in Douglas A. Vakoch and Albert A. Harrison (eds.), *Civilizations Beyond Earth: Extraterrestrial Life and Society*. New York and Oxford: Berghan Books, 2011, pp. 141–155.

139 Bullard, *The Myth and Mystery of UFOs*, pp. 8–9.

140 Kean, *UFOs: Generals, Pilots, and Government Officials Go on the Record*, p. 265.

141 Bullard, *The Myth and Mystery of UFOs*, pp. 312–313.

142 Ibid., p. 84.

chapter 9

1 Nick Cook, *The Hunt for Zero Point: Inside the Classified World of Antigravity Technology*. New York: Broadway Books, 2003, p. 35.

2 Swords and Powell (eds.), "Chapter 3: The Flying Disks and the United States," p. 42.

3 Steven J. Dick, *Life on Other World: The 20th-Century Extraterrestrial Life Debate*. New York: Cambridge University Press, 1998, p. 141.

4 George Garrett, "Flying Disks," July 30, 1947. http://www.bibliotecapleyades.net/sociopolitica/sign/sign.htm.

5 Michael Swords and Robert Powell (eds.), "Chapter 3: The Flying Disks and the United States," in *UFOs and Government: A Historical Inquiry*. San Antonio and Charlottesville: Anomalist Books, 2012, pp. 34–44.

6 Stanton T. Friedman, *Flying Saucers and Science: A Scientist Investigates the Mysteries of UFOs*. The Career Press: Pompton Plains, NJ, 2008, p. 46.

7 Swords and Powell (eds.), "Chapter 3: The Flying Disks and the United States, pp. 34–44.

8 Maccabee, "Forewords" p. 15.

9 Kevin D. Randle, *Case MJ-12: The True Story Behind the Government's UFO Conspiracies*. New York: HarperTorch, 2002, pp. 92–94.

10 After the end of the Cold War, it transpired that Joseph Stalin had instructed his rocketry expert, Sergei Korolve, to look into the UFO sightings. Michael Swords and Robert Powell (eds.), "Chapter 4: A Formalized UFO Project," in *UFOs and Government: A Historical Inquiry*. San Antonio and Charlottesville: Anomalist Books, 2012, p. 57.

11 Philip J. Corso with William J. Birnes, *The Day After Roswell*. New York: Pocket Books, 1997, p. 285.

12 Randle, *Case MJ-12*, pp. 96–97.

13 Michael Kurland, *The Complete Idiot's Guide to Extraterrestrial Intelligence*. New York: alpha books, 1999, p. 110.

14 Dick, *Life on Other Worlds*, p. 143.

15 Seth Shostak, *Confessions of an Alien Hunter: A Scientist's Search for Extraterrestrial Intelligence*. Washington, D.C.: National Geographic, 2009, p. 120.

16 Swords and Powell (eds.), "Chapter 4: A Formalized UFO Project," p. 65.

17 Michael Swords and Robert Powell (eds.), "Chapter 5: Grudge," in *UFOs and Government: A Historical Inquiry*. San Antonio and Charlottesville: Anomalist Books, 2012, p. 73, 87.

18 J. Allen Hynek, "Twenty-one Years of UFO Reports," in Carl Sagan and Thornton Page (eds.), *UFOs A Scientific Debate*. New York: W.W. Norton & Company, 1972, p. 37.

19 Swords and Powell (eds.), "Chapter 4: A Formalized UFO Project," p. 56.

20 Brad Steiger and Sherry Hansen Steiger, *Real Aliens, Space Beings, and Creatures from Other Worlds*. Canton, MI: Visible Ink Press, 2011, p. 204.

21 *The UFO Phenomenon*. Alexandria, VA: Time Life Books, 1987, p. 44; Swords and Powell (eds.), "Chapter 4: A Formalized UFO Project," p. 65.

22 Leslie Kean, *UFOs: Generals, Pilots, and Government Officials Go on the Record*. New York: Harmony Books, 2010, p. 137.

23 Nick Pope, *Open Skies, Closed Minds: For the First Time A Government UFO Expert Speaks Out*. London and New York: Simon & Schuster, 1996, p. 167.

24 Michael Kurland, *The Complete Idiot's Guide to Extraterrestrial Intelligence*. New York: alpha books, 1999, p. 111.

25 Randle, *Case MJ-12*, p. 101.

26 Kurland, *The Complete Idiot's Guide to Extraterrestrial Intelligence*, p. 111.

27 *The UFO Phenomenon*, p. 46.

28 Michael Swords and Robert Powell (eds.), "Chapter 8: Tacking Against the Winds," in *UFOs and Government: A Historical Inquiry*. San Antonio and Charlottesville: Anomalist Books, 2012, pp. 139–140.

29 Michael Swords and Robert Powell (eds.), "Chapter 6: Duck and Cover," in *UFOs and Government: A Historical Inquiry*. San Antonio and Charlottesville: Anomalist Books, 2012, pp. 99–100.

30 Swords and Powell (eds.), "Chapter 8: Tacking Against the Winds," p. 139.

31 Jim Marrs, *Alien Agenda: Investigating the Extraterrestrial Presence Among Us*. New York: Harper Paperbacks, 1997, p. 189.

32 Shostak, *Confessions of an Alien Hunter*, p. 121.

33 Mack Maloney, *UFOs in Wartime: What They Don't Want You to Know*. New York: Berkley Books, 2011, p. 165.

34 Randle, *Case MJ-12*, pp. 99–100 and *The UFO Phenomenon*, p. 49.

35 Kean, *UFOs: Generals, Pilots, and Government Officials Go on the Record*, pp. 103–104.

36 Steiger and Steiger, *Real Aliens, Space Beings, and Creatures from Other Worlds*, p. 208.

37 Kurland, *The Complete Idiot's Guide to Extraterrestrial Intelligence*, pp. 111–113.

38 Maloney, *UFOs in Wartime*, p. 164.

39 Birnes, *The Everything UFO Book*, pp. 44–45.

40 Maloney, *UFOs in Wartime*, p. 168.

41 Steiger and Steiger, *Real Aliens, Space Beings, and Creatures from Other Worlds*, p. 206.

42 *The UFO Phenomenon*, p. 55.

43 Ibid., p. 55.

44 Swords and Powell (eds.), "Chapter 8: Tacking Against the Winds," p. 159.

45 Jenny Randles, *The Truth Behind Men in Black: Government Agents—Or Visitor From Beyond*. New York: St. Martin's Press, 1997, p. 43.

46 Michael Swords and Robert Powell (eds.), "Chapter 12: Something Closer This Way Comes," in *UFOs and Government: A Historical Inquiry*. San Antonio and Charlottesville: Anomalist Books, 2012, pp. 238–239.

47 Kean, *UFOs: Generals, Pilots, and Government Officials Go on the Record*, pp. 105–107 and Nick Redfern, *Contactees: A History of Alien-Human Interaction*. Franklin Lakes, NJ: The Career Press, Inc., 2010, pp. 34–35.

48 Swords and Powell (eds), "Chapter 6: Duck and Cover," p. 103.

49 Ibid., pp. 103–104.

50 Michael Swords and Robert Powell (eds.), "Chapter 11: A Cold War," in *UFOs and Government: A Historical Inquiry*. San Antonio and Charlottesville: Anomalist Books, 2012, pp. 208–212.

51 Ibid., p. 219.

52 Michael Swords and Robert Powell (eds.), "Chapter 10: Intermission," in *UFOs and Government: A Historical Inquiry*. San Antonio and Charlottesville: Anomalist Books, 2012, pp. 204–205.

53 Michael Swords and Robert Powell (eds.), "Chapter 13: Battle in the Desert," in *UFOs and Government: A Historical Inquiry*. San Antonio and Charlottesville: Anomalist Books, 2012, p. 279.

54 Swords and Powell (eds.), "Chapter 13: Battle in the Desert," p. 276.

55 Air Force, *Special Report No. 14: Analysis of Reports of Unidentified Flying Objects* (Project No. 10073), May 5, 1955.

56 Friedman, *Flying Saucers and Science*, p. 36.

57 Menzel, "UFOs—The Modern Myth," p. 133.

58 Air Force, *Special Report No. 14: Analysis of Reports of Unidentified Flying Objects* (Project No. 10073), p. 94.

59 *The UFO Phenomenon*, p. 56.

60 Randles, *The Truth Behind Men in Black*, pp. 44–45.

61 Kean, *UFOs: Generals, Pilots, and Government Officials Go on the Record*, p. 109.

62 Kurland, *The Complete Idiot's Guide to Extraterrestrial Intelligence*, p. 105.

63 McDonald, "Twenty-two Years of Inadequate UFO Investigations," p. 54.

64 Michael Swords and Robert Powell (eds.), "Chapter 14: The Colorado Project," in *UFOs*

and Government: A Historical Inquiry. San Antonio and Charlottesville: Anomalist Books, 2012, pp. 308–309.

65 Swords and Powell (eds.), "Chapter 13: Battle in the Desert," pp. 220–221, 272.

66 Marrs, *Alien Agenda*, p. 202.

67 Swords and Powell (eds.), "Chapter 13: Battle in the Desert," p. 301.

68 Ibid., pp. 272–273.

69 One strange footnote to the story: Ruppelt died allegedly soon after reading a borrowed copy of the Varo edition of Morris K. Jessup's book, *The Case for the UFO*. According to legend, a tradition of bad luck or strange circumstances is connected with those people who possessed the Varo edition of the book. Morris K. Jessup, *The Allende Letters and the Varo Edition of The Case for the UFO*. New Brunswick, NJ: Conspiracy Journal/Global Communications, 2007, p. xiii.

70 Swords and Powell (eds.), "Chapter 12: Something Closer This Way Comes," p. 238.

71 Michael Swords and Robert Powell (eds.), "Chapter 9: The CIA Solution," in *UFOs and Government: A Historical Inquiry*. San Antonio and Charlottesville: Anomalist Books, 2012, p. 185.

72 Swords and Powell (eds.), "Chapter 11: A Cold War," p. 208.

73 Kean, *UFOs: Generals, Pilots, and Government Officials Go on the Record*, p. 109.

74 Swords and Powell (eds.), "Chapter 9: The CIA Solution," p. 196.

75 Dick, *Life on Other Worlds*, p. 151.

76 *The UFO Phenomenon*, pp. 99–103.

77 Ibid., p. 104.

78 Ibid., pp. 106–107.

79 Dick, *Life on Other Worlds*, p. 150.

80 Swords and Powell (eds.), "Chapter 14: The Colorado Project," pp. 306–307.

81 *The UFO Phenomenon*, p. 109.

82 Dick, *Life on Other Worlds*, p. 152.

83 Ibid., p. 152.

84 Ibid., p. 152.

85 Swords and Powell (eds), "Chapter 14: The Colorado Project," pp. 307–308.

86 Ibid., pp. 309–310.

87 Ibid., pp. 310.

88 Ibid., pp. 310.

89 Ibid., pp. 312.

90 Thomas E. Bullard, *The Myth and Mystery of UFOs*. Lawrence, KS: University Press of Kansas, 2010, p. 66.

91 *The UFO Phenomenon*, p. 115.

92 Quoted in Kean, *UFOs: Generals, Pilots, and Government Officials Go on the Record*, pp. 110–111.

93 Swords and Powell (eds.), "Chapter 14: The Colorado Project," p. 322.

94 *The UFO Phenomenon*, p. 116.

95 Swords and Powell (eds.), "Chapter 14: The Colorado Project," p. 322.

96 Kean, *UFOs: Generals, Pilots, and Government Officials Go on the Record*, p. 111.

97 Those scientists who testified included Dr. J. Allen Hynek, the chairman of the astronomy department at Northwestern University; Dr. Carl Sagan, professor of astronomy at Cornell University; Dr. James Harder, professor of civil engineering at the University of California, Berkeley; Dr. Robert L. Hall, head of the department of sociology at the University of Illinois, Chicago; and Dr. Robert M.L. Baker, senior scientist for System Sciences Corp. in El Segundo, California. In addition, the 247-page proceedings included written submissions from six more scientists: Dr. Donald Menzel, astronomer at Harvard University; Dr. R. Leo Sprinkle, psychologist at the University of Wyoming; Dr. Garry C. Henderson, senior research scientist for Space Sciences at General Dynamics in Forth Worth, Texas; Dr. Roger N. Shepard, department of psychology at Stanford University; Dr. Frank Salisbury, head of plant science department at Utah State University; and Mr. Stanton T. Friedman, nuclear physicist at Westinghouse Astronuclear Laboratory in Large, Pennsylvania. Friedman, *Flying Saucers and Science*, pp. 50–51.

98 John B. Alexander, *UFOs: Myths, Conspiracies, and Realities*. New York: St. Martin's Press, 2011, p. 55.

99 Bullard, *The Myth and Mystery of UFOs*, p. 67.

100 Alexander, *UFOs: Myths, Conspiracies, and Realities*, p. 57.

101 Thornton Page, "Education and the UFO Phenomenon," in Carl Sagan and Thornton Page (eds.), *UFO's A Scientific Debate*. New York: W.W. Norton & Company, 1972, p.4.

102 Michael E. Salla, *Exopolitics: Political Implications of the Extraterrestrial Presence*. Temple, AZ: Danelion Books, 2004, p. 11.

103 Shostak, *Confessions of an Alien Hunter*, p. 122.

104 William F. Hamilton III, *Cosmic Top Secret: America's Secret UFO Program*. New Brunswick, NJ: Inner Light Publications, 1991, p. 12.

105 Kean, *UFOs: Generals, Pilots, and Government Officials Go on the Record*, pp. 107–114.

106 Bullard, *The Myth and Mystery of UFOs*, p. 67.

107 Alexander, *UFOs: Myths, Conspiracies, and Realities*, p. 61.

108 Friedman, *Flying Saucers and Science*, pp. 52–53.

109 Pope, *Open Skies, Closed Minds*, p. 57.

110 As Hynek explained, the original UFO reports had to pass through a "narrow band-pass filter" before it qualified as worthy material for scientific study. Only those reports that survived this "running of the gauntlet" could qualify. Hynek, "Twenty-one Years of UFO Reports," p. 40.

111 Pope, *Open Skies, Closed Minds*, p. 171.

112 Alexander Wendt and Raymond Duvall, "Militant Agnosticism and the UFO Taboo," in Kean, *UFOs: Generals, Pilots, and Government Officials Go on the Record*, pp. 17-40.

113 Alexander, *UFOs: Myths, Conspiracies, and Realities*, p. 102.

114 Kean, *UFOs: Generals, Pilots, and Government Officials Go on the Record*, p. 214.

115 Ibid., p. 243.

116 Steiger and Steiger, *Real Aliens, Space Beings, and Creatures from Other Worlds*, p. 222.

117 Quoted in Brad Steiger, *The Philadelphia Experiment & Other UFO Conspiracies*. New Brunswick, NJ: Inner Light Publications, 1990, p. 144.

118 *The UFO Phenomenon*, p. 119.

119 Randle, *Case MJ-12*, pp. 102–106.

120 The General Accounting Office released a report on the hearings titled *Homeland Security: Agency Resources Address Violations of Restricted Airspace, but Management Improvements are Needed*. Alexander, *UFOs: Myths, Conspiracies, and Realities*, p. 73.

121 Kean, *UFOs: Generals, Pilots, and Government Officials Go on the Record*, p. 115.

122 Kean, *UFOs: Generals, Pilots, and Government Officials Go on the Record*, pp. 116–117.

123 Federal Aviation Administration, Air Traffic Organization Policy, Order JO 7110.65U Section 8. Unidentified Flying Object (UFO) Reports. February 9, 2012. http://www.faa.gov/air_traffic/publications/atpubs/atc/atc0908.html.

124 Moon and Back Interview with Robert Bigelow, part 1, http://www.youtube.com/watch?v=DceaEUUcXpI. Accessed August 23, 2013.

125 Robert Sheaffer, "Bigelow's Aerospace and Saucer Emporium," *Skeptical Inquirer*, Vol.

33, Issue 4, (July/August 2009), http://www.csicop.org/si/show/bigelows_aerospace_and_saucer_emporium.

126 Jesse Ventura's Conspiracy Theory, "Skinwalker," http://www.youtube.com/watch?v=KpkTTNKh26Q, accessed August 23, 2013.

127 Michaud, *Contact with Alien Civilizations*, p. 159.

128 Pope, *Open Skies, Closed Minds*, p. 132.

129 Kean, *UFOs: Generals, Pilots, and Government Officials Go on the Record*, pp. 17–40.

130 Michael Swords and Robert Powell (eds.), "Chapter 19: UFOs and France—Beginnings of a Scientific Investigation," in *UFOs and Government: A Historical Inquiry*. San Antonio and Charlottesville: Anomalist Books, 2012, p. 440.

131 Kean, *UFOs: Generals, Pilots, and Government Officials Go on the Record*, pp. 119–127.

132 Alexander, *UFOs: Myths, Conspiracies, and Realities*, p. 218.

133 Ibid., p. 219.

134 Nick Pope, "The Real X-Files," In Michael Pye and Kirsten Dalley (eds.), *UFOs & Aliens: Is There Anybody Out There?* Pompton Plains, NJ: The Career Press, 2011, p. 27.

135 Ibid., pp. 27–28.

136 Pope, *Open Skies, Closed Minds*, p. 44.

137 Ibid., p. 55.

138 Ibid., p. 43.

139 Ibid., p. 212.

140 Ibid., p. 214.

141 Ibid., p. 220.

142 Ibid., p. 222.

143 As Pope noted, the relationship between the MoD and the UFO lobby is full of ironies. The UFO lobby presses for disclosure, but when the MoD complies, the release of information is often dismissed as part of the cover-up. Pope, *Open Skies, Closed Minds*, p. 44, 180.

144 Alexander, *UFOs: Myths, Conspiracies, and Realities*, pp. 98–99.

145 Pope, "The Real X-Files," p. 30.

chapter 10

1 Marie D. Jones and Larry Flaxman, "Identity Crisis: When is a UFO Not a UFO?" In Michael Pye and Kirsten Dalley (eds.), *UFOs & Aliens: Is There Anybody Out There?*

Pompton Plains, NJ: The Career Press, 2011, pp. 221–222.

2 Nick Redfern, *Contactees: A History of Alien-Human Interaction*. Franklin Lakes, NJ: The Career Press, 2010, pp. 25-31.

3 Jenny Randles, *The Truth Behind Men in Black: Government Agents—Or Visitor From Beyond*. New York: St. Martin's Press, 1997, p. 61.

4 Kathleen Marden, "Alien Abduction: Fact or Fiction?" In Michael Pye and Kirsten Dalley (eds.), *UFOs & Aliens: Is There Anybody Out There?* Pompton Plains, NJ: The Career Press, 2011, p. 86.

5 Whitley Strieber, *Communion: A True Story*. New York: Avon, 1987, p. 249.

6 *The UFO Phenomenon*. Alexandria, VA: Time Life Books, 1987, p. 86.

7 Nick Pope, *Open Skies, Closed Minds: For the First Time A Government UFO Expert Speaks Out*. London and New York: Simon & Schuster, 1996, p. 25.

8 Marc Davenport, *Visitors From Time: The Secret of the UFOs*. Tuscaloosa, AL: Greenleaf Publications, 1994, p. 200.

9 Randles, *The Truth Behind Men in Black*, p. 50.

10 Brad Steiger, *The Philadelphia Experiment & Other UFO Conspiracies*. New Brunswick, NJ: Inner Light Publications, 1990, p. 20.

11 In the 1930s, Elijah Muhammad announced that the Japanese had built an enormous "Mother Plane" that had the capability to carry a fleet of smaller planes that would destroy all White people in America with poison bombs. As the theology of the Nation of Islam later evolved, the Japanese were replaced with a council of 24 imams consisting of Black scientists who built the great ship. Thomas E. Bullard, *The Myth and Mystery of UFOs*. Lawrence, KS: University Press of Kansas, 2010, p. 29.

12 Arthur G. Magida, *Prophet of Rage: A Life of Louis Farrakhan and His Nation*. New York: Basic Books, 1996, p. 190; and Louis Farrakhan, "The Great Atonement," in Adam Parfrey, *Extreme Islam: Anti-American Propaganda of Muslim Fundamentalism*. Los Angeles, CA: Feral House, 2001, pp. 238–246. U.S. authorities did take notice of Farrakhan's rapport with the Libyan leader. The U.S. Department of Justice once sent a letter to Farrakhan warning him to register as a foreign agent for Libya if he planned to booster Ghadafi. Magida, *Prophet of Rage*, p. 200.

13 Marden, "Alien Abduction: Fact or Fiction?", pp. 82–83.

14 William J. Birnes, *The Everything UFO Book: An investigation of sightings, cover-ups, and the quest for extraterrestrial life*. Avon, MA: Adams Media, 2012, pp. 56–61.

15 Brad Steiger and Sherry Hansen Steiger, *Real Aliens, Space Beings, and Creatures from Other Worlds*. Canton, MI: Visible Ink Press, 2011, p. 135.

16 Ibid., p. 135.

17 Pope, *Open Skies, Closed Minds*, pp. 22–23.

18 Stanton T. Friedman, *Flying Saucers and Science: A Scientist Investigates the Mysteries of UFOs*. The Career Press: Pompton Plains, NJ, 2008, p. 149.

19 Steiger and Steiger, *Real Aliens, Space Beings, and Creatures from Other Worlds*, p. 135.

20 Birnes, *The Everything UFO Book*, p. 59.

21 John White, "The UFO Problem: Toward a Theory of Everything?" In Michael Pye and Kirsten Dalley (eds.), *UFOs & Aliens: Is There Anybody Out There?* Pompton Plains, NJ: The Career Press, 2011, p. 171.

22 Steiger and Steiger, *Real Aliens, Space Beings, and Creatures from Other Worlds*, p. 136.

23 Birnes, *The Everything UFO Book*, pp. 58–59.

24 John Fuller, *The Interrupted Journey: The Last Two Hours Aboard a Flying Saucer*. London: Dial Press, 1966.

25 Jim Marrs, *Alien Agenda: Investigating the Extraterrestrial Presence Among Us*. New York: Harper Paperbacks, 1997, p. 296.

26 Friedman, *Flying Saucers and Science*, pp. 99–102.

27 Marden, "Alien Abduction: Fact or Fiction?", p. 92.

28 Travis Walton, *The Walton Experience*. New York: Berkley, 1978.

29 Marden, "Alien Abduction: Fact or Fiction?", p. 93.

30 Marrs, *Alien Agenda*, pp. 325–330.

31 Strieber, *Communion*, p. 3.

32 Ibid., pp. 11–24.

33 Ibid., p. 142.

34 Ibid., p. 91.

35 Strieber, *Communion*, pp. 226–228.

36 Pope, *Open Skies, Closed Minds*, p. 58.

37 Neil de Grasse Tyson, *Space Chronicles: Facing the Ultimate Frontier*. New York: W.W. Norton & Company Ltd., 2012, p. 184.

38 Albert A. Harrison, *After Contact: The Human Response to Extraterrestrial Life*. New York: Plenum Press, 1997, p. 74.

39 Harrison, *After Contact*, p. 80.

40 Ibid., p. 79.

41 Michael Talbot, *The Holographic Universe: The Revolutionary Theory of Reality*. New York: Harper/Perennial, 2001, p. 278.

42 Marden, "Alien Abduction: Fact or Fiction?", p. 101.

43 Jim Moroney, "An Alien Intervention," in Michael Pye and Kirsten Dalley (eds.), *UFOs & Aliens: Is There Anybody Out There?*, Pompton Plains, NJ: The Career Press, 2011, p. 59.

44 Marrs, *Alien Agenda*, p. 340.

45 Ibid., p. 337.

46 Steiger and Steiger, *Real Aliens, Space Beings, and Creatures from Other Worlds*, p. 44.

47 Marrs, *Alien Agenda*, p. 355.

48 Rick Strassman, *DMT: The Spirit Molecule: A Doctor's Revolutionary Research into the Biology of Near-Death and Mystical Experiences*. Rochester, VT: Part Street Press, 2000, pp. 217–219.

49 Michael E. Salla, *Exopolitics: Political Implications of the Extraterrestrial Presence*. Temple, AZ: Dandelion Books, 2004, p. 21.

50 Slawek Wojtowicz, "Hypnosis, Past Life Regression, Meditation, and More," in Rick Strassman, Slawek Wojtowicz, Luis Eduardo Luna, and Ede Frecska, *Innter Paths to Outer Space: Journeys to Alien Worlds though Psychadelics and Other Spritual Technologies*. Rochester, VT: Park Street Press, 2008, p. 256.

51 Marrs, *Alien Agenda*, p. 343.

52 Salla, *Exopolitics*, p. 10.

53 Birnes, *The Everything UFO Book*, p. 246.

54 White, "The UFO Problem: Toward a Theory of Everything?", p. 181.

55 Talbot, *The Holographic Universe*, p. 282.

56 Jones and Flaxman, "Identity Crisis: When is a UFO Not a UFO?", pp. 221–222.

57 Randles once wondered if a person caught in the grip of this phenomenon might be able to actually photograph something that he sees even though the entity does not exist in a completely objective form. *The UFO Phenomenon*, p. 71.

58 Marrs, *Alien Agenda*, p. 483.

59 Ibid., p. 467.

60 Ibid., p. 491.

61 Brian Greene, *The Elegant Universe: Superstrings, Hidden Dimensions, and the Quest for*

the Ultimate Reality. New York and London: W.W. Norton & Company, 2003, pp. 3–5.

62 Kaku and Thompson, *Beyond Einstein*, pp. 162–163.

63 Michio Kaku, *Hyperspace: A Scientific Odyssey Through Parallel Universes, Time Warps, and the 10th Dimension*. New York: Doubleday, 1994, p. 8.

64 Kaku, *Hyperspace*, p. 107.

65 Kaku, *Introduction to Superstrings*, p. vii.

66 Kaku and Thompson, *Beyond Einstein*, p. 87.

67 Michio Kaku, *Parallel Worlds: A Journey through Creation, Higher Dimensions, and the Future of the Cosmos*. New York: Anchor Books, 2005, pp. 189–190.

68 Kaku and Thompson, *Beyond Einstein*, p. 119.

69 Michio Kaku, *Einstein's Cosmos: How Albert Einstein's Vision Transformed Our Understanding of Space and Time*. New York and London: W.W. Norton & Company, 2004, p. 226.

70 Minutely small, measuring at the Planck's length, it is estimated that these strings are 100 billion billion times smaller than a proton in an atom. Kaku and Thompson, *Beyond Einstein*, pp. 4–5 and Brian Greene, *The Hidden Reality: Parallel Universes and the Deep Laws of the Cosmos*. New York: Alfred A. Knopf, 2011, pp. 78–79.

71 Michio Kaku, *Parallel Worlds: A Journey through Creation, Higher Dimensions, and the Future of the Cosmos*. New York: Anchor Books, 2005, p. 212.

72 Kaku and Thompson, *Beyond Einstein*, pp. 162–163.

73 Richard Panek, *The 4 Percent Universe: Dark Matter, Dark Energy, and the Race to Discover the Rest of Reality*. Boston and New York: Houghton Mifflin Harcourt, 2011.

74 Rick Strassman, Slawek Wojtowicz, Luis Eduardo Luna, and Ede Frecska, *Innter Paths to Outer Space: Journeys to Alien Worlds though Psychadelics and Other Spritual Technologies*. Rochester, VT: Park Street Press, 2008, p. 5.

75 Rick Strassman, *DMT: The Spirit Molecule: A Doctor's Revolutionary Research into the Biology of Near-Death and Mystical Experiences*. Rochester, VT: Part Street Press, 2000, p. 301.

76 Ibid., p. 201.

77 In 1965, German scientists discovered that DMT was present in human blood and urnine. Rick Strassman, "DMT: The Brain's Own Psychadelic," in Rick Strassman, Slawek Wojtowicz, Luis Eduardo Luna, and Ede Frecska, *Innter Paths to Outer Space:*

Journeys to Alien Worlds though Psychadelics and Other Spritual Technologies. Rochester, VT: Park Street Press, 2008, p. 35.

78 Strassman, *DMT: The Spirit Molecule.*

79 Nevertheless, Strassman observed that some of his subjects displayed rapid eye movement, though they were fully awake. He wondered if the DMT had induced a wakeful dream state. Strassman, *DMT: The Spirit Molecule*, p. 200.

80 Rick Strassman, "The Varieties of the DMT Experience," in Rick Strassman, Slawek Wojtowicz, Luis Eduardo Luna, and Ede Frecska, *Innter Paths to Outer Space: Journeys to Alien Worlds though Psychadelics and Other Spritual Technologies.* Rochester, VT: Park Street Press, 2008, p. 73.

81 Strassman, "The Varieties of the DMT Experience," p. 73.

82 Strassman, *DMT: The Spirit Molecule*, p. 199.

83 Ibid., p. 352, note 7.

84 Slawek Wojtowicz, "Magic Mushrooms," in Rick Strassman, Slawek Wojtowicz, Luis Eduardo Luna, and Ede Frecska, *Innter Paths to Outer Space: Journeys to Alien Worlds though Psychadelics and Other Spritual Technologies.* Rochester, VT: Park Street Press, 2008, pp. 142–161.

85 Strassman, *DMT: The Spirit Molecule*, pp. 314–322.

86 Marden, "Alien Abduction: Fact or Fiction?", pp. 99–100.

87 Strassman, *DMT: The Spirit Molecule*, p. 315.

88 David Jay Brown, "Can Psychedelic Drugs Help us Speak to Aliens?" *Santa Cruz Patch*, June 15, 2012, http://santacruz.patch.com/articles/can-psychedelic-drugs-help-us-speak-to-aliens.

chapter 11

1 Shostak, *Confessions of an Alien Hunter*, p. 112. Likewise, Nearly 80 percent of those Americans surveyed—according to a CNN poll conducted in 1997—believed that the government was hiding knowledge of the existence of alien life forms. "Poll U.S. hiding knowledge of aliens," CNN, June 15, 1997. http://articles.cnn.com/1997-06-15/us/9706_15_ufo.poll_1_ufo-aliens-crash-site?_s=PM:US.

2 Jenny Randles, *The Truth Behind Men in Black: Government Agents—Or Visitor From Beyond.* New York: St. Martin's Press, 1997, p. 43.

3 *The UFO Phenomenon.* Alexandria, VA: Time Life Books, 1987, p. 46.

4 William J. Birnes, *The Everything UFO Book: An investigation of sightings, cover-ups, and the quest for extraterrestrial life*. Avon, MA: Adams Media, 2012, p. 28.

5 Brad Steiger, *The Philadelphia Experiment & Other UFO Conspiracies*. New Brunswick, NJ: Inner Light Publications, 1990, p. 75.

6 Brad Steiger and Sherry Hansen Steiger, *Real Aliens, Space Beings, and Creatures from Other Worlds*. Canton, MI: Visible Ink Press, 2011, p. 286.

7 William J. Birnes, *The Everything UFO Book: An investigation of sightings, cover-ups, and the quest for extraterrestrial life*. Avon, MA: Adams Media, 2012, p. 160.

8 Another scientist who allegedly worked on the flying saucer project was an electrical engineer and SS Colonel by the name of Kurt Debus. He ultimately ended up as a launch control officer at Cape Kennedy in the 1960s. Birnes, *The Everything UFO Book*, p. 31.

9 Michael A.G. Michaud, *Contact with Alien Civilizations: Our Hopes and Fears about Encountering Extraterrestrials*. New York: Copernicus Books, 2007, p. 10.

10 According to Wernher von Braun, Kammler planned to bargain with the Allies for his life in exchange for handing over rocket scientists and engineers. Luftwaffe Major General Walter Dornberger suggested that Kammler committed suicide when Czech resistace fighters overcame SS troops in Prague at the end of the war. Nick Cook, *The Hunt for Zero Point: Inside the Classified World of Antigravity Technology*. New York: Broadway Books, 2003 and Jim Marrs, *Alien Agenda: Investigating the Extraterrestrial Presence Among Us*. New York: Harper Paperbacks, 1997, pp. 104–105.

11 Nick Pope, "The Real X-Files," in Michael Pye and Kirsten Dalley (eds.), *UFOs & Aliens: Is There Anybody Out There?* Pompton Plains, NJ: The Career Press, 2011, pp. 31–32.

12 Thomas J. Carey and Donald R. Schmitt, *Witness to Roswell: Unmasking the Government's Biggest Cover-Up* (Revised and Expanded Edition). Franklin Lakes, NJ: New Page Books 2009, p. 21.

13 Ibid., p. 21.

14 Ibid., pp. 48, 52–53.

15 Ibid., p. 49.

16 According to some versions of the Roswell incident, that same day, a group of civilian archeologists telephoned the sheriff's office in Roswell that they had discovered a set of "little bodies" at a site 35 miles northwest of Roswell in Chaves County. Carey and Schmitt, *Witness to Roswell*, p. 88.

17 Mack Maloney, *UFOs in Wartime: What They Don't Want You to Know*. New York: Berkley Books, 2011, p. 127.

18 Kevin D. Randle, *Case MJ-12: The True Story Behind the Government's UFO Conspiracies*. New York: HarperTorch, 2002, pp. 26–27.

19 Quoted in Whitley Strieber, *Communion: A True Story*. New York: Avon, 1987, p. 232.

20 Randle, *Case MJ-12*, pp. 26–27.

21 Shostak, *Confessions of an Alien Hunter*, pp. 116–117 and Strieber, *Communion*, p. 233.

22 Carey and Schmitt, *Witness to Roswell*, p. 29.

23 Birnes, *The Everything UFO Book*, pp. 35–36.

24 Carey and Schmitt, *Witness to Roswell*, p. 126.

25 *The UFO Phenomenon*, pp. 39–40.

26 Marrs, *Alien Agenda*, p. 135.

27 United States Air Force, *The Roswell Report: Case Closed*. New York: Skyhorse Publishing, 2013, p. 11.

28 William J. Birnes, *The Everything UFO Book: An investigation of sightings, cover-ups, and the quest for extraterrestrial life*. Avon, MA: Adams Media, 2012, pp. 35–36.

29 Carey and Schmitt, *Witness to Roswell*, p. 217.

30 William F. Hamilton III, *Cosmic Top Secret: America's Secret UFO Program*. New Brunswick, NJ: Inner Light Publications, 1991, p. 23.

31 Carey and Schmitt, *Witness to Roswell*, p. 54.

32 Whitley Strieber, *Communion: A True Story*. New York: Avon, 1987, p. 233.

33 Strieber, *Communion*, p. 233.

34 Carey and Schmitt, *Witness to Roswell*, pp. 54–55, 61.

35 Quoted in Carey and Schmitt, *Witness to Roswell*, p. 69.

36 Carey and Schmitt, *Witness to Roswell*, p. 76.

37 Just months after returning home from serving in the Navy, Mack Brazel's son Vernon disappeared. His grandson, William R. Brazel, was shot to death while hunting with two companions in 1964. Another hunter was also killed by a second bullet. State police investigators concluded that both were shot accidentally. Carey and Schmitt, *Witness to Roswell*, p. 73.

38 Carey and Schmitt, *Witness to Roswell*, p. 203.

39 Philip J. Corso with William J. Birnes, *The Day After Roswell*. New York: Pocket Books, 1997, pp. 22–23.

40 Steiger and Steiger, *Real Aliens, Space Beings, and Creatures from Other Worlds*, p. 4.

41 Randle, *Case MJ-12*, p. 29.

42 Michael Swords and Robert Powell (eds.), "Chapter 3: The Flying Disks and the United States," in *UFOs and Government: A Historical Inquiry*. San Antonio and Charlottesville: Anomalist Books, 2012, p. 45.

43 Steiger and Steiger, *Real Aliens, Space Beings, and Creatures from Other Worlds*, p. 2.

44 Carey and Schmitt, *Witness to Roswell*, p. 81.

45 Ibid., p. 29.

46 Steiger and Steiger, *Real Aliens, Space Beings, and Creatures from Other Worlds*, p. 4.

47 Carey and Schmitt, *Witness to Roswell*, pp. 85–86.

48 Marrs, *Alien Agenda*, p. 135.

49 Lydialye Gibson, "Science? Fiction?" *The University of Chicago Magazine*, (September/October 2011).

50 Carey and Schmitt, *Witness to Roswell*, p. 42.

51 Ibid., p. 186.

52 Thomas J. Carey, "Killing the Roswell Story," in Michael Pye and Kirsten Dalley (eds.), *UFOs & Aliens: Is There Anybody Out There?* Pompton Plains, NJ: The Career Press, 2011, p. 194.

53 United States Air Force, *The Roswell Report: Case Closed*. New York: Skyhorse Publishing, 2013, p. 77.

54 United States Air Force, *The Roswell Report: Case Closed*, p. 86.

55 Carey and Schmitt, *Witness to Roswell*, pp. 143–144.

56 For many years, Kittinger held the world records for the highest parachute jump and lengh of a free fall. At one time, he was the only person to exceed the speed of sound without an aircraft or spacecraft. In September 1984, he became the first person to make a solo crossing of the Atlantic by balloon. United States Air Force, *The Roswell Report: Case Closed*, pp. 109–112.

57 United States Air Force, *The Roswell Report: Case Closed*, pp. 117–118.

58 Carey and Schmitt, *Witness to Roswell*, p. 31.

59 Ibid., p. 40.

60 Ibid. pp. 245–254.

61 Ibid., pp. 207–208.

62 Marrs, *Alien Agenda*, pp. 144–145.

63 Carey and Schmitt, *Witness to Roswell*, p. 26.

64 Ibid., p. 223.

65 Michael Swords and Robert Powell (eds.), "Chapter 15: After the Close of Blue Book," in *UFOs and Government: A Historical Inquiry*. San Antonio and Charlottesville: Anomalist Books, 2012, pp. 351–354.

66 Carey and Schmitt, *Witness to Roswell*, pp. 218–219.

67 Ibid., pp. 33–34.

68 Thomas Printy, "The Ramey Document: Smoking gun or empty water pistol?" August 2003, http://home.comcast.net/~tprinty/UFO/Ramey.htm.

69 Carey and Schmitt, *Witness to Roswell*, p. 152.

70 Shostak, *Confessions of an Alien Hunter*, p. 119 and United States Air Force, *The Roswell Report: Case Closed*, p. 5.

71 Marrs, *Alien Agenda*, p. 157.

72 United States Air Force, *The Roswell Report: Case Closed*, pp. 41–42.

73 Carey and Schmitt, *Witness to Roswell*, p. 39.

74 Swords and Powell (eds), "Chapter 15: After the Close of Blue Book," p. 353.

75 United States Air Force, *The Roswell Report: Case Closed*, p. 3.

76 Swords and Powell (eds.), "Chapter 15: After the Close of Blue Book," p. 356.

77 Ibid., pp. 351–352.

78 United States Air Force, *The Roswell Report: Case Closed*, p. 119.

79 Ibid., p. 23.

80 Birnes, *The Everything UFO Book*, p. 37.

81 Carey and Schmitt, *Witness to Roswell*, p. 30.

82 The Horten brothers' flying wing was similar in appearance to the U.S. Air Force's B-2 Stealth bomber. However, the craft was made out of wood, not exotic material. The Luftwaffe never put the flying wing into development and never deployed it. It was only a test aircraft, but by 1945, was powered by jet engines. The craft was inherently unstable and did not fly well. Birnes, *The Everything UFO Book*, pp. 244–245.

83 John B. Alexander, *UFOs: Myths, Conspiracies, and Realities*. New York: St. Martin's Press, 2011, p. 276.

84 Shostak, *Confessions of an Alien Hunter*, p. 117.

85 Lydialye Gibson, "Science? Fiction?" *The University of Chicago Magazine*, (September/October 2011).

86 Steiger, *The Philadelphia Experiment & Other UFO Conspiracies*, p. 146.

87 Randle, *Case MJ-12*, p. 2.

88 "Briefing Document: Operation Majestic 12," November 18, 1952 in Stanton T. Friedman, *Flying Saucers and Science: A Scientist Investigates the Mysteries of UFOs*. The Career Press: Pompton Plains, NJ, 2008, pp. 266–270.

89 Nick Pope, *Open Skies, Closed Minds: For the First Time A Government UFO Expert Speaks Out*. London and New York: Simon & Schuster, 1996, p. 159. Lieutenant Philip J. Corso gave this same roster for MJ-12 as well. Corso and Birnes, *The Day After Roswell*, p. 67.

90 Hamilton, *Cosmic Top Secret: America's Secret UFO Program*, p. 16.

91 Michael E. Salla, *Exopolitics: Political Implications of the Extraterrestrial Presence*. Temple, AZ: Dandelion Books, 2004, p. 175.

92 See for example, Salla, *Exopolitics*, pp. 5–6.

93 Salla, *Exopolitics*, p. 66.

94 Ibid., pp. 70–71.

95 Alexander, *UFOs: Myths, Conspiracies, and Realities*, p. 106.

96 Ibid., p. 106.

97 Salla, *Exopolitics*, pp. 22–23.

98 Birnes, *The Everything UFO Book*, p. 49.

99 Alexander, *UFOs: Myths, Conspiracies, and Realities*, p. 106.

100 Ibid., p. 107.

101 Hamilton, *Cosmic Top Secret: America's Secret UFO Program*, p. 74.

102 Randle, *Case MJ-12*, p. 2.

103 Marrs, *Alien Agenda*, p. 177.

104 Pope, *Open Skies, Closed Minds*, p. 159.

105 Randle, *Case MJ-12*, p. 20.

106 Ibid., p. 263.

107 Ibid., pp. 258–259.

108 Ibid., p. 260.

109 Corso and Birnes, *The Day After Roswell*, p. 67.

110 Ibid.

111 Corso provided testimony that nearly 900 American POWs had been left behind in North Korea after the exchange of prisoners took place. Alexander, *UFOs: Myths, Conspiracies, and Realities*, p. 42.

112 To bolster his case, Corso claimed that General Douglas MacArthur once counseled that the army might be called upon to fight a war in space. "The next war," he predicted in 1955, would "be an interplanetary war...The nations of Earth must make a common front against [an] attack by people from other planents." Philip J. Corso with William J. Birnes, *The Day After Roswell.* New York: Pocket Books, 1997, p. 162. Furthermore, speaking before the graduating class at West Point in 1962, MacArthur stated: "We deal now not with things of the world alone. We deal now with the ultimate conflict between a unified human race and the sinister forces of some other planetary galaxy." Quoted in Steiger and Steiger, *Real Aliens, Space Beings, and Creatures from Other Worlds*, p. 197.

113 Corso and Birnes, *The Day After Roswell*, p. 132.

114 Ibid., pp. 156–171. Corso includes a reprint of a U.S. Army report on Project Horizon in an appendix in his book. See pp. 306–366.

115 Salla, *Exopolitics*, p. 89.

116 Corso and Birnes, *The Day After Roswell*, p. 155.

117 Corso claimed that an alien craft was shot down over Ramstein Air Force Base in Germany in 1974. Corso and Birnes, *The Day After Roswell*, p. 293.

118 Corso and Birnes, *The Day After Roswell*, p. 101.

119 Alexander, *UFOs: Myths, Conspiracies, and Realities*, p. 49.

120 Shostak, *Confessions of an Alien Hunter*, pp. 150–153.

121 Alexander, *UFOs: Myths, Conspiracies, and Realities*, p. 49.

122 Marc Davenport, *Visitors From Time: The Secret of the UFOs*. (Tuscaloosa, AL: Greenleaf Publications, 1994), p. 189.

123 Marrs, *Alien Agenda*, p. 166. According to Alan Bielek, so-called "Gray" aliens invaded Earth in 1954, whereupon they entered into a secret concord with the major world leaders. Supposedly, they reside in underground bases that are officially designated as "emergency shelters" for the president and other government officials in the event of a nuclear war. Steiger and Steiger, *Real Aliens, Space Beings, and Creatures from Other Worlds*, pp. 43, 102.

124 Alexander, *UFOs: Myths, Conspiracies, and Realities*, p. 106.

125 Steiger and Steiger, *Real Aliens, Space Beings, and Creatures from Other Worlds*, p. 23.

126 Salla, *Exopolitics*, p. x.

127 Steiger, *The Philadelphia Experiment & Other UFO Conspiracies*, p. 148.

128 Salla, *Exopolitics*, p. 159.

129 Maloney, *UFOs in Wartime*, pp. 283–285.

130 Salla, *Exopolitics*, pp. 191–226.

131 Len Kasten, *Secret Journey to Planet Serpro*. Rochester, VT and Toronto, Canada: Bear & Company, 2013.

132 *The UFO Phenomenon*, p. 77.

133 For example, mysterious men-in-black were said to have visited witnesses to an alleged UFO sighting in Texas in 1897. Steiger, *The Philadelphia Experiment & Other UFO Conspiracies*, p. 25.

134 Nick Redfern, *The Real Men in Black: Evidence, Famous Cases, and True Stories of these Mysterious Men and Their Connection to UFO Phenomena*. Pompton Plains, NJ: New Page Books, 2011, pp. 27–30.

135 Redfern, *The Real Men in Black*, p. 146.

136 Ibid., p. 155.

137 Randles, *The Truth Behind Men in Black*, p. 42.

138 Redfern, *The Real Men in Black*, p. 17.

139 Davenport, *Visitors From Time*, p. 198.

140 Morris K. Jessup was an astronomer who received an MS degree from the Hussey Observatory in 1926. He was described as having been an instructor in astronomy and mathematics at the University of Michigan and Drake University. He claimed to have completed a doctoral dissertation in astrophysics at the University of Michigan, but it is unclear if he ever received a Ph.D. Carlos Miguel Allende, who did much to establish the legend of the Philadelphia Experiment, corresponded with Jessup. Shortly after the publication of *The Case for the UFO*, Jessup received a letter from Allende. Intrigued with Jessup's *The Case for the UFO*, he made numerous annotations to the volume and sent a copy to the Office of Naval Research. The annotations were written in a style so as to appear that alien beings with advanced knowledge of physics were the true authors. At times the annotations suggest a considerable knowledge of theoretical physics, yet at the same time, numerous elementary grammatical and spelling errors are common. A copy of the annotated manuscript addressed to Admiral N. Furth came in a manila envelope postmarked Seminole, Texas, 1955. Intrigued by his annotations, members of the Office of Naval Research, including Major Darrel L. Ritter, USMC, Captain Sidney Sherby, USN, and Commander George W. Hoover, USN, contracted with a private firm—Varo—to print 500 copies of the annotated version.

At the time, Varo was involved in in aerospace design and manufacturing for the U.S. military. Morris K. Jessup, *The Allende Letters and the Varo Edition of The Case for the UFO*. New Brunswick, NJ: Conspiracy Journal/Global Communications, 2007, p. xii.

141 Davenport, *Visitors From Time*, p. 202.

142 MacDonald's efforts to get to the truth about UFOs exacted a heavy toll on him, as he became professionally isolated and his marriage faltered. Under severe depression, he attempted suicide in April of 1971 by shooting himself in the head. Amazingly, he survived, but was admitted to a psychiatric ward at a Tuscon, Arizona hospitial. On June 13, 1971, a family walking along a creek close to a bridge in Tuscon, discovered a body that was later identified as McDonald's, along with a suicide note and a .38 caliber revolver. Steiger and Steiger, *Real Aliens, Space Beings, and Creatures from Other Worlds*, p. 297.

143 Maloney, *UFOs in Wartime*, pp. 251–254.

144 Redfern, *The Real Men in Black*, pp. 53–59.

145 Michael Shermer, "UFOs, UAPs and CRAPs," *Scientific American*, (April 2011), p. 89.

146 Lydialye Gibson, "Science? Fiction?" *The University of Chicago Magazine*, (September/October 2011).

147 Stanton T. Friedman, *Flying Saucers and Science: A Scientist Investigates the Mysteries of UFOs*. Pompton Plains, NJ: The Career Press, 2008, p. 23.

148 Over the course of his many lectures, Friedmann claims to have encountered only eleven hecklers, two of whom were drunk. Gibson, "Science? Fiction?" http://mag.uchicago.edu/science-medicine/science-fiction.

149 Friedman, *Flying Saucers and Science*, p. 132.

150 Stanton T. Friedman, "Star Travel: How Realistic Is It?" in Michael Pye and Kirsten Dalley (eds.), *UFOs & Aliens: Is There Anybody Out There?* Pompton Plains, NJ: The Career Press, 2011), p. 162.

151 Author interview with Stanton Friedman, July 17, 2012.

152 Friedman, *Flying Saucers and Science*, pp. 162–164.

153 Lydialye Gibson, "Science? Fiction?" *The University of Chicago Magazine*, (September/October 2011).

154 Subsequent historical inquiry suggests that the panic stemming from the broadcast was greatly overstated and became something of an urban legend concerning the influence of the media.

155 Friedman, *Flying Saucers and Science*, pp. 153–158.

156 Stanton T. Friedman, "A Cosmic Watergate: UFO Secrecy," in Michael Pye and Kirsten Dalley (eds.), *UFOs & Aliens: Is There Anybody Out There?* Pompton Plains, NJ: The Career Press, 2011, p. 11.

157 Richard Rhodes, *Dark Sun: The Making of the Hydrogen Bomb*. New York: Simon & Schuster, 2005, pp. 83–93.

158 Friedman, *Flying Saucers and Science*, p. 107.

159 Ibid., pp. 103–128.

160 Stanton T. Friedman, "A Cosmic Watergate: UFO Secrecy," in Michael Pye and Kirsten Dalley (eds.), *UFOs & Aliens: Is There Anybody Out There?* Pompton Plains, NJ: The Career Press, 2011, p. 24.

161 Friedman, *Flying Saucers and Science*, p. 136.

162 Lydialye Gibson, "Science? Fiction?" *The University of Chicago Magazine*, (September/October 2011).

163 Friedman, *Flying Saucers and Science*, pp. 129–152.

164 Author interview with Stanton Friedman, July 17, 2012.

165 Michael Talbot, *The Holographic Universe: The Revolutionary Theory of Reality*. (New York: Harper/Perennial, 2001), pp. 276–277.

166 Shostak, *Confessions of an Alien Hunter*, pp. 146–149.

167 Friedman, *Flying Saucers and Science*, p. 183.

168 Quoted in Ronald N. Bracewell, *The Galactic Club: Intelligent Life in Outer Space*. San Francisco, CA: W.W. Freeman and Company, 1975, p. 75.

169 George Basalla, *Civilized Life in the Universe: Scientists on Intelligent Extraterrestrials*. Oxford and New York: Oxford University Press, 2006, p. 121.

170 Alexander, *UFOs: Myths, Conspiracies, and Realities*, pp. 16–17.

171 Ibid., p. xx.

172 Ibid., pp. 130–131.

173 Alexander, *UFOs: Myths, Conspiracies, and Realities*, p. 98.

174 Alfred Lambremont Webre, "Best-selling author Leslie Kean will approach Congress, Obama on UFO disclosure," Examiner.com, September 22, 2010. http://www.examiner.com/article/best-selling-author-leslie-kean-will-approach-congress-obama-on-ufo-disclosure.

175 Paul Davies, *The Eerie Silence: Renewing Our Search for Alien Intelligence*. Boston and

New York: Houghton Mifflin Harcourt, 2010, pp. 61–62. Clinton was said to have had an interest in aliens and UFOs. In 1995, at a press conference in Belfast, he answered a letter from an Irish boy named Ryan who inquired about his knowledge of Roswell, to which Clinton responded "if the U.S. Air Force did recover alien bodies, they didn't tell me about it either, and I want to know." Alexander, *UFOs: Myths, Conspiracies, and Realities*, p. 108.

chapter 12

1 Jonathan Leake, "Don't talk to aliens, warns Stephen Hawking," *The Sunday Times* (London), April, 25, 2010. Interestingly, in May 2011, Hawkings' daughter Lucy, along with Paul Davies, sent a message into space from an Arizona schoolchild to extraterrestrial aliens if they should be there to receive it. Claudia Dreifus, "Life and the Cosmos, Word by Painstaking Word," *The New York Times*, May 9, 2011.

2 Stephen Hawking, *A Brief History of Time*. New York: Bantam, 1998, p. 166.

3 David Brin, "Shouting at the Cosmos. . . Or How SETI has Taken a Worrisome Turn into Dangerous Territory," September 2006 and Jared M. Diamond, *The Third Chimpanzee: The Evolution and Future of the Human Animal*. New York: Harper Perennial, 2006.

4 Michaud, *Contact with Alien Civilizations*, p. 246.

5 Quoted in Michael A.G. Michaud, *Contact with Alien Civilizations: Our Hopes and Fears about Encountering Extraterrestrials*. New York: Copernicus Books, 2007, p. 241.

6 Brin, "Shouting at the Cosmos. . . Or How SETI has Taken a Worrisome Turn into Dangerous Territory."

7 Adam Korbitz, "The Precautionary Principle: Egoism, Altruism, and the Active SETI Debate," in Douglas A. *Vakoch* (ed.), *Extraterrestrial Altruim: Evolution and Ethics in the Cosmos*. Heidelberg, New York, Dordrecht, and London: Springer, 2013, p. 111.

8 Michio Kaku, "The Physics of Interstellar Travel," November 29, 2002.

9 Davies, *The Eerie Silence*, pp. 171–172.

10 Michaud, *Contact with Alien Civilizations*, p. 241.

11 Ibid., pp. 244–245.

12 Travis S. Taylor and Bon Boan, *Alien Invasion: The Ultimate Survival Guide for the Ultimate Attack*. Riverdale, NY: Baen Publishing Enterprises, 2011, pp. 18–19.

13 Quoted in Seth Shostak, *Confessions of an Alien Hunter: A Scientist's Search for Extraterrestrial Intelligence.* Washington, D.C.: National Geographic, 2009, p. 57.

14 Michael T. Klare *Resource Wars: The New Landscape of Global Conflict.* New York: Holt Paperbacks, 2002.

15 Writing in 2000, Robert Park pointed out that the cost for NASA to launch an ounce of gold into orbit would be about $830. It would cost about the same amount to bring that same ounce of gold back to Earth. At the time, the price of gold was $311 an ounce. Robert Park, *Voodoo Science: The road from foolishness to fraud.* Oxford, UK: Oxford University Press, 2000, p. 70. Presumably, those shipping costs would increase as the distance from Earth increases, although, it is conceivable that they could decrease with new advances in technology. Nevertheless, the point of this example is to illustrate is that it does not seem practical to ship raw materials across interstellar distances. If it would not be practical for a high-valued commodity such as gold, it would be even less practical for commodities such as water, oil, coal, food, etc.

16 I.A. Crawford, "Interstellar Travel: A Review for Astronomers," *Quarterly Journal of the Royal Astronomical Society* (1990), 31, pp. 377–400. Quoted in Harold A. Geller, "Stephen Hawking is Wrong. Earth Would Not Be a Target For Alien Conquest," *The Journal of Cosmology*, (2010), Vol 7.

17 Michaud, *Contact with Alien Civilizations*, pp. 298, 305–306.

18 Ibid., p. 300.

19 As Michio Kaku points out, predators are usually smarter than prey. It follows that an advanced civilization would be descended from carnivores. Michio Kaku, *Physics of the Impossible: A Scientific Exploration into the World of Phasers, Force Fields, Teleportation, and Time Travel.* New York: Anchor Books, 2008, p. 142. Likewise, Gregg Easterbrook observed that "The most disquieting aspect of natural selection as observed on Earth is that it channels intellect to predators." Quoted in Michaud, *Contact with Alien Civilizations*, p. 305.

20 Michaud, *Contact with Alien Civilizations*, p. 305.

21 "Overweight 'top world's hungry,'" BBC News, August 15, 2006. http://news.bbc.co.uk/2/hi/health/4793455.stm.

22 Carl Sagan, in Jerome Agel (ed.). *Cosmic Connection: An extraterrestrial Perspective.* Cambridge: Cambridge University Press, 1973, p. 215.

23 Tim Evans, "Foreword" in Travis S. Taylor and Bon Boan, *Alien Invasion: The Ultimate Sur-*

vival Guide for the Ultimate Attack. Riverdale, NY: Baen Publishing Enterprises, 2011, p. x.

24 Davies, *The Eerie Silence*, p. 198.

25 F.H.C. Crick and L.E. Orgel, "Directed Panspermia," *Icarus* No. 19 (1973), pp. 341–346.

26 Kaku, *Physics of the Impossible*, p. 142.

27 Stephen Webb, *Where is Everybody? Fifty Solutions to the Fermi Paradox and the Problem of Extraterrestrial Life*. New York, Copernicus Books, 2000, p. 2.

28 Kathryn Coe, Craig T. Palmer, and Christina Pomianek, "ET Phone Darwin: What Can an Evolutionary Understanding of Animal Communication and Art Contribute to Our Understanding of Methods for Interstellar Communication?" in Douglas A. Vakoch and Albert A. Harrison (eds.), *Civilizations Beyond Earth: Extraterrestrial Life and Society*. New York and Oxford: Berghan Books, 2011, pp. 214–225.

29 Davies, *The Eerie Silence*, pp. 8–9.

30 Webb, *Where is Everybody?* p. 24.

31 Davies, *The Eerie Silence*, p. 141.

32 Michaud, *Contact with Alien Civilizations*, pp. 312–313.

33 Gian Carlo Ghirardi, "Why Should Hawking's Aliens Wish to Destroy," *The Journal of Cosmology*, (2010), Vol 7.

34 Geller, "Stephen Hawking is Wrong. Earth Would Not Be a Target For Alien Conquest." http://journalofcosmology.com/Aliens100.html#12.

35 Michael Archer, "Slime Monsters Will Be Human Too," *Australian Natural History* 22, (1989), pp. 546–547.

36 "The First Atomic Test," Trinity Atomic Web Site, accessed November 3, 2010. Richard Rhodes, *The Making of the Atomic Bomb*. New York: Simon & Schuster, 1986, p. 664.

37 Webb, *Where is Everybody?*, pp. 128–129.

38 Ibid., pp. 79–84.

39 Michaud, *Contact with Alien Civilizations*, p. 170.

40 Davies, *The Eerie Silence*, pp. 127–128.

41 Taylor and Boan, *Alien Invasion*, pp. 53–114.

42 Ibid., pp. 155–209.

43 George Basalla, *Civilized Life in the Universe: Scientists on Intelligent Extraterrestrials*. Oxford and New York: Oxford University Press, 2006, p. 47.

44 Steven J. Dick, *Life on Other World: The 20th-Century Extraterrestrial Life Debate*. New York: Cambridge University Press, 1998, p. 54.

45 Toney Allman, *Are Extraterrestrials a Threat to Humankind?* San Diego, CA: Reference Point Press, 2012, pp. 24–25.

46 By 1972, when Apollo 16 landed on the Moon, virtually all scientists agreed that the Moon was lifeless. Sagan was one of the few to dispute this conclusion, maintaining that microorganisms might live deep beneath the lunar surface. Basalla, *Civilized Life in the Universe*, pp. 108–109.

47 Allman, *Are Extraterrestrials a Threat to Humankind?*, pp. 24–25.

48 Regulations for this system assign a category for each NASA mission. Destinations for a Category I mission do not require special protection from possible terrestrial contamination. A Category II mission involves a solar system body where the possibility of life is expected to be remote. Venus would fit into this category. For those missions with a spacecraft that orbit or fly by a body where life is possible, Category III is reserved. Finally, Category IV are for missions that actually land where life is possible. This category would include missions to Mars, the moons of Jupiter, and the moons of Saturn. Allman, *Are Extraterrestrials a Threat to Humankind?*, pp. 34–35.

49 Davies, *The Eerie Silence*, p. 129.

50 Allman, *Are Extraterrestrials a Threat to Humankind?*, pp. 38–39.

51 Ibid., pp. 30–32.

52 Shostak, *Confessions of an Alien Hunter*, pp. 267–272.

53 Ibid., p. 275.

54 Michaud, *Contact with Alien Civilizations*, p. 314.

55 Albert A. Harison, "Cosmic Evolution, Reciprocity, and Interstellar Tit for Tat," in Douglas A. Vakoch (ed.), *Extraterrestrial Altruim: Evolution and Ethics in the Cosmos.* Heidelberg, New York, Dordrecht, and London: Springer, 2013, p. 14.

56 Although humans have a natural propensity for violence, they do not necessarily thrive on it. In his book, *Violence: A Micro-sociological theory*, Randall Collins argues that humans are inherently passive. Most humans shirk from violent encounters. Violence is more of a group process than an individual choice. Lary J. Siegel, *Criminology: Theories, Patterns, and Typologies* Eleventh Edition. Belmont, CA: Wadsworth, 2013, pp. 335–336.

57 Quoted in Allman, *Are Extraterrestrials a Threat to Humankind?*, p. 66.

58 Davies, *The Eerie Silence*, p. 199.

59 Seth D. Baum, Jacob D. Haqq-Misra, Shawn D. Domagal-Goldman, "Would Contact with Extraterrestrials Benefit for Harm Humanity? A Scenario Analysis," Acta Astro-

nutica, vol 68 no. 11–12, pp. 21–26. http://sethbaum.com/ac/2011_ET-Scenarios.pdf.

60 Author Interview with Stanton Friedman, July 17, 2012.

61 Baum, Haqq-Misra, Domagal-Goldman, "Would Contact with Extraterrestrials Benefit for Harm Humanity? A Scenario Analysis."

62 Author interview with Michael G. Michaud, July 8, 2012.

63 Webb, *Where is Everybody?* pp. 46–47.

64 Baum, Haqq-Misra, Domagal-Goldman, "Would Contact with Extraterrestrials Benefit for Harm Humanity? A Scenario Analysis," p. 19.

65 Michaud, *Contact with Alien Civilizations*, p. 314.

66 Frank Drake and Dava Sobel, *Is Anyone Out There? The Scientific Search for Extraterrestrial Intelligence*. New York: Delacorte Press, 1992, pp. 160–161.

67 Jill Tarter, "Should we fear space aliens?" http://www.cnn.com/2010/OPINION/04/27/tarter.space.life.fears/. Tarter was the prototype for the character, Ellie Arroway, in Carl Sagan's novel *Contact*. Shostak, *Confessions of an Alien Hunter*, p. 12. For her doctoral thesis, Tarter modeled brown dwarfs (tiny, failed stars that took another 25 years to discover). During a postdoc at NASA Ames, she decided to devote her research to SETI because of her computer skills and because she was already comfortable with the idea that she might have to wait a long time to obtain results. Esther Inglis-Arkell, "Jill Tarter, inspiration for the movie *Contact* tells us about her journey to SETI," io9.com, August 7, 2012, http://io9.com/5932368/jill-tarter-inspiration-for-the-movie-contact-tells-us-about-her-journey-to-seti.

68 Webb, *Where is Everybody?* p. 75.

69 I.A. Crawford, "Where are They? Maybe we are alone in the galaxy after all," *Scientific American*, July 2000, pp. 38–43.

70 Drake and Sobel, *Is Anyone Out There? The Scientific Search for Extraterrestrial Intelligence*, p. 131.

71 Michaud, *Contact with Alien Civilizations*, p. 322.

72 Ibid., p. 304.

73 Ibid., p. 242.

74 Tomislav Janović, "Other Minds, Empathy, and Interstellar Communication," in Douglas A. Vakoch (ed.), *Extraterrestrial Altruim: Evolution and Ethics in the Cosmos*. Heidelberg, New York, Dordrecht, and London: Springer, 2013, p. 167.

75 Michaud, *Contact with Alien Civilizations*, p. 300.

76 From 1970 to 2010, there was an enormous upsurge in the number of democracies around the world. In 1973, only 45 of the world's 151 countries were designated as democracies by Freedom House, a U.S.-based non-governmental organization that conducts research and advocacy on democracy, political freedom, and human rights. But by the late 1990s, some 120 countries—more than 60 percent of the world's independent states—were electoral democracies. The political scientist Samuel Huntington referred to this development as the "third wave" of democratization, which began to crest soon thereafter. In the first decade of the twenty-first century, a "democratic recession" commenced in which approximately one in five countries that had been part of the third wave either reverted to authoritarianism or saw a significant erosion of democratic institutions. Francis Fukuyama, *The Origins of Political Order: From Prehuman Times to the French Revolution*. New York: Farrar, Straus, and Giroux, 2011, pp. 3–4.

77 Francis Fukuyama, *The End of History and the Last Man*. New York: Free Press, 1992.

78 Harison, "Cosmic Evolution, Reciprocity, and Interstellar Tit for Tat," p. 1.

79 Albert A. Harrison, *After Contact: The Human Response to Extraterrestrial Life*. New York: Plenum Press, 1997, pp. 182–184.

80 Baum, Haqq-Misra, Domagal-Goldman, "Would Contact with Extraterrestrials Benefit for Harm Humanity? A Scenario Analysis," p. 12.

81 Kardashev, "Cosmology and Civilizations," p. 35.

82 Ronald N. Bracewell, *The Galactic Club: Intelligent Life in Outer Space*. San Francisco, CA: W.W. Freeman and Company, 1975, p. 33.

83 Bracewell, *The Galactic Club*, pp. 80–81.

84 Adam Korbitz, "Altruism, Metalaw, and Celegistics: An Extraterrestrial Prespective on Universal Law-Making," in Douglas A. Vakoch (ed.), *Extraterrestrial Altruim: Evolution and Ethics in the Cosmos*. Heidelberg, New York, Dordrecht, and London: Springer, 2013, p. 229.

85 Korbitz, "Altruism, Metalaw, and Celegistics: An Extraterrestrial Prespective on Universal Law-Making," p. 231.

86 Ibid., pp. 232–233.

87 Ibid., p. 233.

88 Taylor and Boan, *Alien Invasion*, p 18.

89 Harrison, *After Contact*, p. 294.

90 Carl Sagan, "The Number of Advanced Galactic Civilizations," in Carl Sagan (ed.), *Com-*

munication with Extraterrestrial Intelligence*. Cambridge, MA: MIT Press, 1973, p. 170.

91 Quoted in Michio Kaku, *Hyperspace: A Scientific Odyssey Through Parallel Universes, Time Warps, and the 10th Dimension*. New York: Doubleday, 1994, p. 283.

92 Stephen Webb, *Where is Everybody? Fifty Solutions to the Fermi Paradox and the Problem of Extraterrestrial Life*. New York, Copernicus Books, 2000, pp. 113–114.

93 Some scientists have even speculated that our solar system has been cordoned off and functions as a "zoo" in that the natural development of our planet is allowed to occur. Webb, *Where is Everybody?* pp. 46–48.

94 Webb, *Where is Everybody?* pp. 113–114.

95 Kardashev speculated that advanced alien civilization might have developed sophisticated telescopes that could closely observe other planets. He referred to "Cosmic Ethnography" as a branch of science in which primitive forms of life and civilization would be studied. Kardashev, "On the Inevitability and the Possible Structures of Supercivilizations," p. 501.

96 Brin, "Shouting at the Cosmos. . . Or How SETI has Taken a Worrisome Turn into Dangerous Territory." http://lifeboat.com/ex/shouting.at.the.cosmos.

97 Harrison, *After Contact*, pp. 267–268.

98 George F. R. Ellis, "Kenotic Ethics and SETI: A Present-Day View," in Douglas A. Vakoch (ed.), *Extraterrestrial Altruim: Evolution and Ethics in the Cosmos*. Heidelberg, New York, Dordrecht, and London: Springer, 2013, p. 222.

99 Harrison, *After Contact*, p. 258.

100 Seth Shostak, "A Planet Someone Might Call Home," *The Huffington Post*, October 4, 2010, http://www.huffingtonpost.com/seth-shostak/a-planet-someone-might-ca_1_b_745228.html.

101 Michaud, *Contact with Alien Civilizations*, pp. 279–323.

chapter 13

1 Stanton T. Friedman, *Flying Saucers and Science: A Scientist Investigates the Mysteries of UFOs*. Pompton Plains, NJ: he Career Press, 2008, p. 197.

2 Carl Sagan, "The Lifetimes of Technical Civilizations," in Carl Sagan (ed.), *Communication with Extraterrestrial Intelligence*. Cambridge, MA: MIT Press, 1973, pp. 155–158.

3 Yoji Kondo, "Interstellar Travel and Multi-Generation Space Ships: An Overview," in Yoji Kondo, Frederick Bruhweiler, John Moore, and Charles Sheffield (eds.), *Interstel-*

lar Travel and Multi-Generation Space Ships. Burlington, Ontario, Canada: Apogee Books, 2003, p. 8.

4 Robert Forward, "Ad Astra!" in Yoji Kondo, Frederick Bruhweiler, John Moore, and Charles Sheffield (eds.), *Interstellar Travel and Multi-Generation Space Ships*. Burlington, Ontario, Canada: Apogee Books, 2003, p. 29.

5 Charles Sheffield, "Fly Me to the Stars: Interstellar Travel in Fact and Fiction," in Yoji Kondo, Frederick Bruhweiler, John Moore, and Charles Sheffield (eds.), *Interstellar Travel and Multi-Generation Space Ships*. Burlington, Ontario, Canada: Apogee Books, 2003, p. 21.

6 Paul Gilster, *Centauri Dreams: Imagining and Planning Interstellar Exploration*. New York: Copernicus Books, 2004, p. 6.

7 In his 1941 story, *Far Centaurus*, A.E. van Vogt chronicled the voyage of a spacecraft whose crew traveled in suspended animation en route to its destination. Meanwhile, scientists on Earth invented a method of faster-than-light-travel. The original crew awoke from its long sleep, only to discover that subsequent travelers had beaten them to their destination (Sheffield, "Fly Me to the Stars," p. 26). Marc Millis refereed to this conundrum as "Zeno's Paradox in reverse." According to the fable, Zeno of Eleca (circa 450 B.C.) wrote of a race between the Greek hero Achilles and a tortoise. The tortoise told Achilles that it could win the race if given a head start with the proviso that each time Achilles halved the distance between the two, the tortoise would have advanced a bit further still, and so on ad infinitum. Thus it would be a mathematical impossibility for Achilles to win the race under said circumstances. Applied to insterstellar travel, Zeno's Paradox in reverse implies that no matter how fast a spacecraft can be sent into space, with the passage of time, an even faster method will be discovered. At some point during the journey, the initial probe will be passed by a subsequent probe, which in turn will be passed by another (Gilster, *Centauri Dreams*, pp. 156–157).

8 Joe Haldeman, "Colonizing Other Worlds," in Yoji Kondo, Frederick Bruhweiler, John Moore, and Charles Sheffield (eds.), *Interstellar Travel and Multi-Generation Space Ships*. Burlington, Ontario, Canada: Apogee Books, 2003, p. 62.

9 Forward, "Ad Astra!", p. 31.

10 Carl Sagan, in Jerome Agel (ed.). *Cosmic Connection: An extraterrestrial Perspective.* Cambridge: Cambridge University Press, 1973, pp. 160–161.

11 Ben Austen, "After Earth: Why? Where? How? When? *Popular Science* (March 2012), p. 49.

12 Isaac Asimov conjectured that insofar as our civilization is endowed with such a practical satellite, we might even be able to develop spaceflight capabilities ahead of other civilizations that are much older than ours. In that sense, we could eventually be the heir the universe thanks to the Moon. Isaac Asimov, *Extraterrestrial Civilizations*. New York: Crown Publishers, Inc., 1979, p. 213.

13 Harvey Wichman "Near-Term Extended Solar System Exploration," in Douglas A. Vakoch (ed.), *On Orbit and Beyond: Psychological Perspectives on Human Spaceflight*. (Heidelberg, New York, Dordrecht, and London: Springer, 2013), pp. 267–283.

14 Neil de Grasse Tyson, *Space Chronicles: Facing the Ultimate Frontier*. New York: W.W. Norton & Company Ltd., 2012, p. 77.

15 Mike Wall, Manned Mission to Mars: Is the Moon Really a Stepping Stone?" August 19, 2013. http://news.yahoo.com/manned-missions-mars-moon-really-stepping-stone-102145232.html.

16 "Tens of thousands apply for one-way ticket to Mars," MSN news, April 25, 2013, http://news.msn.com/us/tens-of-thousands-apply-for-one-way-ticket-to-mars.

17 Carl Sagan, *Pale Blue Dot: A Vision of the Human Future in Space*. New York: Ballantine Books, 1994, pp. 279–281.

18 Sagan, *Pale Blue Dot*, p. 332.

19 Austen, "After Earth: Why? Where? How? When?", p. 50.

20 Asimov, *Extraterrestrial Civilizations*, p. 57.

21 Albert A. Harrison, *After Contact: The Human Response to Extraterrestrial Life*. New York: Plenum Press, 1997, p. 12.

22 Sagan, *Pale Blue Dot*, p. 274.

23 Carl Sagan, "Astroengineering Activity: The Possibility of ETI in Present Astrophysical Phenomena," in Carl Sagan (ed.), *Communication with Extraterrestrial Intelligence*. Cambridge, MA: MIT Press, 1973, pp. 188–189.

24 Freeman J. Dyson, "The World, the Flesh, and the Devil," in Carl Sagan (ed.), *Communication with Extraterrestrial Intelligence*. Cambridge, MA: MIT Press, 1973, pp. 371–389.

25 de Grasse Tyson, *Space Chronicles*, p. 163.

26 Michio Kaku, *Physics of the Impossible: A Scientific Exploration into the World of Phasers, Force Fields, Teleportation, and Time Travel*. New York: Anchor Books, 2008, pp. 157–158.

27 Gilster, *Centauri Dreams*, p. 71.

28 Ibid., p. 66.

29 Kaku, *Physics of the Impossible*, p. 163.

30 Forward, "Ad Astra!", p. 38.

31 Geoffrey A. Landis, "The Ultimate Exploration: A Review of Propulsion Concepts for Interstellar Flight," in Yoji Kondo, Frederick Bruhweiler, John Moore, and Charles Sheffield (eds.), *Interstellar Travel and Multi-Generation Space Ships*. Burlington, Ontario, Canada: Apogee Books, 2003, p. 54.

32 Gilster, *Centauri Dreams*, p. x.

33 To date, the United States has operated only one reactor in space. All totaled, the Soviet Union launched around three dozen nuclear reactors into space. The Soviets dabbled with the idea of using nuclear energy for their space program as well. On January 24, 1978, a Soviet satellite carrying a nuclear reactor reentered in the Northwest Territories of Canada. It was the thirteenth Soviet nuclear reactor to operate in space. Friedman, *Flying Saucers and Science*, pp. 47–48.

34 Kaku, *Physics of the Impossible*, pp. 161–162.

35 Landis, "The Ultimate Exploration: A Review of Propulsion Concepts for Interstellar Flight," p. 55.

36 Michael Kurland, *The Complete Idiot's Guide to Extraterrestrial Intelligence*. New York: alpha books, 1999, p. 246.

37 de Grasse Tyson, *Space Chronicles*, p. 159.

38 Stanton T. Friedman, "Star Travel: How Realistic Is It?" in Michael Pye and Kirsten Dalley (eds.), *UFOs & Aliens: Is There Anybody Out There?* Pompton Plains, NJ: The Career Press, 2011, p. 162.

39 Friedman, *Flying Saucers and Science*, pp. 69–81.

40 Kaku, *Physics of the Impossible*, pp. 159–160.

41 Forward, "Ad Astra!", p. 42.

42 Frank Drake and Dava Sobel, *Is Anyone Out There? The Scientific Search for Extraterrestrial Intelligence*. New York: Delacorte Press, 1992, p. 123.

43 Frank Drake, "The search for extra-terrestrial intelligence," *Philosophical Transactions of the Royal Society*, no. 369 (2011), pp. 637–638.

44 Kaku, *Physics of the Impossible*, p. 158 and Gilster, *Centauri Dreams*, p. 102.

45 Gilster, *Centauri Dreams*, p. 105.

46 de Grasse Tyson, *Space Chronicles*, pp. 165–166.

47 Gilster, *Centauri Dreams*, p. 108.

48 Ibid., p. ix.

49 Ibid., p. 97.

50 Ibid., pp. 123.

51 Michio Kaku, "Breakthrough: Anti-Matter Trapped for the First Time," bigthink.com, November 19, 2010, http://bigthink.com/dr-kakus-universe/breakthrough-anti-matter-trapped-for-the-first-time.

52 Sagan, *Pale Blue Dot*, p. 271.

53 Gilster, *Centauri Dreams*, p. 77.

54 As they explained, the trick was to mix cold clouds of trapped positrons and antiprotons. Gilster, *Centauri Dreams*, p. 88.

55 Gilster, *Centauri Dreams*, p. 2.

56 Ibid., p. 79.

57 Paul A. LaViolette, *Secrets of Antigravity Propulsion: Tesla, UFOs, and Classified Aerospace Technology*. Rochester, VT: Bear & Company, 2008, pp. 1–2.

58 Jon Mooallern, "A Curious Attraction: On the quest for antigravity," *Harper's Magazine*, (October 2007), pp. 84–91.

59 Nick Cook, *The Hunt for Zero Point: Inside the Classified World of Antigravity Technology*. New York: Broadway Books, 2003, pp. 2–23.

60 See "DARPA History" at http://www.bibliotecapleyades.net/sociopolitica/sociopol_DARPA01.htm. Accessed October 20, 2013.

61 For more on the Tau Zero Foundation, see its website at http://www.tauzero.aero/, accessed August 30, 2013.

62 Mooallern, "A Curious Attraction," pp. 84–91.

63 Nick Cook, *The Hunt for Zero Point: Inside the Classified World of Antigravity Technology*. New York: Broadway Books, 2003, p. 100.

64 Cook, *The Hunt for Zero Point*, pp. 109–111.

65 For example, Paul Davies believes that a spaceship designed to "scoop" up this dark or vacuum energy would probably not be viable because it would violate the second law of thermodynamics. As he sees it, there would have to be a temperature differential somewhere in the space from which the energy is to be extracted. Paul Davies, *The Eerie Silence: Renewing Our Search for Alien Intelligence*. (Boston and New York: Houghton Mifflin Harcourt, 2010), pp. 150–151.

66 Gilster, *Centauri Dreams*, p. 55.

67 Asimov, *Extraterrestrial Civilizations*, pp. 218–220.

68 Kaku, *Physics of the Impossible*, pp. 53–69.

69 "Seeing Light in a New Light: Scientists Create Never-Before-Seen Form of Matter," *Science Daily*, September 25, 2013, http://www.sciencedaily.com/releases/2013/09/130925132323.htm.

70 Gilster, *Centauri Dreams*, p. 161.

71 Michio Kaku, "The Physics of Interstellar Travel," November 29, 2002.

72 Lynne McTaggart, *The Field: The Quest for the Secret Force of the Universe*. New York: Harper Collins, 2008, pp. 220–221.

73 Jesus Diaz, "NASA Scientists Reveal Warp Drive Ship Capable Of Reaching Alpha Centuri in 2 Weeks!" Higher Perspective, http://higherperspective.com/2014/06/nasa-scientists-reveal-warp-drive-ship-capable-reaching-alpha-centuri-2-weeks.html?utm_source=HP, accessed June 18, 2014.

74 Gilster, *Centauri Dreams*, pp. 166–167 and Richard Gray, NASA researches Star Trek warp drive for future space, travel," *The Telegraph*, August 20, 2013, http://www.telegraph.co.uk/science/space/10254857/Nasa-researches-Star-Trek-warp-drive-for-future-space-travel.html.

75 Michio Kaku, *Einstein's Cosmos: How Albert Einstein's Vision Transformed Our Understanding of Space and Time*. New York and London: W.W. Norton & Company, 2004, p. 175.

76 Stephen Webb, *Where is Everybody? Fifty Solutions to the Fermi Paradox and the Problem of Extraterrestrial Life*. New York, Copernicus Books, 2000, pp. 68–69 and Carl Sagan, *Contact*. New York: Simon & Schuster Trade, 1985, pp. 406–407

77 Davies, *The Eerie Silence*, p. 4.

78 Harrison, *After Contact*, p. 37.

79 Brian Greene, *The Elegant Universe: Superstrings, Hidden Dimensions, and the Quest for the Ultimate Theory*. New York and London: W.W. Norton & Company, 2003, pp. 264–265. Some astronomers suspect that wormholes migh lurk inside black holes. Gilster, *Centauri Dreams*, p. 179. Stephen Hawking even speculated that the number of black holes might even exceed the number of visible stars. Stephen Hawking, *A Brief History of Time: From the Big Bang to Black Holes*. New York: Bantam Books, 1990, p. 95.

80 Paul A.M. Dirac first speculated on the concept of negative energy. Negative energy would be below the conventionally-defined ground state of energy. The so-called Casimir effect has demonstrated negative energy. In 1948, Henrik Casimir demonstrated in a laboratory that when two parallel metal plates are placed side by side, they will attract each other by a nonzero force, Kaku, *Einstein's Cosmos*, pp. 219–221.

81 Webb, *Where is Everybody?* p. 69.

82 Stephen Hawking, *Black Holes and Baby Universes and Other Essays*. New York: Bantam Books, 1993, pp. 115–125.

83 Hawking argued that black holes do indeed emit particles (so-called Hawking radiation) due to the Uncertainty Principle. He found that over short distances, particles and radiation are able to move faster than the speed of light. By doing so, they can escape the event horizon of the black hole. Hawking, *A Brief History of Time*, p. 112.

84 Kaku, *Einstein's Cosmos*, pp. 218–219.

85 Stephen Hawking, *Black Holes and Baby Universes and Other Essays*. New York: Bantam Books, 1993, pp. 115–125.

86 The Higgs particle is thought to be a tiny subatomic particle that weights roughly 130 times as much as an atom of hydrogen, the lightest gas.

87 In 2012, Rolf Heuer, the chief of CERN, announced that a 0.000057 percent statistical chance that the detection of the Higgs boson was erroneous. Dan Vergano, "Physicists break down concept of 'God particle,'" *USA Today*, July 5, 2012.

88 "Higgs boson find could make light-speed travel possible, scientists say," National Post Wire Services, July 5, 2012, http://news.nationalpost.com/2012/07/05/higgs-boson-find-could-make-light-speed-travel-possible-scientists-hope/.

89 Mary Roach, *Packing for Mars: The Curious Science of Life in the Void*. New York and London: W.W. Norton & Company, 2013, pp. 156–166.

90 Roach, *Packing for Mars*, p. 58.

91 Ibid., p. 117.

92 Ibid., p. 135.

93 Nick Kanas, Gro Mjeldheim Sandal, Jennifer E. Boyd, et al., "Psychology and Culture During Long-Duration Space Missions," in Douglas A. Vakoch (ed.), *On Orbit and Beyond: Psychological Perspectives on Human Spaceflight*. Heidelberg, New York, Dordrecht, and London: Springer, 2013, p. 169.

94 Roach, *Packing for Mars*, pp. 211–228.

95 Kaku, *Physics of the Impossible*, p. 172.

96 Roach, *Packing for Mars*, p. 213.

97 According to Robert Zubrin, a 1,500-m long tether could be used to connect a space vessel with a small rocket. Once in space, firing the rocket would cause the system to rotate. He estimates that even a slow rotation rate of 2.6 RPM would produce artificial gravity about equal to that on Earth. Wichman "Near-Term Extended Solar System Exploration," p. 273.

98 Roach, *Packing for Mars*, p. 306.

99 Ibid., p. 200.

100 Ibid., p. 271.

101 Albert A. Harrison and Edna Fiedler, "Behavioral Health," in Douglas A. Vakoch (ed.), *On Orbit and Beyond: Psychological Perspectives on Human Spaceflight*. Heidelberg, New York, Dordrecht, and London: Springer, 2013, p. 13.

102 Roach, *Packing for Mars*, pp. 61–62.

103 Ibid., p. 68.

104 Kanas, Sandal, Boyd, et al., "Psychology and Culture During Long-Duration Space Missions," p. 156.

105 Harrison and Fiedler, "Behavioral Health," pp. 13–14.

106 Ibid., p. 7.

107 Roach, *Packing for Mars*, p. 50.

108 Harrison and Fiedler, "Behavioral Health," pp. 8–9.

109 Roach, *Packing for Mars*, p. 32.

110 As Mary Roach opined, as more and more, doctors, biologists, and engineers are among the ranks of the spacefarers, "astronauts these days are as likely to be nerds as heroes." Roach, *Packing for Mars*, pp. 27–28.

111 Wichman "Near-Term Extended Solar System Exploration," p. 279.

112 David S. Portree, "Skylab-Salyut Space Laboratory (1972)," *Wired Science Blogs*, March 26, 2012, http://www.wired.com/wiredscience/2012/03/skylab-salyut-space-laboratory-1972/.

113 Albert A. Harrison and Edna Fiedler, "Introduction: Psychology and Space Exploration," in Douglas A. Vakoch (ed.), *On Orbit and Beyond: Psychological Perspectives on Human Spaceflight*. Heidelberg, New York, Dordrecht, and London: Springer, 2013, p. xxvii.

114 Harrison and Fiedler, "Behavioral Health," p. 5.

115 To minimize misunderstanding that could arise from the interaction of different cul-

tures, Suedfeld and her colleagues recommend that both crew members and ground staff take in situ language training and familiarization in each other's countries. Peter Suedfeld, Kasia E. Wilk, and Lindi Cassel, "Flying with Strangers: Postmission Reflections of Multinational Space Crews," in Douglas A. Vakoch (ed.), *On Orbit and Beyond: Psychological Perspectives on Human Spaceflight*. Heidelberg, New York, Dordrecht, and London: Springer, 2013, pp. 185–209.

116 Juris G. Draguns and Albert A. Harrison, "Cross-Cultural and Spaceflight Psychology: Arenas for Synergistic Research," in Douglas A. Vakoch (ed.), *On Orbit and Beyond: Psychological Perspectives on Human Spaceflight*. Heidelberg, New York, Dordrecht, and London: Springer, 2013, pp. 211–228.

117 Wichman "Near-Term Extended Solar System Exploration," p. 280.

118 Roach, *Packing for Mars*, p. 237.

119 Drawing upon the results of a simulation designed to mimic a single-stage orbit rocket, Harvey Wichman found that not only could minimally trained civilians tolerate the extreme environment of a space simulator, they could also find the experience very pleasing. As a result, he is sanguine about the prospects of the fledging space tourism industry. Rather than emphasizing selection criteria that would restrict the number of prospective tourists, Wichman argues that the space program should focus more attention on environmental design and training so that the prospective pool of spacefarers could be broadened. The primary tool for this effort, he proposes, is the spaceflight simulator. Harvey Wichman, "Managing Negative Interactions in Space Crews: The Role of Simulator Research," in Douglas A. Vakoch (ed.), On Orbit and Beyond: Psychological Perspectives on Human Spaceflight. Heidelberg, New York, Dordrecht, and London: Springer, 2013, pp. 107–122.

120 Peter G. Roma, Steven R. Hursh, Robert D. Hienz, et al., "Effects of Autonomous Mission Management on Crew Performance, Behavior, Phuysiology: Insights from Ground-Based Experiments," in Douglas A. Vakoch (ed.), *On Orbit and Beyond: Psychological Perspectives on Human Spaceflight*. Heidelberg, New York, Dordrecht, and London: Springer, 2013, p. 246.

121 Harrison and Fiedler, "Introduction: Psychology and Space Exploration," p. xxxviii.

122 Walter Sipes, "Foreword," in Douglas A. Vakoch (ed.), *On Orbit and Beyond: Psychological Perspectives on Human Spaceflight*. Heidelberg, New York, Dordrecht, and London: Springer, 2013, p. vii.

123 Nick Kanas, "Human Interactions On-orbit," in Douglas A. Vakoch (ed.), *On Orbit and Beyond: Psychological Perspectives on Human Spaceflight*. Heidelberg, New York, Dordrecht, and London: Springer, 2013, pp. 93–106.

124 Examples include long-duration missions to severe, remote areas, such as Antarctica. Although these settings were quite different than the near zero gravity environment of space, people who worked in them had to deal with stressful situations that might give insights that could be applicable to space crews. Historical exploratory expeditions were typically of long duration and involved many unknown risks. Because of the uncertainty, there was often a high degree of situational decision-making. Studying individuals in analog environments has taken two basic approaches. One involves constructing an environment within a laboratory setting with maximum control over exogenous variables and utilizing research subjects. To that end, several space-analogous settings have been created, including the NEEMO underwater habitat located new Key Largo, Florida, the Haughton-Mars project on Devon Island in Canada, and the Mars500 simulation at the Institute for Biomedical Problems in Moscow. The second approach examines naturally occurring in situ groups in real environments. For example, studies of submarine crews were thought to be applicable to spacefarers insofar as submariners operated in tightly confined environments for missions of long duration. By examining subjects who perform in extreme environments, researchers can gain insight into the challenges astronauts may face in space missions. Sheryl L. Bishop, "From Earth Analogues to Space: Learning How to Boldly Go," in in Douglas A. Vakoch (ed.), *On Orbit and Beyond: Psychological Perspectives on Human Spaceflight*. Heidelberg, New York, Dordrecht, and London: Springer, 2013, pp. 25–50.

125 Nick Kanas, Stephanie Saylor, Matthew Harris, et al., "High Versus Low Crewmember Autonomy in Space Simulation Environemnnts," in Douglas A. Vakoch (ed.), *On Orbit and Beyond: Psychological Perspectives on Human Spaceflight*. Heidelberg, New York, Dordrecht, and London: Springer, 2013, pp. 231–244.

126 Roma, Hursh, Hienz, et al., "Effects of Autonomous Mission Management on Crew Performance, Behavior, Phuysiology: Insights from Ground-Based Experiments," pp. 245–266.

127 Kaku, *Physics of the Impossible*, p. 172.

128 Wichman "Near-Term Extended Solar System Exploration," pp. 268–269.

129 Nick Kanas, "From Earth's Orbit to the Outer Planets and Beyond: Psychological Issues in Space," in Douglas A. Vakoch (ed.), *On Orbit and Beyond: Psychological Perspectives on Human Spaceflight*. Heidelberg, New York, Dordrecht, and London: Springer, 2013), p. 286.

130 Kanas, "From Earth's Orbit to the Outer Planets and Beyond: Psychological Issues in Space," p. 291.

131 Kondo, "Interstellar Travel and Multi-Generation Space Ships: An Overview," p. 11.

132 Kaku, *Physics of the Impossible*, p. 174.

133 I.S. Shklovskii and Carl Sagan, *Intelligent Life in the Universe*. San Francisco, London, and Amsterdam: Holden-Day, Inc., 1966, p. 442.

134 Harrison, *After Contact*, p. 36.

135 John H. Moore, "Kin-Based Crews for Interstellar Multi-Generational Space Travel," in Yoji Kondo, Frederick Bruhweiler, John Moore, and Charles Sheffield (eds), *Interstellar Travel and Multi-Generation Space Ships*. Burlington, Ontario, Canada: Apogee Books, 2003, p. 81.

136 Sheffield, "Fly Me to the Stars," p. 25.

137 Asimov, *Extraterrestrial Civilizations*, p. 235.

138 Drake and Sobel, *Is Anyone Out There? The Scientific Search for Extraterrestrial Intelligence*, pp. 117–119.

139 Asimov, *Extraterrestrial Civilizations*, pp. 203–205.

140 Ben Austen, "After Earth: Why? Where? How? When? *Popular Science*, (March 2012), p. 50.

141 Kondo, "Interstellar Travel and Multi-Generation Space Ships: An Overview," p. 15.

142 Asimov, *Extraterrestrial Civilizations*, pp. 235–238.

143 Kurland, *The Complete Idiot's Guide to Extraterrestrial Intelligence*, p. 248.

144 Jason P. Kring and Megan A Kaminski, "Gender Composition and Crew Cohesion During Long-Duration Space Missions," in Douglas A. Vakoch (ed.), *On Orbit and Beyond: Psychological Perspectives on Human Spaceflight*. Heidelberg, New York, Dordrecht, and London: Springer, 2013, p. 125.

145 For instance, the experience of people working in polar environments suggest that prolonged space travel could actually increase fortitude, perseverance, independence, self-reliance, ingenuity, camaraderie and even decrease tension and depression. Kanas, Sandal, Boyd, et al., "Psychology and Culture During Long-Duration Space Missions,"

pp. 162. Paradoxically, L. Palinkas found that a depressed mood was inversely associated with the severity of station environments—that is, the better the environment, the worse the depression. For example, in Antarctica, the winter-over experience was associated with reduced rates of hospital admissions. Palinkas speculated that the experience of adapting to isolation and confinement may actually improve an individual's self-efficacy and self-reliance thus engendering coping skills that could be applied to other areas of life to counteract stress. Bishop, "From Earth Analogues to Space: Learning How to Boldly Go," p. 42.

146 Roach, *Packing for Mars*, p. 243.

147 Moore, "Kin-Based Crews for Interstellar Multi-Generational Space Travel," p. 80–88.

148 Dennis H. O'Rourke, "Genetic Considerations in Multi-Generational Space Travel," in Yoji Kondo, Frederick Bruhweiler, John Moore, and Charles Sheffield (eds.), *Interstellar Travel and Multi-Generation Space Ships*. Burlington, Ontario, Canada: Apogee Books, 2003, p. 93.

149 Kanas, "From Earth's Orbit to the Outer Planets and Beyond: Psychological Issues in Space," p. 293.

150 John Gribbin, *Alone in the Universe: Why Our Planet is Unique*. Hoboken, NJ: Wiley, 2011, p. 49.

151 Harrison, *After Contact*, p. 167.

152 Webb, *Where is Everybody?* p. 73. Although this process would be time-consuming, Paul Davies noted that it would still constitute only a small fraction of the age of the galaxy. Based on this reasoning, it would take only one such community somewhere in the galaxy to present us with Fermi's awkward conundrum. Davies, *The Eerie Silence*, p. 120.

153 Davies, *The Eerie Silence*, p. 126.

154 Gilster, *Centauri Dreams*, p. 11.

155 de Grasse Tyson, *Space Chronicles*, pp. 130–131.

156 Toney Allman, *Are Extraterrestrials a Threat to Humankind?* San Diego, CA: Reference Point Press, 2012, pp. 4–8.

157 Roach, *Packing for Mars*, p. 171.

158 Quoted in de Grasse Tyson, *Space Chronicles*, p. 131.

159 Gilster, *Centauri Dreams*, p. 217.

160 Davies, *The Eerie Silence*, p. 120. Some researchers suspect that UFOs are piloted by robotic entities. They point out that he zigzagging maneuvering of the UFOs

would produce enormous g forces that biological entities could not survive. If, however, the spacecraft were manned by non-biological entities, it would explain how the craft cold execute these maneuvers and still operate. Kaku, *Physics of the Impossible*, p. 151.

161 Haldeman, "Colonizing Other Worlds," p. 68.

162 Haldeman added a science fiction twist by speculating that the surrogates might one day have the power to change their form and resemble the native animals, or even plants or mineral formations on the planet and patiently wait for something to happen. Haldeman, "Colonizing Other Worlds," p. 67.

163 Mary Roach, *Packing for Mars: The Curious Science of Life in the Void*. New York and London: W.W. Norton & Company, 2013, p. 171.

164 Jason Koebler, "Poll: Americans Overwhelmingly Support Manned Mars Mission," *U.S. News and World Report*, February 11, 2013, http://www.usnews.com/news/articles/2013/02/11/poll-americans-overwhelmingly-support-manned-mars-mission.

165 Davies, *The Eerie Silence*, pp. 122–123.

166 de Grasse Tyson, *Space Chronicles*, p. 136.

167 Ibid., p. 130.

168 Sagan, *Pale Blue Dot*, pp. 226–227.

169 de Grasse Tyson, *Space Chronicles*, p. 219.

170 Doug Beason, "Interstellar Travel: Why We Must Go," in Yoji Kondo, Frederick Bruhweiler, John Moore, and Charles Sheffield (eds.), *Interstellar Travel and Multi-Generation Space Ships*. Burlington, Ontario, Canada: Apogee Books, 2003, pp. 71–72.

171 Austen, "After Earth: Why? Where? How? When?, p. 55.

Conclusion

1 Carl Sagan, *Pale Blue Dot: A Vision of the Human Future in Space*. New York: Ballantine Books, 1994 p. 245.

2 The most powerful nuclear bomb ever detonated was the "Tsar bomb" that exploded on October 30, 1961 over the Russian island of Novaya Zemlya. The yield was estimated to have been between 50 to 58 megatons.

3 Sagan, *Pale Blue Dot*, p. 251.

4 Neil de Grasse Tyson, *Space Chronicles: Facing the Ultimate Frontier*. (New York: W.W. Norton & Company Ltd., 2012), pp. 53–54.

5 Sagan, *Pale Blue Dot*, pp. 261–263.

6 Michio Kaku, *Visions: How Science Will Revolutionize the 21st Century*. (New York: Anchor Books, 1997), p. 316.

7 Ben Austen, "After Earth: Why? Where? How? When? *Popular Science*, (March 2012), p. 46.

8 John Gribbin, *Alone in the Universe: Why Our Planet is Unique*. Hoboken, NJ: Wiley, 2011), p. 97.

9 Sagan, *Pale Blue Dot*, p. 264.

10 Gribbin, *Alone in the Universe*, p. 98.

11 Ibid., pp. 141–145.

12 Isaac Asimov, *Extraterrestrial Civilizations*. New York: Crown Publishers, Inc., 1979, p. 110.

13 I.S. Shklovskii and Carl Sagan, *Intelligent Life in the Universe*. San Francisco, London, and Amsterdam: Holden-Day, Inc., 1966, p. 86 and Michio Kaku, *Physics of the Impossible: A Scientific Exploration into the World of Phasers, Force Fields, Teleportation, and Time Travel*. New York: Anchor Books, 2008, p. 156.

14 Some astronomers believe that life could develop around red giant stars at a distance of several astronomical units. Bruno Lopez, Jean Schneider, and William C. Danchi, "Can Life Develop in Expanded Habitable Zones Around Red Giant Stars?" *The Astrophysical Journal*, No. 627 (July 2005), pp. 974–985. However, as John Gribbin noted red dwarf stars frequently emit flares that would release large amounts of ultraviolet radiation and other harmful particles that would be particularly damaging because of the propinquity of the star to the planet that happened to reside in the habitable region. Gribbin, *Alone in the Universe*.

15 Asimov, *Extraterrestrial Civilizations*, pp. 110–111.

16 Paul Gilster, Habitability Around Red Giants," December 15, 2008.

17 Gribbin, *Alone in the Universe*. 137.

18 Sagan, *Pale Blue Dot*, p. 307.

19 Paul Davies, *The Eerie Silence: Renewing Our Search for Alien Intelligence*. Boston and New York: Houghton Mifflin Harcourt, 2010, p. 91.

20 Sagan, *Pale Blue Dot*, p. 308.

21 Kathryn Denning, "'L' on Earth," in Douglas A. Vakoch and Albert A. Harrison (eds.), *Civilizations Beyond Earth: Extraterrestrial Life and Society*. New York and Oxford: Berghan Books, 2011, p. 80.

22 Stephen Pinker, *The Better Angels of Our Nature: Why Violence has Declined.* New York: Penguin Books, 2012.

23 On April 8, 2010, U.S. President Barack Obama and Russian president Dmitry Medvedev signed the New Strategic Arms Reduction Treaty (New START). Both the U.S. and Russian governments ratified the treaty within a year. Under the terms of New START, the United States and Russia agreed to dramatically reduce their deployed arsenals to 1,550 nuclear warheads each. According to the Federation of American Scientists, as of 2011, the total nuclear inventories for the United States and Russia were respectively 8,500 and 11,000—a substantial reduction from the peak levels of roughly 32,000 and 45,000 nuclear warheads held by Washington and Moscow, respectively, during the Cold War. George Michael, "Avoiding A Third World War—And a Second Holocaust," *The Nonproliferation Review*, Vol. 19, No. 1 (March 2012), p. 146.

24 "World population projected to reach 9.6 billion by 2050—UN Report," UN News Centre, June 13, 2013, http://www.un.org/apps/news/story.asp?NewsID=45165.

25 Jack A. Goldstone, "The New Population Bomb: The Four Megatrends That Will Change the World," *Foreign Affairs*, Vol. 89, No. 1 (January/February 2010), p. 39.

26 Michael T. Klare, *Resource Wars: The New Landscape of Global Conflict.* New York: Owl Books, 2001.

27 John Vidal, "Food shortages could force world into vegetarianism, warn scientists," *The Guardian*, August 26, 2012, http://www.theguardian.com/global-development/2012/aug/26/food-shortages-world-vegetarianism.

28 Francis Fukuyama, *The Origins of Political Order: From Prehuman Times to the French Revolution.* (New York: Farrar, Straus, and Giroux, 2011), p. 5.

29 Daniel Patrick Moynihan, *Pandemonium: Ethnicity in International Politics.* Oxford: Oxford University Press, 1993.

30 Samuel P. Huntington, *The Clash of Civilizations: Remaking of World Order.* New York: Touchstone, 1996, p. 36.

31 Huntington, *The Clash of Civilizations*, pp. 256–257.

32 Peter Almond, "Beware: the new goths are coming," *The* [London] *Sunday Times*, June 11, 2006, http://www.timesonline.co.uk/tol/news/uk/article673612.ece.

33 Robert Kaplan, "The Coming Anarchy," *The Atlantic Monthly*, February 1994. http://www.theatlantic.com/magazine/archive/1994/02/the-coming-anarchy/304670/.

34 Samuel P. Huntington, *Who Are We? The Challenges to America's National Identity*. New York: Simon & Schuster, 2004.

35 See the United Nations website at http://secint24.un.org/en/members/growth.shtml.

36 Barnett's functioning core consists of North America, both Eastern and Western Europe, Russia, Japan, China, India, Australia, New Zealand, South Africa, Argentina, Brazil, and Chile—roughly 4 billion out of a global population of over 6 billion. Thomas P.M. Barnett, *The Pentagon's New Map: War and Peace in the Twenty-first Century*. New York: G.P. Putnam's Sons, 2004, pp. 131–132.

37 Barnett, *The Pentagon's New Map*, p. 56 and Thomas P.M. Barnett, *Blueprint for Action: A Future Worth Creating*. New York: Berkley Books, 2005, p. 302.

38 Kaku, *Visions*, p. 319.

39 Stephen Webb, *Where is Everybody? Fifty Solutions to the Fermi Paradox and the Problem of Extraterrestrial Life*. New York, Copernicus Books, 2000.

40 Michio Kaku, *Physics of the Future: How Science will Shape Human Destiny and our Daily Lives by the Year 2100*. New York: Doubleday, 2011, p. 327.

41 Kaku, *Visions*, p. 19.

42 Michio Kaku and Jennifer Trainer (eds.), *Nuclear Power: Both Sides*. New York and London: W.W. Norton & Company, 1982.

43 During the Cold War proliferation had been mainly vertical, in the sense that the nuclear powers added more and more weapons to their stockpiles. Today, however, that pattern has been reversed, with the major nuclear powers drastically reducing their arsenals. The trend is now toward "horizontal proliferation." Over the past two decadees, more countries are aspiring to acquire nuclear weapons—for example, India, Pakistan, North Korea, and Iran—and the diffusion of nuclear technology enables this pattern. A steady dissemination of dual-use equipment could enable rogue states or terrorist organizations to produce chemical and biological weapons. Although the hurdles to producing and acquiring nuclear weapons are still formidable, the design and construction problems are being mitigated by advances in computers and commercial equipment. Furthermore, the spread of centrifuge technology and advances in centrifuge designs provide for potentially greater capacity to produce nuclear weapons. Finally, the technology used to produce delivery systems is becoming more readily available as well. Anthony Cordesman, *The War After the War: Strategic Lessons of Iraq and Afghanistan*. Washington, D.C.: Center for Strategic and International Studies, 2004, pp. 21–23.

44 Kaku, *Visions*, p. 326.

45 Ibid., p. 329–332.

46 Kaku, *Physics of the Future*, pp. 331–338.

47 Ibid., pp. 338–339.

48 Ibid., p. 347.

49 Kaku, *Visions*, p. 18.

50 Gribbin, *Alone in the Universe*, p. 414.

51 Davies, *The Eerie Silence*, p. 207.

52 Ibid., p. 208.

53 Donald E. Tarter, "Can SETI Fulfill the Value Agenda of Cultural Anthropology?" in Douglas A. Vakoch and Albert A. Harrison (eds.), *Civilizations Beyond Earth: Extraterrestrial Life and Society*. New York and Oxford: Berghan Books, 2011.

54 Sagan, *Pale Blue Dot*, p. 330.

55 Quoted in Kaku, *Physics of the Impossible*, p. 155.

56 According to astrophysicists, if omega—a property used by astrophysics to measure the density of the universe—is greater than one then there will be enough matter to reverse the expansion of the Big Bang. Kaku and Thompson, *Beyond Einstein*, pp. 152–153.

57 Kaku, *Physics of the Impossible*, p. 293.

58 In 2010, data from the WMAP satellite determined that the universe consisted of 72.8 percent dark energy, 22.7 percent dark matter, and 4.56 baryonic matter (of which we are made). Panek, *The 4 Percent Universe*, pp, 242.

59 Kaku defines Planck energy as consisting of "1019 billion electron volts, the energy at which space-time itself becomes unstable." Kaku, *Physics of the Future*, p. 340.

60 Kaku, *Physics of the Impossible*, p. 249.

61 *Parallel Worlds*, p. 21.

62 Kaku, *Parallel Worlds*, pp. 339–341.

63 Kaku, *Visions*, p. 344. Physicists are divided on the feasibility of time travel. The so-called "Hawking's paradox," named after the legendary physicist Stephen Hawking, suggests that if time travel were possible, then we would be visited by tourists from the future. One explanation advanced for UFOs is that they are visitors from the future. See Marc Davenport, *Visitors from Time: The Secret of the UFOs*. Tuscaloosa, AL: Greenleaf Publications, 1994. For more on the feasibility of time travel see Paul Davies, *About Time: Einstein's Unfinished Revolution*. New York: Simon & Schuster, 1995; Paul

Davies, *How to Build a Time Machine*. New York: Penguin Group, 2001; and Jenny Randles, *Breaking the Time Barrier: The Race to Build the First Time Machine*. New York: Paraview Pocket Books, 2005.

64 Kaku, *Parallel Worlds*, p. 306. On that note, it is worth mentioning the aphorism of Arthur C. Clarke which noted, there are three stages in the way the scientific establishment reacts to a new big idea. First, the idea is dismissed as impossible. Second, it is considered possible, but not practical. Finally, the idea is said to have been good all along. Michaud, *Contact with Alien Civilizations*, p. 125.

65 Sagan, *Pale Blue Dot*, p. 331.

66 Michaud, *Contact with Alien Civilizations*, p. 59.

67 Holman W. Jenkins, Jr., "Will Google's Ray Kurzweil Live Forever?" *The Wall Street Journal*, April 12, 2013, http://online.wsj.com/article/SB10001424127887324504704578412581386515510.html.

CPSIA information can be obtained
at www.ICGtesting.com
Printed in the USA
BVHW041813080620
581123BV00015B/627